Arjen Y. Hoekstra

Perspectives on Water

An Integrated Model-based Exploration of the Future

INTERNATIONAL BOOKS, 1998

This book has been written in the context of the research programme of the School of Systems Engineering, Policy Analysis and Management of the Delft University of Technology. The research has been financed by and carried out at the National Institute of Public Health and the Environment in Bilthoven. The J.E. Jurriaanse Stichting gave a financial contribution to the publishing expenses.

© Arjen Y. Hoekstra, 1998

ISBN 90 5727 018 8

Key words: water assessment, sustainability, future studies, computer modelling

Copy editor: Annemarie Weitsel
Cover design: Marjo Starink
Cover photo (New Delhi, 1990): Arjen Y. Hoekstra
Printed by: Haasbeek

Published by: International Books
Alexander Numankade 17, 3572 KP Utrecht, the Netherlands
Telephone + 31 30 2731840, fax + 31 30 2733614, email i-books@antenna.nl

Contents

Preface		7
1	**Introduction**	**9**
	1.1 Water and sustainable development	*9*
	1.2 Research objective and approach	*15*
	1.3 Outline of the book	*21*
2	**Water in the 21st century: questions and uncertainties**	**23**
	2.1 Introduction	*23*
	2.2 Water demand	*25*
	2.3 Water availability	*32*
	2.4 Water scarcity	*44*
	2.5 Alteration of the hydrological cycle	*49*
	2.6 Sea-level rise	*57*
	2.7 Critical water issues	*62*
3	**AQUA: a tool for integrated water assessment**	**65**
	3.1 The field of integrated water assessment	*65*
	3.2 Tools for integrated water assessment	*68*
	3.3 Introduction to AQUA	*72*
	3.4 Concepts and model schematization	*76*
	3.5 The pressure sub-model	*86*
	3.6 The state sub-model	*91*
	3.7 The impact sub-model	*102*
	3.8 The response sub-model	*106*
	3.9 Discussion	*111*
4	**Water indicators**	**113**
	4.1 The link between indicators and simulation model	*113*
	4.2 Pressure indicators	*115*
	4.3 State indicators	*116*
	4.4 Impact indicators	*118*
	4.5 Response indicators	*120*

5	**Perspectives on water**	**123**
	5.1 Uncertainties: a perspective approach	*123*
	5.2 Cultural theory	*129*
	5.3 Perspectives on water	*133*
	5.4 Perspective-based model formulations	*140*
6	**The AQUA World Model**	**147**
	6.1 Introduction	*147*
	6.2 Exogenous developments and model initialization	*150*
	6.3 Model parameters: sensitivities and calibration	*157*
	6.4 Discussion of results	*184*
7	**A global water assessment**	**185**
	7.1 Introduction	*185*
	7.2 The context	*187*
	7.3 Three water utopias	*191*
	7.4 Three water dystopias	*204*
	7.5 Possible water futures: the full range	*208*
	7.6 Risk assessment	*210*
	7.7 Water and development: the water transition	*212*
	7.8 Conclusion	*216*
8	**The AQUA Zambezi Model**	**221**
	8.1 Introduction	*221*
	8.2 Input data	*225*
	8.3 Calibration of hydrological parameters	*242*
	8.4 Discussion of results	*254*
9	**A water assessment for the Zambezi basin**	**259**
	9.1 Introduction	*259*
	9.2 The Zambezi basin: current issues and prospects	*261*
	9.3 Three water utopias	*266*
	9.4 Dystopias and risks	*277*
	9.5 Reflection on the 'Harare priorities'	*282*
	9.6 Conclusion	*285*

10	**Discussion**	**289**
	10.1 Validity of the AQUA tool	*289*
	10.2 Perspectives as a rationale for composing scenarios	*299*
	10.3 Future research	*307*

References — 311

Summary — 327

Summary in Dutch - Samenvatting — 333

About the author — 339

List of symbols — 341

Abbreviations — 349

Glossary — 351

Preface

Think globally, act locally. Who does not know this environmental slogan today? But who knows precisely what it means? It seems so reasonable: what can we, tiny people, do other than act locally? And of course, we should think about how we affect our environment in a wider context and about how the environment affects us. We know that a lot of small actions can have large effects. We also know that some processes are slow and that some effects become visible only after decades. Nevertheless, despite our good intentions, it seems difficult to really think on a global and long-term level. Because most of our daily business concerns problems and wishes that refer to where we are and the day, week, month or year in which we live, it is not surprising that we are not used to think globally or long term. As a result, our knowledge in this area is comparatively small. My interest in the environment and development and my inclination towards the incomprehensible have made me write this thesis.

The thesis is about water and development. Central questions addressed are: how do water scarcity and socio-economic development interact and how might people anticipate water problems in the 21st century? Although this broad subject seems to require further limitation before it can reasonably be handled, any demarcation would clash with the message of the above slogan. I do not want to exclude anything in advance, because of the aim of a comprehensive approach. By considering only one country or one river basin, one would for instance disregard the fact that many areas import food from far-off irrigated regions, thus calling on water resources elsewhere. Besides, because of their global nature, phenomena such as climate change and sea-level rise cannot be considered at lower than global level. By considering only the physical aspects of water availability, one would ignore important socio-economic processes which determine water demand and which constitute the actual pressure on the physical system. However, an analysis of the dynamics of the globe and its population in all its facets is likely to become such an abstract exercise, that any indicator for a concrete regional issue would be lacking. To bridge the gap between comprehensive thinking and more down-to-earth questions, one needs the eyes of a fish, which can simultaneously look sharply at a specific subject and look at its broad context. The underlying aim of this study is to develop a method that can be used to put the 'fish-eye' methodology into practice.

In this work, I introduce the AQUA tool for integrated water assessment and apply it at both global and river-basin level. The first application, the AQUA World Model, forms an integral part of the more comprehensive model TARGETS, which also addresses demography, human health, economics, energy, food, land and elements cycles. The TARGETS model provides us with a tool for global thinking. By taking the AQUA World Model to the front, one can focus on water. The river-basin application of AQUA, the AQUA Zambezi Model, is intended to illustrate how one can concentrate on a specific area, without losing the broad outlook.

The research for this thesis has been carried out as part of the Global Dynamics and Sustainable Development Programme at the National Institute of Public Health and the Environment (RIVM) in Bilthoven, the Netherlands. As a researcher at the School of Systems Engineering, Policy Analysis and Management at the Delft University of Technology, I have been on detachment to RIVM for four years. An account of the activities carried out within the Global Dynamics and Sustainable Development Programme is given in Rotmans and De Vries (1997).

There are a number of people who directly or indirectly contributed to this book. I am grateful to Wil Thissen and Jan Rotmans, who made it possible for me to do this research and who stimulated me in different ways: Jan through his innovative ideas and Wil through his down-to-earth criticism throughout. I wish to thank all 'globos' for being pleasant colleagues. Some deserve special mention: Bert de Vries, whose comments on earlier drafts were very useful to me; Henk Hilderink, Arthur Beusen and Marco Janssen, for their support in the world of computers; my roommate Bart Strengers, for the stimulating discussions we have had and his comments on a draft of this book; Frank Geels, for the necessary reflection on the use of computer models; and my trainees Renzo van Rijswijk and Jasper Vis, for their work on the Ganges-Brahmaputra and Zambezi respectively. Parts of Jasper's work have been used in the chapter on the AQUA Zambezi Model. Outside the 'globos', my thanks to Bert Bannink, Ton de Nijs and Theo Traas, for their enthusiastic involvement in the Zambezi study; Jaap Kwadijk and Willem van Deursen, for their work on the Ganges-Brahmaputra study; and the SEIS-team of Resource Analysis, for their supportive model studies. I am grateful to Annemarie Weitsel for editing the English text. Finally, my thanks to my dearest Daniëlle, for her love, and to my parents, Jaap and Wik, for their continuing stimulation and interest throughout the years. Mamma and dad, to you I dedicate this work.

1 Introduction

This chapter gives an account of the incentives for and the objectives and methods of the research which is recorded in this book. It starts with a review of the relatively short history of thinking about water and sustainable development. It has become evident that the possible role of water policy in sustainable development can only be understood if people adopt a long-term view, on different levels, from the river-basin to the global. In addition, analysing the long-term interaction between water and humankind requires an integrated assessment, considering quantitative and qualitative aspects of water in relation to human development and environmental change. Such a comprehensive approach should have an explorative character, because large and inherent uncertainties make predictions impossible. As proper means for explorative, long-term water policy analyses were not readily available at the start of this research, one aim has been to develop a generic tool that suits the requirements of such analyses. Another aim was to apply this tool both to a specific river basin and to the globe as a whole. The chapter concludes with an explanation of the approach followed and an outline of the book.

1.1 Water and sustainable development

Water is essential for life. There are few studies of water and development that do not start with this or a similar sentence. A relevant question is therefore not *whether* water plays an important role in development, but *how* water relates to development. Water has traditionally been perceived merely as a resource, to be exploited in order to support development. However, people are becoming increasingly aware of the impact of human activities on water, limitations to water supply and the tension between intensive water use and the functioning of natural ecosystems. This illustrates that the relationship between water and development is changing. Several observers have warned of an approaching worldwide water crisis and argued for sustainable management of the global water resources. This changing attitude among scientists, water policy analysts and water policy makers is a process that started in the 1960s. One of the first books to express awareness of the impact of human beings on the environment was *Silent Spring* by Rachel Carson, published in 1962. Carson's major concern was the spread of pesticides and other chemicals throughout the environment and their lethal effect. She stated that the problem of water pollution can be understood only in context, as part of the whole to which it belongs, namely the pollution of the entire environment by mankind. Worth mentioning also is the book *Le Problème de l'Eau dans le Monde* by the Frenchman

Raymond Furon, published in 1963. In this comprehensive study, Furon already broaches most of the subjects that are considered highly relevant today: the increasing demand for water from agriculture, industries and households, the limits to water supply, erosion and flooding, water pollution, the costs and benefits of dams, hydropower generation, the use of fossil ground water, desalination, and the effects of climate fluctuations.

Increasing interest in water led to the creation in 1965 of the International Hydrological Decade (IHD, 1965-1974), under the auspices of UNESCO, with the purpose of advancing hydrological knowledge through promoting international co-operation and training specialists and technicians. An important product of this at the end of the decade was a study on the world water balance carried out by the USSR Committee for the IHD. The results of this study, consisting of an atlas and a comprehensive monograph, were published in Russian in 1974 and translated into English a few years later (Korzun et al., 1977, 1978). The work was based on new material received from various countries as a result of the implementation of the IHD programme. In the same period, two other major studies on the global water balance were published: by Baumgartner and Reichel (1975) and by L'vovich (1979), the latter an English translation of another Russian publication dating from 1974. For global water studies today, people still rely heavily on these three studies. Although L'vovich and Korzun pay some attention to the socio-economic aspects of water demand, the emphasis in all these studies lies on the hydrological aspects of water availability.

An important event raising global political awareness of the environment was the United Nations Conference on the Human Environment in Stockholm, Sweden, in 1972. At this conference, the foundation was laid of the United Nations Environment Programme (UNEP). The same year also saw the publication of *The Limits to Growth* (Meadows et al., 1972), a report on one of the first efforts to explore the world of the future with the help of a computer model. The emphasis was on population and economic growth, food production, depletion of non-renewable resources and pollution. Water was not an explicit part of the model.

The first global conference specifically dedicated to water was the United Nations Water Conference in 1977, in Mar del Plata, Argentina. The Mar del Plata Action Plan stimulated a number of activities, including the International Drinking Water Supply and Sanitation Decade (1981-1990). This decade, proclaimed by the UN General Assembly at the end of 1980, had as its primary goal the achievement of full access to water supply and sanitation for all inhabitants of developing countries. Although this goal was far from achieved by the end of the decade, it was successful in creating awareness of the importance of clean water and sanitation and in developing workable strategies for further improvements (Christmas and De Rooy, 1991).

The concept of sustainable development was first launched in *World Conservation Strategy*, a joint report by the IUCN, UNEP and WWF (IUCN *et al.*, 1980). Although the strategy received wide attention, there was little political commitment. The concept really entered the political arena a few years later, when at the end of 1983 the General Assembly of the United Nations decided to form a World Commission on Environment and Development with the task of formulating 'a global agenda for change'. In 1987, the commission published its - soon well-known - report *Our Common Future* (WCED, 1987), which set out the global challenge of sustainable development, to meet not only our current needs but also those of future generations. Surprisingly, the availability of fresh water received only marginal attention in this influential report. But through *Our Common Future* the concept of sustainable development reached the wider public and it became a catalyst, bringing together nations from the North and the South, politicians from the left and the right and researchers from different fields in a common endeavour. Since the publication of the report, the number of meetings, books, reports and articles on the subject of sustainability has grown exponentially.

The most important political event since then has been the United Nations Conference on Environment and Development (UNCED) in June 1992 in Rio de Janeiro, Brazil, which produced *Agenda 21*, an action plan for the 21st century. As a preparation for UNCED, the International Conference on Water and the Environment (ICWE) was held in Dublin, Ireland, in January of the same year. This conference, the most comprehensive water meeting since the one in Mar del Plata, produced a *Report of the Conference* and the so-called *Dublin Statement* (ICWE, 1992a, 1992b), the latter containing four 'guiding principles' which should give direction to future water policies at local, national and international level. Briefly, the first principle states that - as fresh water sustains life, development and the environment - a holistic approach to water management is needed, linking social and economic development with the protection of natural ecosystems. According to the second principle, water management should be based on a participatory approach, involving users, planners and policy makers at all levels. The third principle emphasizes the central role of women in the provision, management and safeguarding of water. The final principle states that water should be considered an economic good, adding that it is vital to recognize first of all the basic right of all human beings to have access to clean water and sanitation at an affordable price.

In addition to these four principles, I cite here the seven sustainability criteria for water as provided by Gleick (1995), which give further insight into the key issues today. First, a minimum water requirement should be guaranteed to all humans to maintain health. Second, sufficient water should be guaranteed to restore and maintain the health of ecosystems. Third, data on water resources availability, use, and quality should be collected and made accessible to all parties. Fourth, water quality should be maintained to certain minimum standards. Fifth, human actions should not damage the long-term renewability of freshwater stocks and flows. Sixth, institutional mechanisms should be set up to prevent and resolve conflicts over water. Finally water planning and decision-making should be democratic, ensuring representation of all affected parties and fostering direct participation of affected interests.

One of the key elements in the concept of sustainable development is the recognition that large time-scales are involved in the relationship between development and environment. It is therefore argued that, in addition to five- or ten-year planning studies, long-term studies are needed, to anticipate future problems and to be able to respond accurately before it is too late for preventive policy. In the case of water, for instance, phenomena such as groundwater depletion and groundwater-level decline are slow processes, which often become manifest and problematic only after it has already become difficult to reverse the activities that are responsible for the problems. Even if the activities could be reversed, it might be too late to solve acute problems, due to the slow response of the natural processes. The same applies to global phenomena such as climate change and sea-level rise. Many researchers also warn of the exponentially growing demand for water in many areas of the world. If the water demand cannot be stabilized, they maintain, many regions will face serious water crises, with a disastrous impact on society, agriculture and economics in those regions which lack an escape route through costly measures such as importing water or desalination. Long-term analyses are required to study the future 'water self-sufficiency' of water-poor countries or countries where demand is growing rapidly, as is already done with self-sufficiency and interdependencies in energy and food supply. In many cases water conservation policy has to be formulated now, to prevent water shortages in the long-term future.

Discussions about sustainable development have brought an awareness not only of the large time-scales involved in environmental change, but also of the enormous spatial dimensions of some phenomena and problems. Before the concept of sustainable development had become a common notion, water researchers and policy analysts had already concluded that the national level is often too restricted for the analysis of water-related issues. The appropriate level for studying issues of freshwater availability and pollution would be that of the river basin, because river basins are the natural entities for water flows (UN, 1970). Many of the world's river basins are situated in more than one country. As these international river basins encompass nearly fifty per cent of the global land area (Gleick, 1993a), a large part of the world's population depends on water resources shared by neighbouring countries. In a number of basins this has led to political tensions. In the Jordan basin, for example, the scarcity of fresh water was one of the factors in the numerous hostilities between Israel and the Arab countries Jordan, Syria and the Lebanon, and the problem of water rights in the Middle East has still not been solved (Lowi and Rothman, 1993; Murakami and Musiake, 1994). Another international water dispute has arisen over Turkey's Southeast Anatolia Project, which aims to exploit the upstream waters of the Euphrates and Tigris for irrigation and hydroelectric power generation, thereby threatening the future water supply of Syria and Iraq (Slim, 1993; Kolars, 1994). Recently, Hungary and Slovakia were quarreling over the use of the Danube for hydroelectric power generation. Other large international basins which are often mentioned as vulnerable to water conflicts are the Nile, Niger and Zambezi in Africa, the Indus, Ganges and Mekong in Asia, and the Paraná, Colorado, Rio Grande and Columbia in the Americas (Clarke, 1991; Pearce, 1992; McCaffrey, 1993; Knoppers and Van Hulst, 1995).

The increasingly international character of many water-related problems has brought an awareness that water policy is not merely a matter of national concern. It has become clear that several aspects of water policy require a river-basin approach. However, there is a growing recognition that even the river-basin level is not sufficiently large. Some phenomena have a distinct global dimension. According to the Intergovernmental Panel on Climate Change (IPCC), human-induced global climate change and sea-level rise are phenomena which have become inescapable if no drastic measures are taken (Houghton et al., 1996). Another phenomenon with an unmistakably global dimension is deforestation. At first sight this only affects the conditions in certain regions, which is true if one considers it from the perspective of many people's everyday life (which plays on a scale of days to years), but in the long term, through its influence on global climate and biodiversity, it concerns mankind as a whole. Yet another such phenomenon is the worldwide loss of wetlands, which are among the most valuable ecosystems on earth. It is estimated that since the beginning of the 20th century more than half the world's wetlands have disappeared (Öquist and Svensson, 1996). This means that for the availability of water, which is closely linked to phenomena such as those mentioned above, it is not necessarily enough to take the river basin as the basic entity for analyses and decision-making, especially not if one is interested in a long term view. In addition to river-basin studies, global water studies have become a necessity. Even a seemingly regional issue such as the interaction between water demand and supply has a global dimension. Whereas international trade in *real* water might be small,[1] this does not apply to international trade in *virtual* water, water 'hidden' in agricultural and industrial products. Virtual water refers to water used in the production of products such as food or cotton. Importing water-intensive products means an indirect use of water from elsewhere in the world.[2] Although this international trade in virtual water has no meaning in a hydrological sense, it should be recognized as an important issue in a socio-economic sense, closely connected to the issue of the water self-sufficiency of nations.

As embodied in the first principle of the *Dublin Statement*, sustainable development requires a *holistic* approach towards water management. Considering large-scale and long-term effects are just two elements of such a holistic approach. Another element is to regard water as an integral part of the environment as a whole and to examine the interaction between water and development in a comprehensive way (Figure 1.1). This aspect of the holistic approach has become known as the need for *integration*. It has to be said that the need for an integrated approach to water management has not arisen exclusively from the call for sustainable development. In water management practice, this need has been recognized since the early 1980s, due to the increasing complexity of problems which had to be solved and the growing number of parties involved. Later in this book I will elaborate on the emergence of the field of integrated water assessment, and discuss present insights into integrated water assessment and the requirements for integrated assessment tools. Here, I would like to confine myself to saying that integrated water assessment studies typically require an explorative approach, where predictions give way to 'what if' propositions. Tools for integrated water assessment should be able to explore the implications of various assumptions and hypotheses, give an insight into system behaviour, help identify problems in the long term and assist in formulating water policy priorities as part of a long-term strategy for sustainable development. As will be shown later, explorative models for integrated water assessment which

---

[1] Water is generally used in the basin where it is naturally available and *if* it is taken out of the basin, it is transported to a neighbouring basin and not to the other end of the world. Some water is traded at global level in the form of bottled water, but the amounts are insignificant.

[2] As far as I am aware there have been no studies into this subject, and I can therefore give no data. I know of only one study into a similar subject, namely the *indirect use of land*: Harjono *et al.* (1996) find that the Netherlands actually use 23 million hectares abroad (an area nearly seven times as large as the country itself), three quarters of which is due to the import of food products. Undoubtedly, many of these products have been cultivated on irrigated lands. Although this estimate is probably an overestimate (many of the 23 million hectares are used only partly and the country exports as well as imports), this study is at least illustrative of the global socio-economic dimension of natural resources use.

Introduction

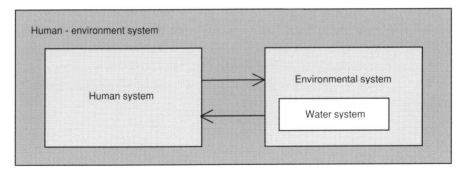

Figure 1.1. Context of the water system.

satisfy such requirements were not readily available at the start of this research. To develop such a model has been the main incentive for the research reported in this book.

1.2 Research objective and approach

The problem with integrated water assessment is not a lack of appropriate tools in any of the separate research fields, but rather the lack of integration of these tools and the difficulty of translating analytical results into policy-relevant information. The first objective of this research is therefore to improve the methodology of integrated water assessment through a better integration of instruments. The second objective is to search for methods to translate information about future developments into simple indicators which can be easily understood by policy makers and the wider public. Another major obstacle in integrated water assessment is a lack of appropriate methods to handle uncertainties and risks in an explicit way. For this reason, the third objective of this research is to develop a proper method of dealing with uncertainties. The final objective is to apply the new integrated methodology at different spatial levels and see whether understanding of the long-term interaction between water and development can indeed be improved. Below, I explain which approaches have been chosen to accomplish each of these objectives within the overall framework of the research.

Research framework
The research consists of four phases: problem statement, design, application and evaluation. In the first phase it is explored what types of water problems might be encountered in the 21st century. Several critical water issues are formulated and translated into design criteria for the development of a generic computer tool for

integrated water assessment. In the second phase, this generic tool is designed, together with a method to handle uncertainties. In the third phase, the tool and the method of uncertainty analysis are applied to two case studies, a global water study and a study concerning the catchment area of the Zambezi river in Southern Africa. The final phase of the research is to evaluate the tool and the method of uncertainty analysis which have developed.

The systems approach

As the central approach in this research I have adopted the systems approach, the foundations of which were laid in the 1940s and 1950s (Von Bertalanffy *et al.*, 1951). Some of its roots, however, go back much further, to classical times: a basic notion of the system concept can be found in Aristotle's observation that 'the whole is more than the sum of its parts'. Nowadays, a system is generally defined as 'a coherent whole of interacting entities'. A central idea in systems theory is that some system properties cannot be inferred from the individual properties of the constituting entities, but emerge only at the system level. Through its holistic nature, systems theory typically responds to the quest for an integrated approach as formulated in the environmental sciences. A particularly useful approach in modelling the evolution of systems is system dynamics, developed in the 1960s and 1970s by people such as Jay Forrester (1961, 1968). The mathematical structure of system dynamics is used in this research as a formal method of describing a system's changes over time. The technique of meta-modelling is used to simplify specialist models into components that fit within an integrated framework. A meta-model is a simplified representation of an original expert model and it thus describes the modelled system at a higher level of abstraction than the original (Rotmans, 1990). Techniques of simplification are for example: combination of functions through parameterization, approximation of non-linear functions, and omission of redundant parts and relatively small lag times (Thissen, 1978). Other ways of simplification are to take a lower spatial or temporal resolution or to combine separate categories into a few aggregated classes. As part of this research, a system-dynamics-based integrated water assessment tool has been developed, hereafter referred to as AQUA, which can be understood as a set of interlinked meta-models. The modelled system has been schematized into four subsystems: a pressure, state, impact and response subsystem. This approach has been adapted from the pressure-state-response schematization proposed by the OECD (1993) and corresponds to the approach followed by Rotmans and De Vries (1997). The advantage of this schematization is that it frames some of the main interests of policy makers: what makes things change (pressure), what are the actual changes in the environmental system (state), how are things affected (impacts) and how do or can people react (response). At the same time, the schematization roughly represents a major causal feedback loop. Socio-economic developments exert pressure on the environment and, as a consequence, the state of the environmental system changes. These changes have impacts on socio-economic activities and ecological processes which depend on the environment. These impacts in turn will evoke societal responses, in the form of actions that feed back upon the pressures (e.g. preventive measures), or directly upon the state or impacts (curative or adaptive measures).

Indicators
The second objective of this research is to search for methods to translate information about water developments into indicators that can easily be understood. Policy analysts and policy makers have increasingly become used to the concept of indicators. An indicator can be defined as an instrument for communicating key information about a system in a simplified way, for instance to policy makers, but also to the public. The aim of an indicator is to provide information in a comprehensive and quantitative form, in order to facilitate understanding of what is happening and to support the policy-making process. An indicator is often not a directly observable variable, but a construct made up by aggregating or averaging a number of different variables or by comparing a variable with a reference variable. In the case of highly aggregated measures, they are frequently referred to as indices rather then indicators. The growing interest in the use of indicators and indices is closely connected to the increasing complexity of policy problems and the overwhelming amount of data now available. Economic indicators have already been used for decades. More recently, several social indicators have been developed (OECD, 1976, 1982; UN, 1989). At present, the field of environmental and sustainability indicators is growing rapidly (Kuik and Verbruggen, 1991; Adriaanse, 1993; Bakkes *et al.*, 1994; Gouzee *et al.*, 1995). Within this field there has also been some specific interest in water indicators (Buijs and Dogterom, 1995). A common feature in most current studies is that indicators are regarded mainly as instruments for communicating observed data or trends. However, in the debate about sustainable development it is at least as urgent to communicate projections of future developments to policy makers. Therefore, it is proposed to use indicators not only as vehicles for the communication of observed data and trends, but also as a means of presenting model results. In this way, not only observed data but also model results can be made easily understandable and accessible. In order to implement this idea in practice, a framework of water indicators has been developed as a second

component of the AQUA tool, and linked to the first component, the simulation model.

Uncertainties and risks: the perspective approach
The third objective of this research is to develop and apply an appropriate analytic method to deal with uncertainties and risks. Historically, the most common approach to uncertainties in policy analysis has been to ignore them, but nowadays it is generally accepted that uncertainties should be made explicit. However, there is no general agreement on how uncertainties should be dealt with, for a number of reasons. First, there are many different types of uncertainty. Morgan and Henrion (1990) mention for instance uncertainty about quantities and uncertainty about the structure or functional form of a model. Furthermore, uncertainties might be about what is (or will be), about what we like, about what to do, and about our degree of uncertainty. I will refer to this as different *subjects* of uncertainty. A second reason for the lack of consensus on how to deal with uncertainties is the existence of different *sources* of uncertainty. Morgan and Henrion (1990) mention the following sources: statistical variations in measurements, linguistic imprecision, variability of the measured quantities, inherent randomness, estimating quantities on the basis of simplified models, subjective judgement and disagreement. A third and final reason for lack of consensus is that there are various *methods* to express uncertainties.

One of the most common methods of expressing uncertainties is the *probability method*. In this formal approach, uncertainty is expressed as the probability of a certain event. The concept of risk is defined as the probability that a certain undesirable event will take place. Another formal method - more recently developed and less common - is *fuzzy set theory*, in which an ill-defined concept is formalized by classifying it as partly in one class and partly in another (Kosko, 1993). In addition to these formal methods, there are methods which are completely different in nature. For example, a socio-economic way to measure uncertainties is through *insurance*, expressing risks in financial terms. The more it costs or would cost to insure against a certain event, the greater the risk of this event apparently is (Beck, 1997). Uncertainty is thus reduced to an economic concept. A psychological measure of uncertainty is *fear*, in which personal experience plays an important part. In this approach uncertainty is clearly a subjective concept. Further, some anthropologists point out that each form of society has its own risk portfolio, which means that the perception of risks is closely bound up with the cultural bias of a particular social organization (Douglas and Wildavsky, 1982). In this view risk is a collective construct, to be understood through *cultural analysis*. Finally, a commonly used method among policy analysts is the *scenario approach*, a method of analysing how different futures evolve if basic assumptions are varied.

It is important to distinguish between the several subjects and sources of uncertainties on the one hand and the various methods to handle them on the other, because each method has a limited domain in which it can be applied. Morgan and Henrion (1990) for instance argue that probability is an appropriate way to express uncertainty about empirical quantities, but that it is not applicable to any other subject of uncertainty. Fuzzy sets are particularly useful in cases of linguistic imprecision. In many socio-economic settings insurance has appeared to be a useful tool, because all kinds of non-numerical or subjective risk elements can be taken into account in the insurance costs. As Beck (1997) observes, the economic realism of insurance companies forbids them to insure the risks of nuclear power plants and several types of industries, while engineers maintain - based on probability analyses - that low risks are guaranteed. According to Beck (1997), modern societies operate on the borders and even beyond the limits of the insurable, due to the use of incorrect risk assessment tools. However, it may be questioned whether the insurance method can in practice be applied to express the risks of large projects or public policy, because one reaches a level at which it becomes impossible to answer the question of responsibility. As everyone is somehow involved and responsible for future developments, who is to blame if something goes wrong? Who has to insure against what? The less frequently a certain type of event occurs or the more inclusive a certain 'event' is, the more difficult it is to apply the insurance method for estimating risks. To deal with complex issues such as risks of public policy, the scenario approach might be used to advantage. This approach is particularly useful if structural uncertainties dominate uncertainties about quantities. In the scenario approach, structural uncertainties are analysed by varying the basic assumptions of an analysis.[1] Although the scenario approach has been proven to be a useful tool, an important shortcoming is the usually *ad hoc* character of scenarios. There seems to be no method to determine which scenarios to present and which not to, so that scenario composition appears to be a process left to the personal insights of the researchers and analysts involved. This brings us to the following question: can a certain set of scenarios really tell something about the uncertainties involved, or does the choice of a certain set of scenarios imply a prior choice of the uncertainties one wishes to face or to ignore? Douglas and Wildavsky (1982), who pose a similar question, argue that any assessment of uncertainty and risk is socially biased and

---

[1] In practice, scenarios are sometimes composed by varying one or two important parameter values, but I do not consider this as the original idea behind the construction of scenarios.

that uncertainties and risks should therefore be studied within a cultural context. According to these authors, cultural analysis is an essential tool in really understanding the different points of view in current debates on technological and environmental dangers, where structural uncertainties clearly play a dominant role.

Assuming that each of the different subjects and sources of uncertainty requires its own approach, the question arises what kind of uncertainties pertain to the type of long-term policy research reported in this book. According to Morgan and Henrion (1990), policy analysts often argue that uncertainty about structure is usually more important, and more likely to have a substantial effect on the results of the analysis, than uncertainty about quantities. In my opinion this is indeed the case, particularly in integrated assessment, where the formulation of the complex interrelations between different systems is an important element of the analysis and where disagreement between different disciplines is likely to occur. In this light, of the methods mentioned above the scenario approach seems to be best qualified for my purpose, while the cultural analysis method can be used as a rationale for composing different scenarios. For reason which will be stated later, the cultural theory of Thompson *et al.* (1990) has been chosen to provide a framework for the scenario analysis. Four of the five ways of life which are described in this theory are used here: the hierarchist, egalitarian, individualist and fatalist. Each way of life is closely connected to certain values and beliefs and to a particular world-view. Because these ways of life actually encompass more than just a way of life, I will refer to them as 'cultural perspectives', or just 'perspectives'. By taking different perspectives, scenarios will be constructed which are fundamentally different, yet coherent.

The generic approach
The final objective of this research is to apply the integrated and perspective-based assessment methodology at different spatial levels, to see whether insight into the long-term interaction between water and development can indeed be improved at these different levels. For this reason, AQUA has been designed as a generic tool, which can be applied from the river-basin to the global level. An obvious reason for applying an integrated water assessment tool at river-basin level is that the physical aspects of water problems are typically river-basin specific. The reason for applying AQUA at the global level is that several relevant phenomena have a distinct global dimension, as argued in the first section of this chapter: e.g. climate change, sea-level rise, deforestation and loss of wetlands. In this study, AQUA is applied to the world as a whole and to one particular river basin. The global water study was an integral part of the global change research carried out within the Global Dynamics

and Sustainable Development Programme at RIVM.[1] The river basin study was also part of this research programme, but not of the core work. I had originally chosen to work on the Ganges-Brahmaputra basin, but later on it was decided to start research on the Zambezi basin in order to contribute to the Zambezi Basin Water Assessment Study, a joint project of UNEP, SADC-ELMS and RIVM.[2] Here, I do not report on the work done on the Ganges-Brahmaputra,[3] but limit myself to a discussion of the AQUA World Model and the AQUA Zambezi Model and the actual use of these two tools in integrated water assessment. The global and river-basin case studies were carried out in parallel, so that experience and insight obtained in one study could be used in the other one and conversely.

1.3 Outline of the book

The main structure of the book reflects the four phases in the overall research framework: problem statement, design, application and evaluation (Figure 1.2). Chapter 2 explores the types of water problems which might be encountered in the 21st century. One of the aims of this chapter is to identify the water policy questions policy makers should pose today if they seriously wish to strive towards sustainable development. Another aim is to trace the main uncertainties in answering these questions and formulating the necessary policy response. As a result several critical water issues are formulated which provide a focus for the remaining chapters.

The design phase is reported in Chapters 3 to 5. Chapter 3 starts with a concise overview of the research field of integrated water assessment and discusses the requirements of integrated water assessment tools. The remaining part of the chapter describes AQUA, the computer tool for integrated water assessment which has been developed as part of this study. Chapter 4 discusses the indicators which have been linked to the simulation model of AQUA in order to communicate model outcomes in a relatively simple way. Chapter 5 explains how uncertainties are dealt with and shows how different perspectives have been included with respect to how the world functions and how people act.

Chapters 6 to 9 report on the two case studies carried out: the global water study and the Zambezi study. In Chapter 6 the AQUA World Model is discussed, an

---

[1] For the main findings of this research programme, the reader is referred to Rotmans and De Vries (1997).

[2] Intermediate findings of the Zambezi study can be found in two workshop reports (UNEP, 1995b, 1996); the final report is currently being written (Bannink, 1998).

[3] Preliminary results of the Ganges-Brahmaputra study were promising (Hoekstra, 1995; Van Rijswijk, 1995), but there was no time to work on two river basin studies simultaneously.

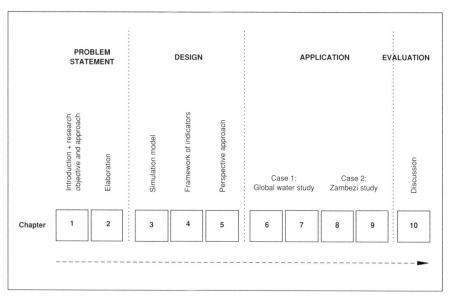

Figure 1.2. Set-up of the book.

application of AQUA to the world as a whole. Chapter 7 consists of a global water assessment, supported by the AQUA World Model. Chapter 8 discusses the application of AQUA to the catchment area of the Zambezi river in Southern Africa. Chapter 9 gives an integrated water assessment for the Zambezi basin on the basis of the AQUA Zambezi Model.

Chapter 10 discusses the validity of the AQUA tool, reflects on the use of perspectives as a rationale for composing scenarios and gives some directions for future research.

2 Water in the 21st century: questions and uncertainties

In this chapter some key questions are identified which have to be posed by water policy makers today if they wish to take the challenge of sustainable development seriously. Will mankind have enough water during the 21st century? Will it be of sufficiently good quality? What are the human abilities needed to manage the future? The chapter reviews the major water problems which humanity can expect to encounter in the next century and which are already being experienced in many parts of the world today. The discussion focuses on long-term developments and aims to trace the main underlying mechanisms. The most important gaps in knowledge are discussed and, where relevant, the controversies which have resulted from them. It is shown that various scholars have different expectations about future developments, due to different perceptions of basic concepts such as water demand, availability and scarcity. The chapter concludes by formulating a number of critical water issues, which provide a focus for the following chapters.

2.1 Introduction

Some people have suggested that, whereas the 20th century was the era in which mankind became aware of limitations to energy use, the 21st century will be the era which confronts the human species with the exhaustibility of water resources (Saeijs and Van Berkel, 1995). Water shortage might even be the factor which ultimately limits the scale of the human enterprise (Ehrlich and Ehrlich, 1991). Will water indeed play a key role in development and does it therefore deserve to be a major focus point in studies on long-term change? Or is water just a secondary issue, ranking behind issues such as economic development, human health, energy and food supply, biodiversity and global warming? Some researchers and institutions addressing long-term change and sustainable development assign comparatively high importance to hydrological processes, water pollution, irrigation and public water supply. Such interest in water has been expressed, for example, in the *Global 2000 Report* (Barney, 1980), *Agenda 21* (UN, 1992a), and various reports from the Worldwatch Institute (Brown *et al.*, 1993, 1996) and the World Resources Institute (WRI, 1992, 1994, 1996). Some observers expect that disputes over fresh water might even escalate into 'water wars'. Prominent politicians, among them the former Secretary-General of the United Nations Boutros Boutros Ghali, have for instance predicted that the next war in the Middle East will be over water (Donkers, 1994).

However, there are also many analysts of long-term change who touch only lightly upon the issue of fresh water. One can see this, for example, in *Limits to Growth* and its sequel *Beyond the Limits* (Meadows *et al.*, 1972, 1991), *Our Common Future* (WCED, 1987) and IIASA's *Sustainable Development of the Biosphere* (Clark and Munn, 1986). Many studies of climate change also pay comparatively little attention to the role of water.

In a discussion about water and sustainable development, the central question might be formulated as follows: *Can human beings, worldwide and in the long term, be provided with sufficient clean fresh water without imposing an unacceptable strain on ecosystems?* And if the answer is yes: *How?* These questions can only be answered by making assumptions about, for example, the physical or economical limits to freshwater supply, the development of water-conserving technology, humanity's ability to adapt its lifestyle to water scarcity and the extent to which people can shape their environment. Many authors point out that, given the expected continuation of population growth, water will become scarcer all over the world, and that many regions in the world experience serious water scarcity already (La Rivière, 1989; Clarke, 1991; Postel, 1992; Kulshreshtha, 1993). Human beings therefore should make considerable efforts to save us from a world water crisis. However, there are also researchers who are very optimistic. Engineers indicate that using water more efficiently and applying re-use techniques can lower demand significantly. In addition, through the construction of reservoirs and desalination of salt or brackish water, we are able to increase the potential water supply. In countries with a high degree of water scarcity, new techniques are indeed being developed and used.[1] Many economists do not expect serious water problems, because, as they argue, considering water as an economic asset will result in far more efficient water supply (Anderson, 1995). It is obvious, therefore, that different authors have different perceptions of water supply in the future.

To answer the central question and identify the various complex problems involved, some more specific questions need to be posed: what level of demand might be expected in the future, what determines the demand and how can this demand be managed? How much fresh water is available, what determines the maximum possible water supply and how can it be increased? What makes clean fresh water scarce and how can this scarcity be overcome? How will the hydrological cycle respond to increasing consumption of water by humans, land use

---

[1] It is no coincidence that 60 per cent of the world's desalination capacity is found in the water-poor but energy-rich nations of the Persian Gulf and that Israel has cut its water use per irrigated hectare substantially by applying more efficient, self-developed techniques.

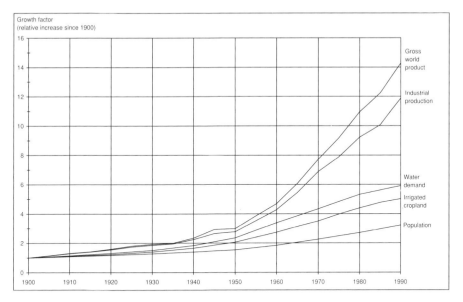

Figure 2.1. Relative growth of world population, gross world product, value added in the industrial sector, global irrigated cropland area and global water demand for the period 1900-1990.

changes and global warming? And how will the sea level be affected by these changes? These questions are discussed in the following sections, which review present understanding and try to uncover the major uncertainties and controversies involved.

2.2 Water demand

At the end of the 20th century, global water demand is six to seven times greater than it was at the beginning of the century (Figure 2.1). The increase in size of areas of irrigated cropland, especially during the second half of the 20th century, has contributed significantly to the so-called Green Revolution. Should people anticipate a similar increase in water demand during the 21st century and expect a continued Green Revolution? If present growth rates are extrapolated, the increase in water demand in the 21st century will be even greater than in the 20th century. But some recent studies foresee a slowdown in growth rates, especially in the developed world. However, a review of water demand projections in the literature shows little agreement. Below I first discuss some types of water demand and the actual demand today, and secondly I present the results of a review of different water demand projections. It will become clear that most projections have been made without paying much attention to the actual mechanisms which determine water demand: in these studies demand is regarded as a requirement which can simply be calculated on the basis of population figures, the area of irrigated cropland, etc. After the review of water demand projections, I will consider the concept of water demand more closely. It will be shown that water demand is perceived in fundamentally different ways, not only with respect to the factors which determine it, but also with regard to its manageability.

Types of water demand
It is useful to distinguish between offstream use of water for domestic, agricultural or industrial purposes and instream use of surface waters for navigation, recreation or hydroelectric power generation (Van der Leeden *et al.*, 1990). Offstream water use is characterized by the withdrawal of water from one place and its transportation to another place where it is needed. In the case of irrigation, most of the water eventually evaporates or filters through into the soil. Water withdrawn for domestic, livestock and industrial purposes also partly evaporates and filters through into the soil, but it is largely discharged into open water as wastewater. For instream use of surface water no diversions are needed, but some minimum water level or minimum flow is required. Apart from human uses such as navigation, recreation and hydroelectric power generation, aquatic and riverine ecosystems can also be regarded as instream 'water users'. The term water demand, however, is generally used (in this study as well) to refer exclusively to offstream water use.

The difference between total and 'consumptive' water use
Many authors distinguish between total and consumptive water use, which is helpful if one considers the effect of water use on the hydrological cycle. Whereas total water use refers to the total withdrawal of water, mostly from surface or ground water, consumptive water use refers to that part of the withdrawal which evaporates. The percentage of consumptive water use is largest in irrigation, where between 50 and 95 per cent of the total water used evaporates, partly through take-up by the crops but also partly due to losses during transport. A type of use with a very low rate of consumption is for example water use for cooling thermoelectric power plants, where only 0.5 to 3 per cent of the water withdrawal evaporates (Shiklomanov, 1997). The non-consumptive or recoverable part of total water use recharges the ground water (drainage of irrigation water) or is discharged into open water (domestic and industrial wastewater), and can therefore possibly be re-used. The picture is more complex when one realizes that the evaporated water will at least in part come back in the form of precipitation, so that it also can possibly be re-

used, but this does not alter the fact that it is useful to at least know the primary fate of water used.

Current water demand
According to Shiklomanov (1997), about 70 per cent of current global water demand comes from the agricultural sector (mainly for irrigation), 20 per cent from the industrial sector and 10 per cent from the domestic sector. Shiklomanov does not mention livestock water demand as a separate category but regards this as a small fraction of the total agricultural demand. Irrigation is of vital importance for the agricultural sector, because irrigated lands are much more productive than rain-fed lands. According to Postel (1992), 36 per cent of the global harvest comes from the 16 per cent of the world's cropland which is irrigated. In economic terms irrigated lands are probably even more important: Shiklomanov (1997) estimates that in dollars more than half of the global agricultural output comes from irrigated lands. A significant amount of irrigation water is used for the production of meat.[1] The main component of industrial water demand is the water demand for thermoelectric power generation, some 60-70 per cent. The remainder is for manufacturing. Domestic water demand is small compared to agricultural and industrial water demand but not less important, because there is a strong connection between public health and access to a proper water supply.

Projections of future water demand
A number of researchers has made forecasts of future water demand. Figure 2.2 shows the results of some major worldwide analyses carried out since the 1970s and also the growth of water demand in the past. Most authors are hesitant to look ahead more than a few decades, due to the great uncertainties. Although the earlier projections are clearly now obsolete, they are presented here nevertheless, to illustrate the extent to which projections can diverge and change over time. It is interesting to see that the projections done in the 1970s for the 1980s and 1990s exceed the actual development in these two decades.[2] Although both Falkenmark and

---

[1] In 1990, about 30 per cent of the global agricultural yield (measured in grain equivalents) was used for feed (FAO, 1996). This is a rough indication of the ratio between water demand for meat production and total irrigation demand, if one supposes that land for feed production is as frequently irrigated as land for food production and that most feed is used to grow animals which are ultimately used for meat production.

[2] The 'real' development during the past two decades is not precisely known, as shown by the divergence of the estimates in some recent publications, but it is clear that the actual development has been below even the lowest projections made in the 1970s.

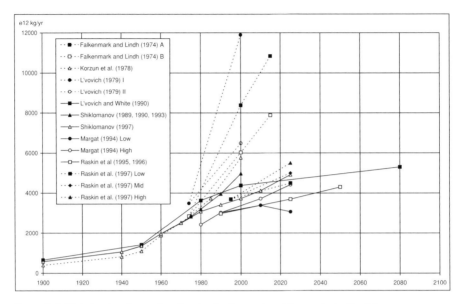

Figure 2.2. Global water demand: estimates for the past and forecasts for the future.

Lindh (1974) and L'vovich (1979) presented two scenarios which they considered to be low and high extremes, their scenario ranges have actually been infringed by the real development. Without detracting from the merits of the scenario approach (as opposed to presenting only one development path), these examples show the danger of this approach: once an uncertainty range has been established, it is suggested that - although the facts are not precisely known - the uncertainties at least are known. In all likelihood nothing is further from the truth in considering the long-term future of a system such as the earth. I do not regard this as an argument against long-term policy analysis, but rather as a serious warning not to underestimate the uncertainties. One should continually realize that the application of given uncertainty ranges to a few seemingly important parameters does not guarantee that the actual development will stay within the projected extremes.

As shown in Figure 2.2, Falkenmark and Lindh (1974) give two projections up to the year 2015. Even their low projection, which includes major reductions in industrial water use, is high if compared to other projections. The projection by Korzun *et al.* (1978) for the year 2000 is about the same as the low projection by Falkenmark and Lindh (1974). L'vovich (1979) gives a high and a low projection, the former possibly the highest projection of water demand ever made: nearly 12×10^{15} kg/yr. More than a decade later, L'vovich and White (1990) have a different view and suggest that with a conservative growth of population and industrial production the global water withdrawal in the year 2080 might be about 5.3×10^{15} kg/yr, but they add that this could be a very low estimate. In 1987 the Russian State Hydrological Institute (SHI) produced a projection of global water use up to the year 2000, which was later also published in English (Shiklomanov, 1989, 1990, 1993). With an irrigated area of 3.17×10^6 km^2, global water use in 2000 was estimated at 5.0×10^{15} kg/yr.[1] Recently, after concluding that the 1987 projection was a gross overestimate, SHI did a completely revised projection, now taking the year 2025 as the time horizon (Shiklomanov, 1997). They now estimate a total water use of 4.9×10^{15} kg/yr by 2025, a value which according to the earlier projection would already have been passed by 2000. The new projection is based on a global irrigated cropland area in 2025 of 3.29×10^6 km^2 and a reduction of water-use intensities in the period 1990-2025 of 10-25 per cent, depending on the region. Margat (1994) gives a low and a high projection of future global water demand, considering the period 1990-2025. The two projections are based on different assumptions with respect to population growth and efficiency improvements. Water demand is basically calculated as a function of population figures. Economic growth has not been distinguished as a separate determinant and the expansion of irrigated areas is supposed to be proportional to population growth. Margat's low projection of future water demand is the lowest in the literature, probably due to a low estimate of industrial water use. Margat's estimates of industrial water use do not include water requirements for cooling thermoelectric power plants and are based on an average efficiency improvement in developed countries of 40 per cent by 2010 and 50 per cent by 2025. In addition, industrial water use is supposed to be linked to population growth and not to economic growth or the growth of industrial production. As it is generally assumed that future global economic growth will be greater than future population growth, this leads to a conservative estimate of industrial water demand.

In 1995, the Stockholm Environment Institute projected global water use in 2025 at 3.7×10^{15} kg/yr and in 2050 at 4.3×10^{15} kg/yr, based on a so-called conventional development scenario, assuming middle-range estimates of population and economic growth and progressive improvements in water-use efficiencies (Raskin *et al.*, 1995, 1996). For domestic water use, they assume a reduced water-use intensity of 20 per cent in North America by 2050, no change in other OECD regions and growing intensities in the developing regions (as a function of economic growth).

---

[1] Korzun *et al.* (1978) and Shiklomanov (1989, 1990, 1993, 1997) include in their figures for total water use the evaporation from artificial surface reservoirs. For comparison with others, I present their figures excluding this component. I agree with Korzun and Shiklomanov that reservoir evaporation should be accounted for, but I do not consider this part of water use but as a change to the hydrological cycle (see Section 2.5).

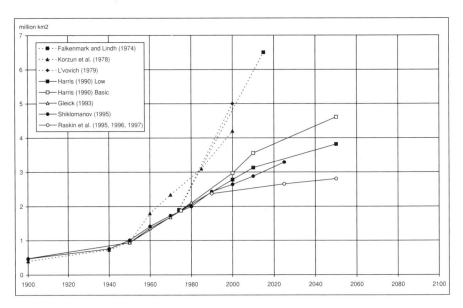

Figure 2.3. Global irrigated cropland area: estimates for the past and forecasts for the future.

For irrigation they assume a global area of 2.81×10^6 km^2 by 2050, with a 10 per cent reduction in water-use intensity. For water-use intensities in manufacturing, they assume an improvement by 2050 of 60 per cent for OECD regions. The improvement in non-OECD countries is assumed to be a function of economic growth, using a convergence algorithm to approach average OECD values. For thermoelectric power generation, they assume a 30 per cent reduction of water-use intensities by 2050. Not only the projected but also the current water use in Raskin *et al.* (1995, 1996) is much lower than for example in Shiklomanov (1997), due to the use of a different set of basic data. They used national data from WRI (1994), which refer to years between 1970 and 1992 but which were taken as initial data for 1990. The fact that the data are on average ten years behind is enough to explain the low estimates, and illustrates how a misinterpretation of figures can easily lead to a radically different assessment of the future situation. The Stockholm Environment Institute has recently published new water projections (Raskin *et al.*, 1997), still based on the conventional development scenario and using the same assumptions, but now distinguishing three cases of efficiency improvements and correcting the initial situation. The correction means that Raskin *et al.* (1997) assume a global water withdrawal in 1995 which in Raskin *et al.* (1995, 1996) was projected for the year 2025. The new low, medium and high projections for global water use by the year 2025 are 4.5, 5.0 and 5.5×10^{15} kg/yr respectively. These projections still only cover a very limited set of possible futures, as they are all based on the same assumptions about population growth and expansion of irrigated cropland.

The most recent assessments of future water demand, by L'vovich and White (1990), Margat (1994), Shiklomanov (1997) and Raskin *et al.* (1997), give lower projections than earlier assessments. According to Shiklomanov (1997), the earlier assessments greatly overestimated future water use because they assumed more progressive population scenarios, a continued expansion of irrigated land and a continued growth of industrial water use. They did not take account of limits to continued expansion of irrigated land, which it has been possible to observe since the 1980s (Postel, 1993). Because irrigation is a major component of human water demand, I give here an overview of projections of the global irrigated area by different authors (Figure 2.3). This figure also shows that the projections done in the 1970s were much more progressive than more recent projections.

A reflection on the concept of water demand
From the above it may appear that large differences between projections of future water demand are mainly due to different assumptions about population growth and agricultural, economic and technological development. However, one should realize that a confusing factor in discussing future water demand is also the existence of radically different perceptions of the *concept* of water demand. Three main schools of thought are discussed below, each having a somewhat stereotypic character. The literature in fact offers a whole range of opinions, which fall somewhere between these stereotypic views. The first common attitude towards water demand is to consider it as a given need which should be met. This approach takes population growth and agricultural and industrial developments as given processes which imply certain water requirements. In fact, all projections of future water demand discussed above have been based on this 'requirement approach'. One can recognize the same approach in a recent report from the World Resources Institute (WRI, 1996) which poses the question: will future water needs be met? This school of thought starts from the premise that there *is* a certain demand, which will or will not be met. An advantage of this approach is that projections of future water demand can be made in a relatively simple way, using growth scenarios for population, agriculture, industry and economics and assuming certain efficiency improvements. A major drawback is that many factors, such as social customs, individual preferences, the price mechanism and water policy, are ignored.

Another view of water demand is that water use is a necessity only if it is related to the fulfilment of 'basic needs', such as for example drinking (Gleick, 1996). Water demand above the minimum requirements is considered a luxury and largely subject to social and political desires. Political allocation priorities are supposed to strongly influence the extent of water use in the different sectors. Large irrigation schemes, for example, have been centrally planned by governments and are still being planned all over the world. A review of public water supply projects financed by the World Bank has shown that about 65 per cent of the average supply costs are covered by public funds; this figure is even larger for irrigation costs (Serageldin, 1995). At community level, water demand is thought to be largely a function of customs and human behaviour, which may change through improvement of environmental awareness or through for example the imposition of water taxes.

A third perception of water demand is the economic view, in which water demand is considered subject to the price charged (Bower *et al.*, 1984; Rogers, 1985). According to this view, water demand and supply achieve (or should achieve) equilibrium through the price mechanism. Increasing water scarcity leads to higher prices, which result in lower demand and incentives to develop more efficient technology. According to Anderson (1995) for example, a 10 per cent increase in price would decrease agricultural water use in California by 6.5 per cent and cut overall water use by 3.7 per cent in the seventeen western states of the USA. Critics consider the economic view of water demand to be an ideal of economists rather than a reflection of the actual world. Anderson (1995) recognizes this when he notes that despite the evidence that most water projects do not make economic sense, political pressure continues to allow these projects to proliferate because the interest groups which capture the benefits constitute a formidable political force.

Conclusion
A major drawback of many individual water demand studies is that water demand is interpreted according to one particular view. In many cases, basic assumptions and hypotheses are not even specified. As a result the outcome of individual water demand studies differs considerably. Additionally, many published water demand projections are the result of a calculation scheme rather than an explicit idea about the basic processes which determine water demand. The result is that it is difficult to *explain* the projections in terms of underlying mechanisms or to show where and how one can actively influence demand through public policy.

2.3 Water availability

Various researchers have addressed the question of how many people can live on earth. The central idea behind this question is that resources are limited and therefore so is the carrying capacity of the earth. In *An Essay on the Principle of*

Population, Malthus (1798) already pointed out the limited availability of land for food production. Much later, in the last few decades and especially as a consequence of the oil crises in the early 1970s, people became aware of the limited energy resources. Global awareness of limited water resources is of a much more recent date. It is true that in the naturally dry regions of the world people have already for thousands of years thought about water in terms of water scarcity, but this was an issue of how to cope with natural variability, rather than one connected to growth. Today it is sometimes argued that it is not land or energy, but water which is the critical resource for the growing human population on earth. Not being able to either endorse or contradict this argument at this point, it is interesting to take the idea seriously and pose some questions which then need to be answered. How much water is actually available for human use? How can a concept such as *potential water supply* be defined, as a measure of the maximum amount of water people can use annually? Is the potential water supply of the earth a natural constant or can human beings interfere? If this last is the case, which activities influence potential water supply and how can people actively enlarge it? And finally: how many people can live on earth from a water resources viewpoint, or at which stage does water become a critical issue for further development? These questions are analysed below, with the intention of reviewing current knowledge and understanding the difficulties which arise in formulating concrete answers.

Water availability on earth

As there are no data which indicate a significant one-way outflow of vapour from the earth's atmosphere, the total volume of water on earth can be considered constant and the different water stores can be seen as a closed system of communicating vessels.[1] Most of the water on earth is saline, stored in the oceans. For purposes such as drinking, washing, irrigation and manufacturing, however, people need *fresh* water. The stock of fresh water on earth comprises only 3.1 per cent of the total water stock, equal to about 43×10^{18} kg (Table 2.1). About three quarters of the fresh water is stored as ice and snow in the Antarctic and Arctic regions and is thus not available for human use. The available fresh water is stored as ground water (nearly one quarter of the total freshwater stock) and surface water (lakes and rivers, only 0.22 per cent of the total). If there were no replenishment of the fresh surface water and groundwater stores, mankind would - at the present level of water withdrawal - empty the fresh surface water store in about twenty-five years

---

[1] Walker (1977) estimates that during the last few billion years, only 0.1 per cent of the water on earth has been lost through photolysis of H_2O in the upper atmosphere.

Table 2.1. Global stocks of water.

	Total water stock (10^{15} kg)[1,2]	Freshwater stock (10^{15} kg)[1,2]	Average renewal time (yr)[3]
Atmospheric water	12.9	12.9	0.022
Salt surface water			
- Oceans	1338000	-	2500
- Saline lakes	85.4	-	not available
Ice sheets			
- Antarctica	29000	29000	~ 15000
- Greenland	2950	2950	~ 5000
Glaciers, ice caps and permanent snow cover	180	180	~ 260
Soil moisture	16.5	16.5	1
Water in freshwater wetlands	11.5	11.5	5
Biological water	1.12	1.12	a few hours
Ground water			
- Upper zone: active water exchange	3600	3600	270
- Middle zone	6200	6200	not available
- Lower zone (below sea level)	13600	730	not available
Ground ice in permafrost regions	300	300	10000
Fresh surface water			
- Rivers	2.12	2.12	0.044
- Freshwater lakes	91	91	2
Total	1394051	43095	

[1] Although hydrologists traditionally express water quantity in litres, cubic metres or cubic kilometres (units of volume), I use kilograms (unit of weight) here, because it is more accurate to consider the invariable weight of water than its variable volume. In converting water volumes given in the literature into weights, it has been assumed that 1 m^3 = 1000 kg, which is actually true only at 4 °C (for comparison: 1 m^3 of water weighs 998 kg at 20 °C).

[2] The data are from Korzun et al. (1978), except those for the ice sheets and glaciers which were taken from Warrick et al. (1996). The figure given for the soil moisture stock refers to the total soil moisture content of all ecosystem types, excluding wetlands. The stock of water in freshwater wetlands represents soil moisture and standing water.

[3] The data are from Korzun et al. (1978) with a few exceptions. Those for the ice sheets are from Warrick and Oerlemans (1990) and for glaciers a calculation has been made based on data from Warrick et al. (1996). For the upper groundwater zone of active water exchange, the renewal time has been calculated on the basis of the global groundwater runoff of 13.32×10^{15} kg/yr. For freshwater lakes, a calculation has been made based on a global river runoff of 44.7×10^{15} kg/yr (Korzun et al., 1978).

and cause a worldwide decline in groundwater tables of two metres on average in another twenty-five years.[1] This, however, is unrealistic, because the freshwater stores are replenished through precipitation. The size of freshwater stocks is therefore often considered an improper measure of the amount of water available for human purposes. Instead, it is argued, one should look at the *freshwater renewal rate*, equal to the precipitation on the continents minus the evaporation, which is equal to total continental runoff, supposing that the amount of water on land does not change. According to the four most thorough studies of the global water balance, the best estimate of total continental runoff is about $40\text{-}47\times10^{15}$ kg/yr (Table 2.2). The present global annual water withdrawal by mankind is about 8 to 9 per cent of this amount (Shiklomanov, 1997). Understanding the nature of runoff requires an understanding of the hydrological cycle, which will therefore now be briefly explained.

The primary driving force of the hydrological cycle is the evaporation of water from land and oceans. Because only negligible amounts of water escape from the atmosphere into space and water does not accumulate in the atmosphere, all evaporation becomes precipitation sooner or later. The average residence time of a water particle in the atmosphere is about eight days. The total precipitation into the oceans does not replenish the loss through evaporation, thus causing a net deficit. In contrast, precipitation onto the continents exceeds evaporation, resulting in a surplus. This surplus, the so-called net precipitation or freshwater renewal rate, flows underground and via rivers back to the oceans. Net precipitation on the continents depends on numerous factors and mechanisms. Total continental precipitation depends on continental evaporation and net advective moisture transport from the oceans. Important natural factors involved in continental evaporation are incoming radiation from the sun, land cover and soil properties. Some anthropogenic factors which influence evaporation from land are consumptive water use, artificial freshwater reservoirs and land use changes such as deforestation and drainage of wetlands. In the longer term, if the globe warms up as a result of increasing atmospheric concentrations of greenhouse gases, global evaporation and precipitation will increase, the so-called intensification of the global hydrological cycle. Because net precipitation is equal to precipitation minus evaporation, one cannot easily say what this intensified water recycling means for the water availability in particular regions, as this depends very much on the change in regional patterns of evaporation and precipitation. It is expected that some areas will

---

[1] The fresh surface water stock is estimated at 93×10^{15} kg (Korzun *et al.*, 1978). Shiklomanov (1997) estimates the global water withdrawal in 1990 at 3.42×10^{15} kg/yr. The global land area is about 133×10^{6} km^2 (excluding Greenland and Antarctica). Average porosity has been assumed at one third.

Table 2.2. Water balance per continent.

Water flow (10^{12} kg/yr)	Europe	Asia	Africa	North/C. America[1]	South America	Australia[2]	Antarctica	World[3]
Baumgartner and Reichel (1975)								
Precipitation	6587	30724	20743	15561	27965	7144	2376	111100
Evaporation	3761	18519	17334	9721	16926	4750	389	71400
Total runoff	2826	12205	3409	5840	11039	2394	1987	39700
L'vovich (1979)								
Precipitation	7165	32690	20780	13910	29355	6405	n.a	110305
Evaporation	4055	19500	16555	7950	18975	4440	n.a.	71475
Total runoff	3110	13190	4225	5960	10380	1965	n.a	38830
Korzun et al. (1977, 1978)								
Precipitation	8290	32200	22300	18300	28400	7080	2310	119000
Evaporation	5320	18100	17700	10100	16200	4570	0	72000
Total runoff	2970	14100	4600	8180	12200	2510	2310	47000
Shiklomanov (1997)								
Total runoff	2900	13508	4047	7770	12030	2400	n.a.	42655

n.a. = not available

[1] In L'vovich and Shiklomanov excluding Greenland; in L'vovich also excluding the Canadian Arctic islands.
[2] Including New Zealand, New Guinea and Tasmania.
[3] In L'vovich and Shiklomanov excluding Antarctica and Greenland; in L'vovich also excluding the Canadian Arctic islands.

become drier and others wetter, but the uncertainties are so great that a decisive assessment cannot be made. Despite the many studies carried out on this subject, there is no agreement yet in the outcomes of the various General Circulation Models regarding the regional hydrological changes which can be expected as a result of global warming.

The availability of fresh water does not only depend on the recharge rate, but also on that part of the recharge which forms *stable runoff*, i.e. runoff available throughout the year. The most important natural factor responsible for stable flow is the amount of the net precipitation which recharges the groundwater stores. Through its high residence time, this store has a strong stabilizing effect on the variable inflow. In many parts of the world, stable runoff has been increased through artificial reservoirs, but reduced through intensified land use (deforestation, urbanization, erosion) and river canalizations. It has been estimated that the original stable runoff in inhabited areas has increased by nearly 40 per cent due to artificial surface reservoirs (Postel *et al.*, 1996). There are no global estimates available on the extent to which intensified land use has reduced stable runoff, but this factor has certainly been of less significance. As to the future, however, the possibility of further growth in the number of reservoirs is being questioned and the effect of intensified land use may become more manifest. At the same time, large-scale artificial groundwater recharge is being put forward by some authors as an important method of enlarging the global amount of stable runoff in the future (e.g. L'vovich, 1979).

A final but not less important aspect of water availability is the quality of the water. Generally, water of a natural, pristine quality requires no or only minor treatment before use for any purpose. However, nowadays a large part of the easily available fresh water is polluted, most heavily in these areas where the need for clean water is greatest. Human activities have increased the concentrations of various substances in both surface and ground water. Total dissolved nitrogen and phosphorus concentrations in surface water, for example, have increased globally by a factor of two and locally - in Western Europe and North America - by factors of ten to fifty (Meybeck, 1982). This increase has manifested itself in many rivers and lakes as the phenomenon of eutrophication: rich in nutrients, excessive plant growth and deprivation of oxygen. The increased nutrient concentrations are due to domestic and industrial wastewater disposal, agricultural fertilizer use, increased erosion and increased atmospheric deposition. Other types of water quality deterioration result from the disposal of wastes containing heavy metals and organic micro-pollutants such as PCBs, and from the use of pesticides in agriculture. Water quality changes are part of a disturbance of total elements cycles. As surface water is usually more polluted than ground water, people increasingly rely on groundwater resources. The polluted surface water still being used requires expensive water purification. On a small scale, people have started to artificially recharge ground water with surface water, in order to improve the quality of the water and at the same time enlarge the stable flow.

To sum up, there is a wide variety of human factors which may reduce the availability of clean water, such as consumptive water use, evaporation from artificial water surfaces, untreated wastewater disposal, fertilizer use in agriculture, and erosion. However, there are also factors which increase the availability of clean water, such as artificial reservoirs, artificial groundwater recharge and wastewater treatment. Finally, there are factors which can either reduce or increase water availability, such as land use changes and climate change. In setting water policy

Table 2.3. Range of estimates of water stocks and flows in the literature.

	Lowest estimate	Highest estimate
Water stocks (10^{15} kg)		
Atmospheric water	10.5	14.0
Oceans	1320000	1370000
Saline lakes	85.4	125
Ice sheets and glaciers	16500	32170
Soil moisture	16.5	150
Biological water	1	50
Ground water	7000	330000
Rivers	1.02	2.12
Freshwater lakes	30	150
Total	1343644	1732661
Water flows (10^{15} kg/yr)		
Evaporation	446	577
- Ocean	383	505
- Land	63	73
Precipitation	446	577
- Ocean	320	458
- Land	99	119
Runoff to oceans	33.5	47
- Streams	27	45
- Ground feed	0	12
- Glacial ice	1.7	4.5

Source: Speidel and Agnew (1988), who used sources from the 1960s and 1970s, including Nace (1967, 1969), Baumgartner and Reichel (1975), L'vovich (1973b, 1977, 1979) and Korzun et al. (1977, 1978). The current position does not differ significantly from the one presented here, as the sources mentioned are still those most cited. However, the highest estimate for ice sheets and glaciers has been increased, in accordance with the most recent estimates by the IPCC (Warrick et al., 1996).

priorities, one would like to have an informed estimate of the relative importance of the different factors, both locally and globally. Well-founded messages along the lines of 'forget about the importance of land use changes but anticipate on climate change' or 'don't expect that the construction of artificial surface reservoirs can still bring much benefit, concentrate now on the possibilities of artificial groundwater recharge' would be extremely useful for policy makers who have to formulate future policy and allocate research budgets. But can such statements be made? There are two kinds of problem which have to be dealt with and which are discussed below: uncertainties about data and uncertainties about how to define the concept of potential water supply.

Uncertainties in the data

Even in respect of simple facts such as the size of water stocks or the intensity of water flows, the literature does not offer broad consensus on definite figures (Table 2.3). However, one can at least select the most reliable sources, which are based on comprehensive analysis of available hydrological data, and cull the sources which are second-hand or not well documented. On hydrological data for the globe as a whole, I consider Baumgartner and Reichel (1975), L'vovich (1977, 1979) and Korzun *et al.* (1977, 1978) the most up-to-date, reliable and primary sources currently available. The Russian State Hydrological Institute is completing a comprehensive monograph, *World Water Resources by the Beginning of the 21st Century*, in Russian, which could be regarded as an update of Korzun *et al.* (1977, 1978), but this work is not yet available in English.[1] At present, I can only use a draft report which contains some of the new results and which will be published by the WMO (Shiklomanov, 1997). Shiklomanov's recent estimate of global runoff (excluding Antarctica) is 4.3 per cent lower than the estimate by Korzun *et al.* (1977, 1978), but his regional estimates differ much more, due to the availability of more field data, especially for Africa, Asia and Latin America. The biggest difference is for Africa, where Shiklomanov (1997) finds a total runoff value which is 12 per cent lower than Korzun's estimate. Baumgartner and Reichel (1975) have a different method of estimating runoff than Korzun, Shiklomanov and L'vovich. They calculate runoff from the continents not by using river runoff measurements, but by taking the difference between precipitation and evaporation, thus depending on methods to estimate evaporation. For this reason, Shiklomanov (1997) considers Baumgartner and Reichel's method to be less accurate, especially in arid regions.

Uncertainties with regard to water quality are even greater than uncertainties over hydrological data. Water quality measurements are relatively rare, so an overall assessment of the water quality in most river basins in the world is difficult. Also, estimates of emissions are subject to many uncertainties and theoretical models which describe the fate of emissions are not unambiguous. Should people expect that the accumulation of pollutants in soils and sediments over the past decades will eventually lead to exploding chemical time bombs at some point in the future, as argued by Stigliani (1991)? Or is this view too pessimistic and are the quality of the environment and emissions related in a more linear way, so that water quality will improve immediately when emissions are lowered?

---

[1] Personal communication from Professor Igor Shiklomanov of the State Hydrological Institute, St. Petersburg, Russia.

The concept of potential water supply

The fact that there are so many factors which somehow influence the amount of water available for human use is probably the reason why there are different approaches to defining the concept of potential water supply. The most common approach is to take the total annual runoff in a river basin as a measure of the maximum possible or potential water supply in that basin, on the basis that fresh water is a renewable resource and the renewal rate is therefore a measure of water availability. Many authors divide the total annual runoff in an area by the number of people in that area, to obtain a measure of the available water resources per capita (L'vovich, 1979; WRI, 1992, 1994, 1996; Kulshreshtha, 1993; Shiklomanov, 1993, 1997). An advantage of this 'total runoff approach' is that potential water supply is defined in an unambiguous way, leaving no room for dissent other than over the runoff data (see above). A major drawback is that this definition of potential supply does not account for losses due to flood-runoff, runoff in remote areas and pollution, thus giving a profound *overestimate*. Another criticism is that one only looks at the possible supply of *fresh* water, ignoring the possibility of desalinating sea water. According to this point of view, the approach yields a *conservative* measure of potential water supply.

Some authors regard the total runoff in a river basin as the upper limit to potential water supply and propose reductions for losses due to flooding and runoff in uninhabited areas (Ambroggi, 1977, 1980; Postel et al., 1996). This approach results in a much lower assessment of potential water supply than if one were to consider total runoff (Figure 2.4). On a global scale, Ambroggi (1977, 1980) for example thus arrives at a figure of 23 per cent of the total runoff and Postel et al. (1996) at a figure of 31 per cent. A further reduction should be made to account for pollution and to guarantee a certain minimum runoff for maintaining aquatic and riverine ecosystems. Both Ambroggi and Postel allocate part of the stable runoff to diluting wastewater disposals, leaving only the remaining part for other purposes. A final factor of importance in potential water supply is climatic variability. It would be better to use a dry year for calculations rather than an average one, to ensure that the measure of potential water supply also applies in dry years. The 'reduced runoff approach' may have the advantage of carrying carefully balanced information on potential water supply, but the definitions used may give rise to many different interpretations and calculations.

In the views discussed above, water is perceived as a renewable resource. Water can however also be regarded as a non-renewable resource, especially in cases where humans have to rely on ground water (Rogers, 1985). A common example is when the water in rivers and lakes is too heavily polluted for human use, so that people are

Water in the 21st century: questions and uncertainties 41

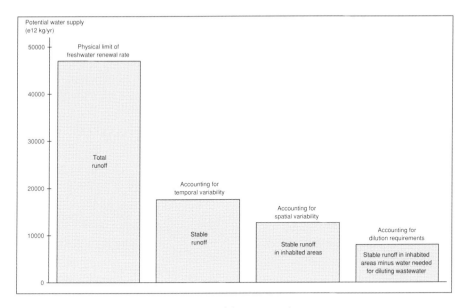

Figure 2.4. Some perceptions of global potential water supply.

entirely dependent on ground water. This leads to falling groundwater tables, most pronounced if withdrawals exceed natural replenishment, and depletion of deeper aquifers. Additional pressures on groundwater availability are saltwater intrusion in coastal areas and other types of groundwater contamination. Two key words in this approach are therefore pollution and depletion, both processes which reduce the stock of clean, fresh water. To calculate potential water supply one now has to use the size of the available freshwater stocks rather than flows. From this perception, the remaining amount of clean fresh water is clearly a better measure of the effect of pollution and intensive water withdrawals, and thus for the capacity for more withdrawals, than the ratio between water use and total (or stable) runoff. For the availability of fresh surface water, a proper measure could be the size of the stock which meets certain quality standards. This is an important measure, as rivers and freshwater lakes form only 0.22 per cent of the total freshwater stock, while providing about 71 per cent of the world water supply (Kulshreshtha, 1993). For the availability of fresh ground water it would make little sense to consider the entire stock, as most ground water is unexploitable. More useful information is for instance the depth of the groundwater table and the type of aquifer. Declining groundwater tables can be regarded as a signal that water withdrawal exceeds water availability.

Although the concepts of potential water supply mentioned above differ considerably, they agree in their recognition of *some form of limitation*. However, as Falkenmark (1989) observes, it is by no means generally accepted that a limit to potential water supply actually exists. Both engineers and economists exhibit a certain amount of opposition to the so-called 'water barrier' concept. Their technological optimism leads them to believe that problems of scarcity will be solved through new technologies which can enlarge supply or make water use more efficient. A confirmation of this view is found in the growing capacity of desalination plants in many water-poor regions. In Saudi Arabia, for example, desalination of salt or brackish water already accounts for about 20 per cent of the total water supply (Gleick, 1993a). Because the oceans can be regarded as both the primary source and the ultimate sink of all water on earth, the possibility of obtaining our water from the sea implies, in principle, that there is no limitation on

---

The effect of consumptive water use on runoff and potential water supply

Consumptive water use can have two kinds of effect on runoff, depending on regional climatic circumstances. In one extreme case, the anthropogenically evaporated water leaves the region and is 'lost' to this region. In the other extreme case, all evaporated water stays within the region and falls down again as precipitation. In the latter case, the local hydrological cycle is intensified as a result of human water use. In both cases, it is an error to take *actual* total or stable runoff as a measure of potential water supply, because one thus excludes water already used and evaporated. If water withdrawals are relatively small, this effect may be ignored, but if people withdraw and evaporate a significant amount of the net precipitation, the effect will be significant. In the Colorado basin, for example, the remaining flow to the ocean has become nil, due to the intensive use of water in the basin (Schwarz *et al.*, 1990). However, the fact that *actual* runoff has become zero due to consumptive water use does not mean that the potential water supply in the basin is zero, only that the potential supply has been fully consumed. To get a more accurate picture of potential water supply, I therefore propose to take the *natural* total or stable runoff, understood as the total or stable runoff which would be measured if there was no consumptive water use. This natural total or stable runoff could be defined as the actual total or stable runoff plus the volume of consumptive water use. If one uses this corrected definition of potential water supply, how does it affect the two extreme cases mentioned above if consumptive water use increases? In the first case, where all consumptive water use is really lost, runoff will decrease by as much as consumptive water use increases and the potential water supply will remain constant. In the second case however, runoff will remain constant and potential supply will increase. This is an interesting phenomenon which has never been studied at global level, but which will become more important if global water use continues to increase.

potential water supply, apart from a possible restriction from an energy perspective. Another possibility is water re-use after treatment (Dean and Lund, 1981) for either the same or a different purpose, thus creating a large new source of fresh water; only actual losses would have to be made up for from outside the recycling system. Other unconventional technologies to extend our resource base, attracting attention in recent decades but still in an experimental and conceptual stage and often regarded as mere fantasies, are weather modification through cloud seeding and towing icebergs to wherever water is needed.

The global carrying capacity from a water resources viewpoint
The uncertainties about water availability data and the different perceptions of potential water supply are the reasons why not much agreement exists on the number of people who can live on earth if water is the limiting factor. Most studies on global limitations to population growth analyse the limitations to *food* and not to water production. Although water is an important input in food production, it is often not seen as a critical factor. Heilig (1994) for example argues that irrigation can become much more efficient and points to the possibility of desalination for water-poor but energy-rich nations in North Africa and West Asia. Furthermore, according to Heilig, it is hard to imagine that mankind can be approaching global limits of freshwater withdrawal when about 92 per cent of the known reserves are still untouched. On a closer study, it appears that Heilig takes the total runoff as a measure of potential water supply, which is clearly on the optimistic side (see Figure 2.4). The same measure is used by for example Luyten (1995), who calculates that the earth can carry 21 to 177 billion people if there is no limitation on water supply and 20 to 152 billion people if water limitations are taken into account, precise numbers further depending on assumptions about fertilizer use and human diet. In his comprehensive study of the global water resources, L'vovich (1979) also takes the total runoff as the measure of potential water supply and suggests that freshwater resources are sufficient for a population in the order of 60 to 100 billion people. But he adds that even if all freshwater resources are exhausted this will not put a limit on further growth, referring to other possibilities such as desalination of sea water. According to L'vovich, "the views of certain authors who believe that population growth and economic development will be limited by the shortage of fresh water can serve as the most vivid manifestation of pessimism". However, taking the situation in Israel as an example, Falkenmark and Lindh (1976) argue that 70 per cent of total runoff is the real upper limit to potential water supply. They calculate that our planet can support a maximum of 25 billion people. La Rivière (1989) takes the stable runoff in inhabited areas as a measure of potential supply and considers this enough

to sustain 20 billion people. Saeijs and Van Berkel (1995) give an even lower estimate of 4.5 to 9 billion people. These estimates are all based on simple assumptions about the minimum water requirement per capita. Falkenmark and Lindh (1976) for example use in their calculation a per capita requirement of 1.1×10^6 kg/yr, while La Rivière (1989) uses a much lower value of 0.45×10^6 kg/yr.[1] But water demand is in fact a complex and controversial concept too, as has been discussed in the previous section.

Conclusion
To return to the question of how many people can live on earth from the viewpoint of water resources, one has to admit that so many disputable assumptions have to be made that a satisfactory answer will never be obtained. However, as Heilig (1994) observes, whatever physical limitations one might believe in, long before the human species will reach these it will be confronted with other kinds of limitations: ecological, technological, economic and social constraints.[2] In addition, reaching upper limits might be possible, but is likely to be highly undesirable. For these reasons, it is probably more useful to study the actual dynamics of water demand and supply and analyse how different conceptions of the upper limits can influence these dynamics, than to focus research on the upper limits as such and explore futures that are possible merely from a theoretical point of view.

2.4 Water scarcity

A recent study by the Stockholm Environment Institute warns that the number of people living under water stress conditions will increase from 1.9 billion in 1995 to 5.3 billion in 2025. The percentage of the total world population in water stress conditions would rise from about 34 to 63 per cent.[3] According to a French study, the number of people in the world living under water shortage conditions will

---

[1] The per capita water supply in 1990 was about 0.66×10^6 kg/yr (Shiklomanov, 1997).

[2] A few researchers who dispute that growth is limited by the availability of natural resources, such as Simon (1981) and Kahn (1982), even question the existence of these types of restriction and assert that the only restriction is the human brain, an argument in favour of population growth (more people = more brain = less restriction).

[3] A country is here supposed to experience water stress if water demand exceeds 20 per cent of total runoff. The figures for 2025 are mid-range estimates for a conventional development scenario (Raskin et al., 1997).

increase from 0.35 billion in 1990 to between 0.9 and 2.8 billion in 2025.[1] It is expected that affected populations will be concentrated in China, Central Asia, the Indian subcontinent, the Middle East and North Africa, all regions where the population is expected to grow rapidly. Although the figures differ, the message is the same: water will become scarcer, due to the increase in demand. However, the two examples also show that estimates can differ considerably. Apparently, water scarcity can be measured in quite different ways. Moreover, and this probably leads to more difficulties, people appear to have different opinions about what causes water scarcity and which solution should be chosen.

Indicators of water scarcity
Most of the water scarcity indicators which have been proposed are based on two basic ingredients: a measure of water demand and a measure of water availability (potential supply). As a measure of water demand, most authors take the total annual demand or the average annual demand per capita. Similarly, potential supply is generally expressed as a total annual volume or an annual volume per capita. Useful reviews of scarcity indicators are given by for example Kulshreshtha (1993), Gleick (1993b) and Raskin *et al.* (1995, 1996, 1997). One of the most common indicators of water scarcity is the ratio between water demand in a certain area and total runoff in that area, called variously the water utilization level (Falkenmark, 1989; Falkenmark *et al.*, 1989), the use-availability ratio (Kulshreshtha, 1993) or the use-to-resource ratio (Raskin *et al.*, 1995, 1996, 1997). Through analysis of the water situation in several European countries, the Polish scientist Balcerski found that the prospects of solving questions of water supply are favourable as long as water demand is less than 10 per cent of total runoff (Falkenmark and Lindh, 1976). If water demand rises to between 10 and 20 per cent, water supply can be said to constitute a problem, which can only be solved through considerable investment and comprehensive planning. According to Balcerski, water supply will become an absolute limiting factor of economic development if water demand exceeds 20 per cent of total runoff. Other authors use different criteria to classify the socio-economic risk of water scarcity. According to Gleick (1993b), serious water shortages may arise if demand exceeds 33 per cent of total runoff. Raskin *et al.* (1995, 1996) take 25 per cent to indicate 'water stress', and Raskin *et al.* (1997) use a value of 20 per cent. Kulshreshtha (1993) makes a distinction between regions with low and regions with high water demand per capita. For the latter, Kulshreshtha

---

[1] A country is supposed to be exposed to water shortages if consumptive water use exceeds 25 per cent of total runoff. The figures for 2025 correspond to a low and a high estimate of future water demand (Margat, 1994).

assumes that there is a water surplus if total demand is below 40 per cent of total runoff, marginal vulnerability between 40 and 60 per cent, water stress between 60 and 80 per cent and absolute water scarcity if total demand exceeds 80 per cent. In regions with low demand per capita, where demand per capita is likely to rise in the future, the demarcations between the different risk categories lie at lower percentages.

Another common indicator of water scarcity is population of an area divided by total runoff in that area, called the water competition level (Falkenmark, 1986, 1989; Falkenmark *et al.*, 1989) or water dependency (Kulshreshtha, 1993). Many authors take the inverse ratio, thus getting a measure of the per capita water availability (Gleick, 1993b). Falkenmark proposes to consider regions with a water competition level below 600 people per 1 million m^3/yr as regions without water stress, where only general water management problems occur. Between 600 and 1000 people for this annual amount of water would indicate stress, 1000 to 2000 people chronic water scarcity and more than 2000 people absolute water scarcity, beyond the 'water barrier' of manageable capability. An obvious shortcoming of this scarcity indicator is that it neglects regional differences in water-use intensities. However, if combined with information on demand per capita, this indicator can be as useful as the use-to-resource ratio.

Although all authors mention complicating factors such as temporal and spatial variability, none of the indicators proposed includes these factors in an explicit way. However, these factors are certainly important, because most of the present water scarcity problems would not occur if water had been distributed more homogeneously over time and space. This problem has largely been solved by implicitly taking these factors into account in defining the criteria for the various risk categories, which has been done on the basis of empirical data. Nevertheless, the criteria should be applied with care, because each particular area has its own peculiarities (e.g. large inter-annual differences). Furthermore, application of the indicators at a relatively small or a large spatial level (below or beyond river-basin level), might require adjustments of the criteria between the risk categories. Another complicating factor - one which many authors do not envisage - is the variable nature of demand. Whereas irrigation water largely evaporates, domestic and industrial water often largely returns to the river from which it was taken. Total water demand, defined as the water withdrawal required, does therefore not represent the pressure on the water system very accurately. As Margat (1994) shows, the ratio between *consumptive* water use and total runoff can be a more useful water scarcity indicator. Particularly in industrialized countries, where industrial water demand constitutes a considerable part of total water demand, the use-to-resource ratio may exaggerate the extent of water scarcity.

A water scarcity indicator with quite a different character is the cost of water supply. Whereas the above-mentioned indicators might be regarded as inappropriate under various specific circumstances and require modification in each particular case, economists argue that the cost of water supply is an indicator which responds to all factors of any importance. According to sceptics, however, the fact that water is a public commodity makes it difficult to determine its actual price. An estimate of the value of water is too easily limited to estimating the costs of transporting the water from where it is found to where it is needed, which is clearly insufficient. In both a theoretical and a practical sense, it is indeed not easy to determine the economic value of water (Gibbons, 1986).

Three views on the nature of water scarcity

Apart from the question of how water scarcity can be measured, there is also the question of what causes it. It seems that the concept of water scarcity can be viewed in fundamentally different ways. Closely connected with the perception of what scarcity *is*, is the perception of how water scarcity can or should be solved. Below, I discuss three extreme points of view. In reality one can of course find all kinds of mixtures of these stereotypes. In the first place, water demand is often regarded as something 'given' and water supply as something which has to meet demand. Water demand is not considered a policy issue, but a fact emanating from population growth and agricultural and economic developments (Shaw, 1994). The actual issue is understood to be the provision of enough water of sufficient quality for the relevant sectors of society, leaving enough to fulfil ecological requirements. Water scarcity is thus a *supply problem*. In this view, water policy should aim at proper management of the physical water system, an approach found all over the world. Attention is given principally to the analysis of available water quantities and qualities and the construction of a water supply infrastructure. If relevant, studies should include possible effects of erosion, consumptive water use and climate change. Water pollution is described in terms of the violation of water quality standards. Wastewater should be treated to bring it up to the required standards. It is perhaps not surprising that this line of thought is often found among natural scientists (hydrologists, climatologists) and engineers.

Another point of view is that potential water supply is limited and that demand cannot continue to increase. Water scarcity is thus a *demand problem*. The augmentative demand is seen as the actual driving force behind growing water scarcity. Underlying forces are population growth and economic development

(Falkenmark *et al.*, 1987). In nearly all parts of the world, the water utilization level increases, which is a signal for action in regions which have reached critical levels. Water quality deterioration is a further consequence of the increasing pressure on the water system and this problem has to be solved at its roots. Wastewater *treatment* is not enough, wastewater *production* should be reduced. Solutions for water scarcity should somehow manage demand and thus human behaviour. As La Rivière (1989) states, a water management project should lean toward increasing the efficiency of water use rather than toward increasing the supply of water. The only exception might be primary needs such as drinking. In this view, minimum water requirements (small but important) should be fully met, while remaining demands (large and of secondary importance) should be reduced. A reduction in water use could be achieved by for example increasing 'water literacy' among the population and charging the full costs of water to the user with - if necessary - an additional amount

---

Dams: a solution to water scarcity?

Building dams has become the pre-eminent engineering solution to water scarcity. The benefit of artificial reservoirs is their stabilizing influence on a variable water inflow. This can greatly increase the stable runoff per year, which is often useful for a further expansion of water supply. The primary purpose of dams, however, is in most cases hydroelectric power generation, which also requires a stable flow. Secondary purposes are often downstream flood control and improved navigation. A beneficial side effect might be the recreational value of the reservoir. Nevertheless, most of the plans for new dams today are heavily criticized. According to the opponents, the benefits of dams far from outweigh the disadvantages: loss of land and valuable ecosystems, forced displacement of people (in some cases hundreds of thousands, up to one million), and - after completion - evaporation losses, sedimentation and water quality problems (Pearce, 1992). Large dams are often said to be prestige projects of governments.

 The different points of view of dam advocates and opponents do not just reflect different results of a simple weighing of pros and cons, but can be brought back to more fundamental differences in the perception of scarcity and how people should interact with their environment. If water scarcity is perceived as a supply problem (see main text), it can easily be understood that dams are considered an important solution to water scarcity. Negative aspects may be serious, but *have to be overcome*. If water scarcity is perceived as a demand problem, however, dams cannot be regarded as a fundamental solution, and it is therefore wise to reject dam construction if there are negative side effects. From the economic point of view, plans for dams should be evaluated on the basis of an economic cost-benefit analysis, which might give different results in each case. In this study I will take all points of view seriously, and analyse how the different views can lead to dissimilar futures, each of which might be desirable from the respective point of view.

in the form of a tax. The price of water for *primary* needs should also reflect the ability of people to pay (Young *et al.*, 1994).

A third view of water scarcity is the economic one. Simon (1980) states that the only meaningful measure of scarcity in peacetime is the cost of the asset in question. A substantial group of scientists applies this view to water as well (e.g. Anderson, 1995). This idea has also caught on in politics, because one of the 'guiding principles for action', embodied in the so-called Dublin Statement (ICWE, 1992b) is that "water has an economic value in all its competing uses and should be recognized as an economic good". In this view the cost of water is the correct indicator for water scarcity, not indicators which contain physical information about water availability and demand. If the price mechanism functions well, factors such as droughts, pollution and increasing demand will automatically and properly be accounted for in the water costs. Solutions to water scarcity are sought through introducing water markets, charging true costs to water users and privatizing water supply companies.

Conclusion

From what is stated above, it follows that trade-offs between water demand and water supply policy options in a country or a river basin do not just depend on hydrological and socio-economic circumstances. Apart from the 'facts', a subjective element plays a role: which view on water scarcity dominates. Following the rationale of those who believe in limits to water supply and who feel that a global water crisis is close, one can understand their arguments for a radical change in demand patterns. However, following the rationale of people who do not really believe in a radical change in demand, one can understand their conviction that new water resources *must* be developed. Adopting the idea of markets as a regulating mechanism, one can understand why some believe that all water problems can be solved if water is treated as an economic good. In other words, it is easier to understand the different opinions in the debate on water and development if the basic attitudes and beliefs of people are taken into consideration.

2.5 Alteration of the hydrological cycle

On a global scale, not much is known about the hydrological changes which have taken or will take place due to human interference. The main reason for a lack of knowledge about changes in the past is the lack of reliable data. Another reason is the difficulty of distinguishing between natural climatic variations and long-term trends. Long-term records of hydrological data exist for only a few areas in the world. Notwithstanding the data problem for the globe as a whole, there is enough evidence of local changes to expect that the global hydrological cycle has also changed and will continue to change. Most of today's research on global hydrological change comes under the heading of *climate research*, well represented these days because of the increased political interest in the phenomenon of climate change. However, this type of research is obviously biased towards the exploration of hydrological changes through climate change, while this is actually only one of the agents which might alter the global water balance. Important driving forces at global level might also be consumptive water use, the damming of river flows, canalizations, and land use changes such as deforestation, drainage of wetlands and urbanization. An interesting question is which phenomena will be most important in the next century, and should therefore deserve major attention in future research and policy.

A classification of pressures
Following Shiklomanov (1990, 1993, 1997), the various human pressures on the hydrological cycle can be divided into four groups: direct diversion of water flows, transformation of the stream network, transformation of drainage basin characteristics, and activity altering regional or global climate. The first group of pressures, direct diversion of water flows, consists of water withdrawals for purposes such as domestic, agricultural and industrial water supply. If the amount of water people withdraw from surface and ground water fully returned to where it was taken, the natural water cycle would only be disturbed to a minimum extent. However, this is generally not the case, which has led to a serious alteration of the water balance in several parts of the world. One of the most famous examples is the Aral Sea: in the period 1960-1990 water withdrawals for irrigation from the rivers Amu and Syr in Central Asia caused a shrinking of this inland sea to about 60 per cent of its original area and one third of its previous volume (Gleick, 1993a). Another example is the Colorado basin, where intensive water use prevents even a drop of water reaching the Pacific Ocean (Schwarz *et al.*, 1990).

With respect to the second group of pressures, transformation of the stream network, Shiklomanov suggests that considering the effects of large reservoirs would be enough. The hydrological effect of reservoirs is twofold: they stabilize runoff and increase evaporation. On a global scale, Shiklomanov (1997) estimates that the additional evaporation as a result of artificial reservoirs at the end of the 20th century will be about 210×10^{12} kg/yr, which constitutes about 0.3 per cent of the total land evaporation. According to L'vovich and White (1990), the stable runoff as a consequence of artificial reservoirs has increased in the period 1900-1980 by about 2.2-3.5×10^{15} kg/yr (Table 2.4). A type of stream network transformation which might be ignored at global level, according to Shiklomanov, is river canalization. Although canalization tends to increase peak flows, because floods are no longer temporarily (partly) stored on the flood plains of a river, the effect of canalization on low flows is probably small. As far as I am aware, there is no author who regards river flow stabilization from flood plains as a significant contributor to the stable runoff of a river (see also e.g. L'vovich, 1979).

The third group of pressures, transformation of drainage basin characteristics, includes agro-technical measures, drainage of wetlands, de- or reforestation and urbanization. According to Shiklomanov, one could ignore the effects of this group of pressures at global level. However, is there indeed enough justification for such disregard? There has not as yet been a comprehensive study showing that the effect of soil and land use changes on the global hydrological cycle is significant, but neither has there been a study proving that the effect is insignificant (L'vovich and White, 1990). Various *local* studies show that land use changes can have considerable effects on evaporation, groundwater recharge and total runoff. Savenije (1995) for instance finds that in the Sahel, where recycling of moisture through evaporation appears to be responsible for 90 per cent of the rainfall, an important feedback mechanism exists between land use and climate. Furthermore, as rainfall in Mali and Burkina Faso, for instance, appears to be influenced by land use in Guinea, the Ivory Coast and Ghana, there are also interesting international aspects to this issue. As well as the possible effect of land use change on evaporation, precipitation and total runoff, there is its possible effect on stable runoff. Changes such as deforestation and urbanization generally enlarge direct surface runoff and thus reduce groundwater recharge and stable runoff. Berry (1990) for example shows how, after urbanization, runoff occurs more rapidly and with a greater peak flow, and flood frequencies increase.

The fourth group of pressures on the hydrological cycle is human activity altering regional or global climate. An example are activities which enhance the greenhouse effect, such as burning fossil fuels. One of the most important consequences of climate change may be the alteration of the terrestrial water balance, including changes in soil moisture, groundwater recharge and river runoff (Rind, 1988; Miller and Russell, 1992). With a doubled concentration of atmospheric CO_2, General Circulation Models estimate that the global mean surface air temperature will increase by 1.7 to 5.3 ^{0}C and global evaporation and precipitation by 2.5 to 10 per cent (Gates et al., 1992).[1] A serious problem, however, is that mankind's ability to predict hydrological changes is even smaller than its ability to predict temperature changes, so figures should be treated with caution, in particular estimates of *regional* hydrological changes. Another reason for increased rates of evaporation and precipitation is consumptive water use. Doubling the present volume of consumptive water use will increase evaporation from land by about 3 per cent. If the evaporated water does not get lost to the oceanic atmosphere, precipitation on land will increase at the same rate. However, if spread over the entire earth, global precipitation will increase by 0.4 per cent only and the continents will, as a result of this, be confronted with a reduced runoff of about 2.5 per cent.[2]

Changes: an interplay of forces
The various pressures which alter the hydrological cycle sometimes reinforce each other, but at other times counteract and neutralize. Changes generally come into being through an interplay of different forces. Earlier in this chapter I discussed how a wide variety of factors acts upon the availability of fresh, clean water (Section 2.3). Here I will discuss in some detail how different factors increase and decrease stable runoff and how groundwater tables may rise or decline. Sea-level change will be discussed in the next section. Stable runoff, defined as that part of the total runoff which is available throughout the year, is often assumed to be equal to groundwater runoff. In a natural situation, groundwater runoff over a year is equal to the groundwater recharge, i.e. the percolating through of net precipitation. Net precipitation, however, may increase or decrease as a result of land use changes or climate change (Figure 2.5). Land cover changes also influence groundwater recharge in another way: processes such as deforestation, drainage of wetlands and urbanization generally result in an increase of the ratio between direct runoff and net precipitation, so that less water will percolate through and recharge the groundwater stock. Soil erosion is a further process which may lead to reduced groundwater recharge. On the other hand, drainage of irrigation water and artificial groundwater recharge are processes which lead to increased recharge. Groundwater runoff is not necessarily equal to groundwater recharge, because groundwater withdrawals can reduce runoff significantly. The groundwater store is not the only stabilizing factor in runoff: natural lakes also stabilize runoff and thus contribute to stable runoff.

---

[1] Today, these estimates are slightly adjusted downwards, due to new insights with respect to negative radiative forcing by aerosols.

[2] For these calculations, current global consumptive water use has been assumed at 2.1×10^{15} kg/yr (Shiklomanov, 1997), land evaporation at 72×10^{15} kg/yr and global evaporation at 577×10^{15} kg/yr (Korzun et al., 1978).

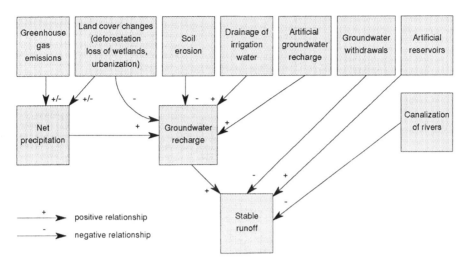

Figure 2.5. Human activities affecting stable runoff.

However, at the global level, this last factor is relatively small. According to L'vovich (1973a, 1973b), natural lakes contribute only 2 per cent to the total stable runoff in the world. Artificial reservoirs, however, are specifically constructed and managed to increase stable runoff and therefore contribute to stable runoff to a much greater extent. L'vovich and White (1990) estimate that, by 1980, the contribution of artificial reservoirs to stable runoff had grown to 20-30 per cent of the natural stable runoff (Table 2.4).

Most of the factors which influence stable runoff also affect groundwater tables. Depending on the local situation, groundwater tables will either rise or fall. Tolba and El-Kholy (1992) give some examples of places where irrigation has led to a significant rise in groundwater tables. In two extreme cases, in Egypt and Morocco, groundwater tables have even risen by two to three metres per year. More often, however, falling water tables are reported, particularly in areas where people depend on ground water for their water supply. According to Shiklomanov (1997), groundwater tables have declined tens of metres, and sometimes even more than a hundred metres, in a number of locations in Europe, North America, China and several Arab countries. Most of the problems currently occur in large cities and in mining areas. As a result of declining groundwater tables, many locations have to cope with significant land subsidence. The proposed solution is to artificially recharge ground water with fresh surface water, an additional advantage being that water is thereby naturally purified (L'vovich, 1979). However, in many regions people rely on ground water because surface water is not available, and therefore the idea of artificial groundwater recharge raises the question of where to get the required water. At present there are several artificial recharge projects around the world, but the total recharge is still relatively small (Van der Leeden, 1990). Information about the effect of land use or climate change on groundwater tables is very scarce.

In the past few decades groundwater tables have probably changed most strongly as a result of groundwater withdrawals and irrigation, especially in places where large changes have been observed. However, land reclamation (deforestation, drainage of wetlands) may have had effects on groundwater tables too, probably fairly small on average but widespread. As a global average, it is generally assumed that, during the past one hundred years, there has been some net loss of fresh ground water, which has resulted in a sea-level rise of 'a few centimetres' (Gornitz et al., 1982). However, there have been few comprehensive studies of this phenomenon. Sahagian et al. (1994) have analysed groundwater loss in some specific cases in North America, North Africa, Arabia and the Aral and Caspian basins and they deduce a conservative estimate of an 8.6 mm sea-level rise since 1900. Korzun et al. (1978) arrive at a larger estimate of 0.8 mm/yr for the period 1900-1964.

Changes in the past

The most comprehensive paper on historical changes to the terrestrial hydrological cycle by humans is that written by L'vovich and White (1990), which considers changes during the period 1680-1980. Their quantitative estimates of the main alterations in this period have been collected in Table 2.4. Evaporation from land has been assumed to increase by 3.5 per cent, mainly due to the increase in consumptive water use. According to L'vovich and White, this has resulted in a decrease of total runoff of 6 per cent, supposing that land precipitation has not been subject to change. Apparently, they assume that increased evaporation from land is completely lost through advective moisture transport to the oceanic atmosphere, where it precipitates into the ocean. This is a disputable assumption, because part of the anthropogenic evaporation is probably recirculated on land, just like natural evaporation, only 30-35 per cent of which is lost through advective moisture transport to the oceanic atmosphere (estimated from Korzun et al., 1978). A major problem is that available hydrological data do not provide a sufficient basis for giving an empirical answer to the question of how the global water balance has changed. To detect trends in the water balance of particular regions, precipitation and runoff measurements can often give an answer, provided that long-term historical records are available. In these cases, evaporation is regarded as the closing entry in the regional water balance. For detecting trends in the global water balance, however, the lack of data takes on such importance that theoretical considerations about evaporation changes may become the most definite component in the water balance, with precipitation and runoff being the less certain factors.[1] According to an analysis of runoff data by Shiklomanov (1997), total runoff on earth in the period 1920-1990 varied between about 40×10^{15} and 45×10^{15} kg/yr, i.e. a variation of ±6 per cent around the middle value. With such a high natural inter-annual variability, a long-term trend of decreasing runoff in the order of a few per cent cannot be detected easily. Indeed, Shiklomanov's data do not indicate any decisive trend. So there can be no conclusive empirical answer to the question of the extent to which average precipitation and total runoff have changed during the past one hundred years. Taking the estimated increase in continental evaporation as a starting point, it is most probable that precipitation has increased and runoff decreased by a few per cent.

A few decades ago, rivers were considered 'the product of climate'. Today, it is common knowledge that soil is an intermediary and that practically all hydrological phenomena on land are considerably influenced by soil and vegetation conditions (L'vovich and White, 1990). For this reason, it might be expected that historical land use changes such as deforestation, drainage of wetlands and urbanization significantly altered regional hydrological circumstances. L'vovich and White (1990) estimate that worldwide urbanization has had the effect of lowering both global evaporation and groundwater runoff. The hydrological effect of deforestation is generally fourfold: evaporation decreases as a result of reduced vegetation cover, total runoff increases due to reduced evaporation, dry season runoff diminishes as a consequence of reduced groundwater recharge, and flood runoff increases due to increased surface runoff. However, this is very much a generalization and the effects in particular regions may differ. Evaporation might for instance increase as a result of management practices which increase soil moisture content. Total runoff might remain unchanged due to a reduction in precipitation (as a result of the reduced evaporation). Groundwater runoff and peak flows might remain unchanged if percolation was already low under undisturbed conditions (e.g. due to an impeding layer at shallow depth). Drainage of wetlands for agricultural purposes can have similar effects to deforestation: reduced soil moisture content and evaporation,

---

[1] For a proper understanding, it is useful to distinguish between the average water balance (e.g. measured over thirty years) and the water balance in a particular year. Speaking about a *trend*, I refer to a change in the *average* water balance. I claim that a global trend might be best estimated on the basis of theoretical considerations of evaporation changes, because precipitation and runoff measurements are too scarce and natural variability is too high to detect a trend empirically. Measurements of precipitation and runoff are only useful in estimating trends in particular regions, where long-term records are available. They might also be useful in estimating historical year-to-year changes.

Table 2.4. Estimated historical changes to the hydrological cycle.

	1680[1]	1900[1]	1980[1]
Water flows (10^{12} kg/yr)			
Precipitation on land	110305	110305	110305
Evaporation from land	69605	70017 [2,5]	72075 [5]
- Increase due to artificial reservoirs	0	0.3 [2]	130 [3]
- Increase due to consumptive water use	86	522	2570 [4]
- Decrease due to urbanization	0	?	137
Total runoff	40700	40288 [2]	38230
Stable runoff	11140	11146 - 11149 [2]	13320 - 14640 [6]
- Increase due to artificial reservoirs	0	6 - 9 [2]	2180 - 3500 [6]
- Decrease due to urbanization	0	?	26
- Artificial groundwater recharge	0	0	0
Water withdrawal	104	654	3640 [4]
Inter-basin water transfer	0	0	100 [7]
Other			
Artificial reservoirs (10^{12} kg)	0	14	5525 [3]
Groundwater-level decline (m)	0	0	0-3
Length of improved waterways (10^3 km)	1.4	20	500 [3]

[1] Unless stated otherwise, the data are taken directly from L'vovich and White (1990).
[2] Interpolation on the basis of data from L'vovich and White (1990).
[3] Data for the year 1985.
[4] L'vovich and White (1990) calculated the 1980 value by adding the 1980 values for the domestic, livestock and industrial sector and the 1985 value for the irrigation sector.
[5] The increase in the total evaporation from land corresponds to the increase due to consumptive water use and artificial reservoirs minus the decrease due to urbanization.
[6] The lower values are taken from the table (p.236) and the higher values from the text (p.239) of L'vovich and White (1990).
[7] Estimate of the order of magnitude, based on L'vovich and White (1990) and Golubev and Vasiliev (1978).

increased flooding and total runoff and decreased dry season flow. As far as I am aware, no quantitative estimates of the changes of the global water balance as a result of deforestation and loss of wetlands have been made.

Changes in the future

Future changes will probably be a continuation of those in the past one hundred years, as most of the underlying forces for change will continue to grow. However, the relative importance of the different forces might change, and thereby their effect. Climate change cannot be said to have had a significant influence on hydrological conditions in the 20th century, something which might change during the 21st century. The construction of artificial surface reservoirs has changed inter-seasonal runoff patterns throughout the world considerably, but this factor might become less important in the future. On the other hand, artificial groundwater recharge could become a relevant factor, while this was not a factor of importance in the past. According to L'vovich (1979), the future contribution of artificial groundwater recharge to stable runoff might even equal the contribution of artificial surface reservoirs. Processes such as growing consumptive water use, deforestation and drainage of wetlands will probably continue in many parts of the world, with a resulting increase in continental evaporation. The fate of the additional evaporation will be of crucial importance to the ultimate effect on regional hydrological circumstances.

Conclusion
Many studies of future global hydrological changes focus on the effects of global warming. Although these effects may be considerable, continued global land cover changes and growth of consumptive water use may play an important role as well, so that it would be better to study the combined effect. One can expect that the combined effect is not equal to the sum of the three separate effects, as evaporation and runoff are typically non-linear processes. Temperature changes influence evaporation through changed potential evaporation, whereas land cover changes influence evaporation through changed interception and soil moisture conditions. In estimating long-term changes to river runoff at global level, consumptive water use, land cover changes and global warming are three mechanisms which might be equally important, so none should be ruled out beforehand.

2.6 Sea-level rise

The oceans are the ultimate sink for water on earth. The amount of water in the oceans depends on the amount of water stored in the polar ice sheets and on the continents.[1] If for whatever reason the polar ice sheets shrink or the volume of water on the continents drops, the sea level will rise. Conversely, increasing water storage in the ice sheets and on land will result in a decline of the sea level. A changing sea

---

[1] In theory, the volume of water in oceans also depends on atmospheric water storage. This can be ignored, however, because atmospheric water storage is small compared to land, ice sheet and oceanic water storage, so that fluctuations in atmospheric water storage can cause only relative small changes in sea level. Supposing we could put the whole atmospheric water content of about 12.9×10^{15} kg (Korzun et al., 1978) into the ocean, and assuming an ocean surface of 361×10^6 km^2, a sea-level rise of only 36 mm would result.

level can be regarded as a symptom of a changing global hydrological cycle, and a special one, because the sea level responds to nearly every change in the global water balance. As a consequence, sea-level change is a symptom for which it is difficult to find the cause. It is hard to distinguish between natural and anthropogenic causes of sea-level change, and within these categories it is difficult to identify the most important mechanisms. This section reviews possible driving mechanisms behind sea-level changes and the knowledge available about the relative importance of the various mechanisms in the past and the future. It further considers existing projections of future sea-level rise and assessments of the coastal impacts to be anticipated.

Historical changes
During the past millions of years, there have been large variations in sea level. The fluctuations were closely connected with changes in the earth's climate. During the last glacial maximum about eighteen thousand years ago, when the global mean surface air temperature was about four degrees centigrade below the present temperature, the sea level was about 120 metres lower than today (Fairbanks, 1989). About a hundred thousand years ago, during a warm interglacial period, the sea level was a few metres higher than today. As then, we are now in an interglacial period and the sea level is comparatively high. Superimposed on these long-term fluctuations in climate and sea level on a geological time scale, scientists now expect an additional change due to human activities. Because the human-induced change plays on a much shorter time scale than the natural change, the anthropogenic component of change might well dominate the natural component in the coming centuries.

Mechanisms behind human-induced sea-level change
There are various mechanisms of anthropogenic sea-level change (Figure 2.6). That which is currently most widely discussed starts with the emission of greenhouse gases. This results in an enhancement of the greenhouse effect and thus an increase in the global mean temperature. This would warm up and expand the ocean and thus result in sea-level rise. Global warming would also increase the melt rates of glaciers, again contributing to sea-level rise. The ice sheets of Greenland and Antarctica would probably both experience increased accumulation. For Greenland, increasing melt rates would counteract the increased accumulation, thus resulting in a negative balance and sea-level rise. The accumulation on the Antarctic ice sheet, however, would cause a drop in sea level. Other anthropogenic mechanisms, not related to climate change, which may contribute to sea-level rise concern water loss

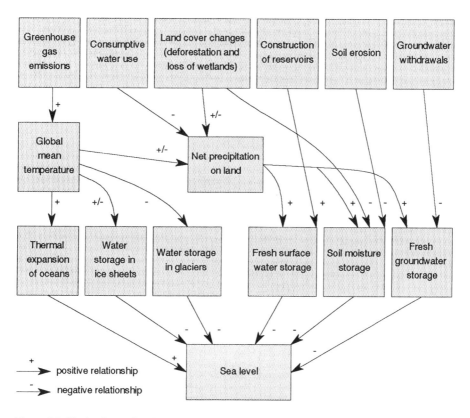

Figure 2.6. Mechanisms of anthropogenic sea-level change.

from the continents. Such loss can be the result of large-scale groundwater withdrawals, surface water diversions and land use changes such as deforestation and drainage of wetlands. By contrast, the construction of freshwater reservoirs increases the amount of water stored on the continents and thus causes a sea-level decline.

According to the 'best estimates' of the Intergovernmental Panel on Climate Change (IPCC), thermal expansion of the oceans and melting of glaciers were the main causes of 20th century sea-level rise (Warrick *et al.*, 1996; see also Table 2.5). However, the uncertainties are great, which is illustrated by the fact that the IPCC can 'explain' a sea-level rise in the past one hundred years of -190 to +370 mm, while the observed rise is 100 to 250 mm. The main reason for uncertainty relates to the mass balances of the polar ice sheets. The present imbalance of the Antarctic ice sheet might be ± 1.4 mm/yr and the imbalance of the Greenland ice sheet ± 0.4 mm/yr (expressed in sea-level rise equivalents). These uncertainty ranges can largely be understood as uncertainties about the natural trend. A source of uncertainty

Table 2.5. Estimates of historical and future sea-level rise.

Sea-level rise (in mm)	Past (1890-1990)[1]			Future (1990-2100)[2]		
	Low	Middle	High	Low	Middle	High
Component contributions						
Thermal expansion	20	40	70	-	280	-
Glaciers	20	35	50	-	160	-
Greenland ice sheet	-40	0	40	-	60	-
Antarctica ice sheet	-140	0	140	-	-10	-
Fresh surface and ground water	-50	5	70	-	0	-
Total sea-level rise explained	-190	80	370	130	490	940
Total sea-level rise observed	100	180	250	-	-	-

[1] Estimates from Warrick *et al.* (1996).
[2] Ibid. The middle estimates are the 'best estimates' for the IS92a-scenario. The low estimate of total future sea-level rise is the low estimate for the IS92c-scenario, and the high one the high estimate for the IS92e-scenario (including aerosol effects).

which gets little attention is the possible contribution of fresh surface and ground water. Without argumentation, the IPCC assumes that the different land components probably balance. However, this might not be the case at all. Sahagian *et al.* (1994) argue for instance that the net effect of groundwater withdrawals, deforestation, wetland reduction and reservoir construction accounts for at least 30 per cent of the 20th-century sea-level rise.

Projections of future sea-level rise
The literature offers a large number of sea-level projections. The big differences between the projections are due to uncertainties about the quantity of future emissions of greenhouse gases and production of aerosols, the climate sensitivity and the response of glaciers and ice sheets. According to the most recent review by the IPCC, one might expect a total rise of 130 to 940 mm over the period 1990-2100, taking account of six scenarios for greenhouse gas emissions and uncertainty ranges for climate sensitivity and ice melt parameters (Warrick *et al.*, 1996; see also Table 2.5). For the 'Business as Usual' scenario IS92a, they expect a sea-level rise of 530 ± 330 mm. Considering all emission scenarios and using the 'best estimates' for the climate sensitivity and ice melt parameters, they arrive at an estimate of 465 ± 85 mm. Apparently, as the latter uncertainty range of ± 85 mm is much smaller than the uncertainty range of ± 330 mm mentioned above, the uncertainties in the human system are supposed to be smaller than the uncertainties in the climate system. The

largest source of uncertainty concerns the ice sheets of Antarctica and Greenland. This uncertainty is such that it is even difficult to assess the order of magnitude. As mentioned above, current methods and data do not allow the detection of an imbalance of Antarctica of up to ± 1.4 mm/yr, but the lower values are unlikely, because they would have led to a substantial sea-level drop and are therefore inconsistent with the observed rise (Warrick *et al.*, 1996).

As shown in Table 2.5, the IPCC can explain a sea-level rise in the period 1890-1990 of -190 to +370 mm, which implies an uncertainty range of ± 280 mm. The sea-level rise during the period 1990-2100 under the IS92a-scenario is estimated to be 200 to 860 mm, which means an uncertainty range of ± 330 mm. One would expect the uncertainty range for the future to be much larger than that for the past, because the former includes uncertainties about climate sensitivity which are not part of the latter (for the past, one can use observed records of temperature change and does not depend on estimates of the climate sensitivity). A source of uncertainty which the IPCC included in its explanation of past sea-level rise but excluded in its projections for the future, is the possible contribution from surface and ground water. In its explanation of past sea-level rise, the IPCC estimates this contribution at -50 to +70 mm, but for the future the IPCC supposes this contribution to be nil, without uncertainty range, which is highly questionable if one considers our poor knowledge on this issue. In fact, according to IPCC's own estimates of past sea-level change, the net contribution of surface and ground water is, after the contribution of Antarctica, the most uncertain component in sea-level change. Considering the fact that the uncertainty on the Antarctica contribution is largely an uncertainty about the *natural trend*, it might well be that the contribution of surface and ground water is the most uncertain component in *anthropogenic* sea-level rise. I therefore consider it of great importance that this component should be analysed in more detail.

Impact and responses
According to Vellinga and Leatherman (1989), the impact of sea-level rise will be most severe for deltas, barrier islands, atolls and marshy coastlines. A rise in sea level will inundate wetlands and other lowlands, accelerate coastal erosion, increase the risk of flood disasters, create problems with respect to drainage and irrigation systems, and increase saltwater penetration into ground water, rivers and farmlands. Hoozemans *et al.* (1993) estimate that a sea-level rise of one metre would increase the number of people subject to annual flooding from about 47 million (in 1990) to 61 million if no adaptive measures were taken. Adding the effect of population growth over a 30-year period, the population at risk would double to about 100 million a year. Furthermore, in combination with human activities, a sea-level rise of

one metre over the next century would threaten half of the world's coastal wetlands of international importance. In some areas, valuable coastal wetlands could be virtually eliminated, because their ability to migrate inland would be limited over such a short time scale. Furthermore, the areas behind the coastal fringe are often intensively used by humans, so the pressure actually comes from two sides. The impact of sea-level rise can be anticipated in three basic ways: preventive (taking measures which will reduce sea-level rise), defensive (building dykes or raising lowlands), and adaptive (allowing the sea to advance and accommodating to the new circumstances). Because it is generally assumed that future sea-level rise will be largely due to climate change, most of the preventive strategies proposed are aimed at reducing greenhouse gas emissions. Many authors have argued that preventive strategies have to be implemented now for them to be effective (Edgerton, 1991).

Conclusion
A mechanism contributing to sea-level rise which has received little attention in recent studies is the loss of ground water as a result of groundwater withdrawals and land use changes. In a comprehensive assessment of sea-level rise, all possible driving forces should be given equal attention. From a policy analytical point of view, equal attention should be paid to the desirability and possibility of preventive strategies, strategies of risk acceptance and defensive or adaptive strategies.

2.7 Critical water issues

In this section I formulate a number of critical issues which provide a focus for the following chapters. From the previous sections it follows that a key issue will be to improve understanding of the interaction between various water-related problems, rather than to improve understanding of the nature of one particular problem.

With respect to future water demand, the focus should not be to develop new, more accurate projections of future water demand as such, but to analyse the different possible determinants of water demand, the dynamics which underlie the establishment of demand and the desirability and possibility of interfering through public policy. With respect to water availability, the next chapters will focus on how a certain concept of potential water supply will act on the dynamics of demand and actual supply. In other words: how will different perceptions of the upper limits influence society long before these limits have been reached? A further issue which deserves attention involves assessing the risks of certain development paths in terms of possible feedback from the natural system to society. Because irrigation is the largest water user, extra attention will be given to this sector. An interesting question is for example how food and water policy strategies can reinforce or counteract each other: importing water-intensive agricultural products in water-poor countries can relieve the pressure on their own water system, but the opposite, cultivating water-intensive products for export, will increase this pressure. With respect to water scarcity, questions will be addressed such as: what happens under a variation of the basic attitude towards water scarcity, what kind of policy fits in what kind of future and, finally, what are the risks of different policy strategies?

It has been shown in this chapter that basic concepts such as water demand, water availability and water scarcity are interpreted in quite different ways. Various examples have been given of individual water studies which were based on one particular interpretation of the concepts used, in many cases even without mentioning the choices made. The problem of working from a single interpretation of concepts is that this strongly biases the results of the analysis in advance. For this reason, it is considered essential to make uncertainties on basic assumptions and hypotheses explicit, to study the coherence between different types of assumptions and to analyse the implications of varying assumptions on the basis of an 'if then' approach.

Regarding changes to the hydrological cycle, a critical issue will be the combined effect of global warming, land cover changes and consumptive water use. This study will therefore focus on an estimation of the relative importance of the different mechanisms, in order to assess the effectiveness of different response strategies. Another issue will be the impact of human activities on groundwater systems, an area which has had relatively little attention until now. In the next chapters ground water will not be regarded as a static system which converts recharge into stable runoff, but as a dynamic system which responds to numerous factors such as withdrawals, drainage of irrigation water, land use changes, and climate change. People increasingly rely on groundwater resources, which adds to the pressure on this system and may lead to further lowering of groundwater tables and depletion of fossil aquifers. Additionally, if groundwater exploitation continues to grow, this may result in a significant rise of the sea level.

With respect to sea-level rise, the issue is not only how much sea-level rise might be expected during the next century, but also what will the dominant mechanisms be. In this study equal attention will be paid to the different possible mechanisms behind sea-level rise, including the loss of water on land. Because I have chosen to adopt a comprehensive point of view, it is impossible to make an in-depth study of the land contribution to sea-level rise, but the various land components have been included in this study to make some first explorations. A final issue is how trade-offs could be made between preventive policy strategies, defensive or adaptive strategies and a strategy of risk acceptance.

3 AQUA: a tool for integrated water assessment

This chapter describes AQUA, a computer tool for integrated water assessment. It starts with a discussion of the field of integrated water assessment and explains that tools within this field should meet particular requirements, which differ from the requirements for specialist tools. AQUA is typically an explorative model, meant to explore the implications of varying assumptions and hypotheses. The model simulates four types of processes: pressures on the water system, changes in the state of the water system, impacts on functions of the water system and societal responses to these impacts. Accordingly, the model is subdivided into four sub-models, each of which is described in a separate section.

3.1 The field of integrated water assessment

Traditionally, water policy issues were relatively simple issues, in the sense that most of the problems had one particular cause and the solution - although sometimes expensive or difficult to implement - was generally straightforward. The typhoid and cholera epidemics which broke out in large cities in the 19th century could be traced back to unclean drinking water and people responded by controlling water intakes, taking water from clean sources outside the cities and improving public water supply and sanitation. Problems with periodical flooding in a number of inhabited deltas around the world were clearly due to the natural variation of rainfall in the upstream areas and had to be solved through river canalization and the building of dykes and dams. Problems of a shortfall in agricultural yield were a consequence of dry years and could be solved through irrigation. Today, problems are often more complex, because they are no longer independent of each other and more interests are involved. In many cases a solution for one problem appears to be the driving force for another. Irrigation is meant to increase food availability and ensure high yields in dry years, but it has turned out to be the cause of soil degradation and the drying up of rivers in many regions of the world. Dams are constructed in order to generate energy, improve water availability and solve periodical flooding of lowlands, but they often cause the destruction of forests and the expulsion of people from their homelands, and then silt up with sediment. Better water supply in cities has greatly improved health conditions in many parts of the world, but the production of wastewater is becoming an increasing pressure on the natural environment. As a result, apparently separate problems do not have a clear beginning and end, but appear to be part of a web of interlinked phenomena. Furthermore, as illustrated in the previous chapter, uncertainties are becoming a major element in the equation. It is difficult to identify the main causes of problems because these are multiple and, as a consequence, solutions can no longer be achieved through single measures. As water problems are embedded in a complex of regional and global changes, they probably can be better analysed by considering the dynamic context in which they occur than by focusing merely on the problem itself and looking for a single solution. Many of the problems of water scarcity and pollution are interrelated with environmental change and socio-economic development to such an extent, that a single discipline or sector approach can no longer provide a satisfactory solution. According to Young et al. (1994), water policy analysts increasingly recognize that managing water resources can no longer be regarded as an independent field of expertise and a separate domain of public policy, and they now consider that particularly the inter-linkages between the water system, the rest of the environmental system and the human system should be studied.

In the actual process of water policy-making, the concept of integration was first used in the 1980s, particularly in industrialized countries (Wisserhof, 1994). However, the concept is still crystallizing, both in theory and in practice. System boundaries in water studies are gradually widening: water quality aspects are being added to the hydrological aspects, water demand aspects to the water supply aspects, and socio-economic aspects to the physical aspects. The ultimate goal of integrated water assessment is seen as analysing the role of water in environmental change and socio-economic development. Integrated water assessment is defined in this study as a process of combining and interpreting knowledge from diverse scientific disciplines to allow a better understanding of the long-term interaction between water and development, explicitly distinguishing the role of water policy. An integrated water assessment may result in recommendations for an integrated water policy.[1]

But what does integration mean? The usual approach is to explain the idea of integration through its historical evolution. One can distinguish three successive stages or levels of integration (Mitchell, 1990). On the first level, integration means the systematic consideration of the various dimensions of water: surface and ground water, quantity and quality. The key aspect here is the recognition that water comprises an ecological system which is formed by a number of interdependent components. The necessity of integrated management emanates from the fact that different components influence one another. On this level, attention is directed at joint consideration of such aspects as water supply, wastewater treatment and

---

[1] Integrated water *assessment* is distinct from integrated water *management* or integrated water *policy*. See the glossary in the back of this book.

disposal, hydroelectric power generation, flooding and water quality. One recognizes that some measures might be counterproductive in a different area of concern, but also that some measures might profit several targets at once (in this way, the concept of 'multi-purpose dams' has emerged). On this first level of integration, one has started to realize the importance of considering river basins as a whole, because upstream measures will change downstream conditions. On a second and broader level of integration, water is seen as a system which interacts with other environmental systems, such as land, soil and climate. Analysts thus take a step from *internal* integration within the water field towards *external* integration with surrounding fields. Management interests focus on issues such as agricultural practices, deforestation, erosion, loss of wetlands, diffuse sources of pollution, preservation of aquatic ecosystems and recreational use of water. On a third and even broader level of integration, water is considered in relation to social and economic development. Here, the concern is to determine the extent to which water is both an opportunity for and a barrier to economic development, and to ascertain how it is possible to ensure that water is managed and used so that development can be sustained in the long term. Currently, most of the 'integrated' water assessment studies are confined to the second level of integration. Even Mitchell (1990), who explicitly mentions the *three* levels of integration, ignores the last one and focuses on the idea of integrating water and land resources management. Neither theoretical nor practical examples of an elaboration on the third level of integration can easily be found. In my opinion, the third level of integration deserves most attention at the moment, not only from a scientific point of view - it is a rather undefined field of research - but also from a societal point of view, because many of today's water problems can probably only be solved through a comprehensive outlook and set of measures. In this book, I use the term integrated water assessment exclusively for the third level of integration. As integrated water assessment in this sense is an area of research which is only just emerging, it may be useful to look at a comparable research area and draw some parallels.

Within environmental research, a useful parallel can be drawn with the field of integrated *climate* assessment, equally a relatively new research area. The two fields have many common properties, as can be seen from what Morgan and Dowlatabadi (1996) call the seven hallmarks of a good integrated assessment of climate change. First, the characterization and analysis of *uncertainty* should be the central focus. Second, the approach should be *iterative*. The focus of attention should be permitted to shift over time, depending on what has been learned and which parts of the problem are found to be critical to answering the questions being asked. Third, parts of the problem of which one has little knowledge must not be ignored. *Order-of-*

magnitude analysis, and carefully elicited expert judgement should be used when formal models are not possible. Fourth, treatment of *values* should be explicit, and when possible parametric, so that many different players can all make use of the results of the same assessment. Fifth, to provide a proper perspective, expected changes and impacts should be placed in the *context* of other natural and human background stochastic variation and secular trends. Where possible, relevant historical data should be used. Sixth, a successful assessment is likely to consist of a set of co-ordinated analyses which span the problem, not a single model. Different parts of this set will probably need to adopt different analytical strategies. Finally, there should be multiple assessments. Different players and problems will require different formulations. No one project will get everything right, nor are results from any one project likely to be persuasive on their own. Because these seven attributes particularly refer to *structural* properties of integrated climate assessment, they can be taken as hallmarks of integrated *water* assessment as well.

But which subjects should be covered by an integrated water assessment? I would like to clear up one misconception I am often confronted with: integrated water assessment means adopting a comprehensive and inclusive approach, but it is not intended to *replace* more detailed water studies. Integrated water assessment is not necessarily a mammoth study which makes an in-depth analysis of all relevant phenomena. Rather, most integrated assessment studies will necessarily lack detail. The aim of an integrated water assessment is to find the right balance between the various phenomena, problems and questions, and study the different issues in their relationship to one another accordingly. The right balance differs in each case, depending on the particular objectives of the individual assessment study. In an integrated assessment, the correct reason for leaving issues out of consideration is because they are regarded as of secondary importance. If issues are omitted because one would just like to focus on some specific subjects, it is no longer possible to speak of an integrated assessment.

3.2 Tools for integrated water assessment

Integrated water assessment requires specific analytical methods and tools, which differ from the methods that can be used for short-term planning studies. Because integrated water assessment requires knowledge from a number of disciplines, the tools which could be useful originate from different research fields. However, most of these tools have not been designed to be used in an integrated assessment, but to support understanding in a specific discipline. For this reason, one needs to simplify or in some way adjust the specialist tools, so that they can be used within one comprehensive and coherent framework. Before considering the current situation with respect to the integration of different specialist tools, I will briefly review the areas covered by integrated water assessment and look at the specialist tools available within these specific areas.

The primary component in all water studies is hydrology. Hydrological models which describe water flows and changes in water storage have a long history, but show progressive development particularly during the last few decades, due to the availability of computers. A variety of computer models has become available for all separate elements of the hydrological cycle, from evaporation, precipitation and atmospheric circulation to river and groundwater flows. In the past, water flow models were often simple calculations on a scrap of paper or physical analogues to scale, to simulate the effect of proposed measures on the flow of water. Nowadays, several kinds of computer model describe processes on such a large spatial or temporal scale and on such a high level of complexity that they can never be equalled by physical models to scale. The best example are the so-called General Circulation Models (GCMs), which describe processes of evaporation, atmospheric moisture transport and precipitation on the global level, at resolution levels of up to $1° \times 1°$ latitude by longitude (Gates *et al.*, 1996). Laboratory experiments to analyse the behaviour of such systems are impossible, unless the real world is regarded as an experiment. The variety of computer models has become so great, that several classifications have been proposed. Relevant classification criteria are for instance: purpose of the model application, type of the system modelled, hydrological processes considered, degree of causality, and time and space discretization (WMO, 1991; Nemec, 1993).

As well as the hydrological component, there is the water quality component. About a century and a half ago, water pollution was largely a problem of contamination through faecal bacteria, especially in large cities in industrialized countries. During the past few decades the concept of water pollution has steadily widened, to include successively organic pollution, salinization, eutrophication, metal pollution, acidification, nitrate pollution, and pollution by radioactive wastes and organic micro-pollutants (UNEP, 1991, 1995a). The realization has also grown that water quality is not merely a matter of physical properties and chemical composition, but that bodies of water can also be regarded as aquatic ecosystems, comprising complex interrelations between physical-chemical processes and processes of life. Although epidemics as a consequence of faecal pollution (such as typhoid, cholera and other gastro-enteric diseases) have largely been eliminated in industrialized countries, they still occur with great frequency in developing countries. Several industrialized countries have made progress in eradicating other types of pollution too, but a large number of countries in the world is facing many of the different types of water quality problems simultaneously. Alongside the growing number of water quality problems during the past decades, the number of water quality indicators has increased exponentially, from about ten at the beginning of the 20th century to more than a hundred now (Meybeck and Helmer, 1989). The demand for appropriate tools to assess complex water quality problems has pushed the science of water quality modelling through a rapid process of development. At present, one-, two- and three-dimensional water quality models are available for rivers, lakes and ground water and for various substances (including the interrelations between different chemicals and sediments).

A third important component in many water studies is water demand. Different methods of estimating water demand have been developed, such as the water requirement approach (Shaw, 1994), the water allocation approach (Pulles, 1985) and the economic approach (Kindler and Russell, 1984). In the first type of calculation, water demand is regarded as a need emanating from population and economic growth, while in the second approach competing demands require a distribution of water according to priorities. In the economic approach, the price of water is the crucial factor in explaining of demand. Furthermore, methods often differ as a result of the water-demanding sector and level at which demand is calculated, for instance individual water-use activity, municipal, regional, national or river-basin level. Water studies may have many more components than the three mentioned. It is not possible to list them all, due to the large diversity. To mention just a few, they include hydroelectric power generation, navigation, fishery, aquaculture, wastewater production, diffuse sources of pollution, erosion and sediment transport, impact of droughts on agricultural yields, salinization of soil, and impact of river or coastal flooding on population. Needless to say, for each of these issues various specialist models have been developed.

A major question in integrated water assessment is how to select the most important components for analysis and how to choose between the various analytic tools. Should half of the time and money be spent on the analysis of hydrology and the other half on water quality? Or do these two physical aspects of water deserve no more than 50 per cent of the total effort, so that the remaining effort can be put into an analysis of socio-economic issues such as water demand, water conservation, wastewater production and wastewater treatment? Or should socio-economic development even be studied in its entirety, so as to be able to put water in the right context? Although obviously all relevant areas have to be considered, most people are biased in one particular direction, influenced by their own background. Traditionally, water studies are the domain of hydrologists and engineers, so it is not surprising that hydrological and constructional aspects still get most attention in many 'integrated' water assessment studies. Here, I will not go further into the question of the correct balance, because the answer will in any case depend on the particular goals of an assessment. However, a general feature of any integrated water assessment is that it covers a broad array of disciplines. Most of the specialist tools developed in the different research fields will prove to be too narrow within the context of integrated assessment, because each specialist area is just one component of the whole study. Although simplification is often unacceptable to specialists who know how complex particular issues really are, such simplification is essential for an integrated study.[1] Because any integrated water assessment study will have a limited budget, details below a certain level have to be left to other, more specific studies. An integrated study which does not take relevant aspects into consideration due to budget constraints is simply not an integrated study. An integrated study which covers all relevant fields but goes into few details due to budget constraints, however, is still an integrated study.

Efforts to integrate specialist water assessment tools have resulted in quite comprehensive computational frameworks, linking (simplified versions of) originally separate models. The Danish for instance have developed the Mike11 framework, and the Dutch have build upon the computational framework which was developed in the so-called PAWN-study (Pulles, 1985). Although these computational frameworks have proved to be very useful, they address a level of detail which makes them useful for five- to ten-year planning studies, but not for the formulation of priorities in the context of sustainable development policy. They are rather elaborate and focus on operational and short-term strategic aspects, rather than on the long-term strategic aspects of water policy. Furthermore, in the words of Bankes (1993), they have a consolidative rather than an explorative character. Consolidative models 'consolidate' known facts into a single package and are used as surrogates for the actual system. They can be powerful in understanding the behaviour of systems which are well known, but they become inappropriate when insufficient knowledge and irresolvable uncertainties are involved. If one looks only a few years ahead, the behaviour of even complex systems can sometimes be sufficiently known and described by consolidative models, but if one aims to look more than ten years into the future, it becomes impossible to uphold the notion that a

---

[1] A fundamental difference between specialists, who are mainly interested in their own speciality, and generalists, who have to manage the process of integration, is that specialists have to limit the object of their study and go into *greater* detail if the object becomes too complex, unmanageable or time consuming, while generalists have to retain the comprehensive point of view and go into *less* detail if things become too complex.

model is a reliable representation of the actual system. In such cases, Bankes argues, one needs explorative models which are based on assumptions in the areas where no definite knowledge is available. Explorative models are specifically meant to explore the implications of varying assumptions and hypotheses. Whereas consolidative models should be able to *predict* system behaviour to some extent of reliability, explorative models should be able to improve *insight* into system behaviour. Furthermore, there is a reversal of the relationship between problem formulation and modelling: whereas consolidative models are specifically developed to support the *solution* of problems (model formulation is directed by the problem to be solved), explorative models are meant to *identify* and *diagnose* problems (problem formulation is directed by model results). Finally, regarding tools to be used in an integrated assessment, a similar statement can be made to the one about the idea of integrated assessment in general: a comprehensive tool for integrated assessment should not be intended to be one big model replacing other tools. Instead, a comprehensive tool should be as simple as possible for its purpose and should not go into details which go beyond the objectives of integrated assessment. At the start of this research explorative models for integrated water assessment which satisfy these requirements were not readily available, which has been the main incentive for the development of the integrated water assessment tool AQUA.

3.3 Introduction to AQUA

AQUA is an integrated water assessment tool which has been designed to analyse the long-term interaction between water and development. The tool should thus be able to address the critical water issues formulated in Chapter 2. It is an explorative tool in the sense explained in the previous section, meant for exploring the implications of varying assumptions and hypotheses. The tool has been developed in five phases. In the first phase, the purposes and requirements of the tool were specified. In the second phase, basic concepts were formulated which would underlie the tool to be developed. This phase focused on issues such as the definition of the boundaries of the system to be considered, the identification of essential elements and relationships within this system, the specification of elements of interest to policy makers, the detection of major uncertainties, and the formulation of a methodology to deal with these uncertainties. In the third phase, the concepts were expanded into a mathematical model, which could be operated on a personal computer. The result of this phase was a generic tool, ready for application in different settings, but not yet provided with particular input data. In the fourth phase, the generic tool was made operational and used in two case studies: for the world as a whole and for the Zambezi basin. In the final phase, the tool was validated on the basis of the functions formulated in the first phase. The five phases were defined beforehand so that they would logically follow one another, but it was recognized that there would and should be a great deal of iteration between the different phases. In retrospect, I can confirm that this iteration has been intensive, which sometimes caused confusion about the current stage of the project, but without doubt it has resulted in a better product. However, this study is not intended to describe the process of development exhaustively. I focus instead on a discussion of the final product. In this section, I briefly discuss the functions and main characteristics of the tool.

Purposes and requirements

The primary objective of the tool is to improve insight into the interaction between long-term socio-economic development and changes in the water system. The tool should be useful in showing how this interaction can be perceived according to different world-views. Furthermore, the tool should be helpful in analysing the effects and risks of different water policy strategies, in supporting agenda setting and in formulating water policy priorities. These are the general objectives of the tool, which should be made specific for each particular application. In addition to these general purposes, a few design criteria were formulated. First, in order to link scientific analysis to the interests of policy makers, the tool should be able to translate model results into policy-relevant information. Second, the tool should be reliable from a scientific point of view. Third, the tool should be generic, so that it is possible to apply the tool at different spatial levels, from the river basin to the global. A last design criterion was user-friendliness, including a short running time, so that interactive use would be possible. Users aimed at are policy analysts and researchers in national or international institutions who are in some way concerned with water and long-term policy. It was decided not to specify particular user groups in more detail, a choice closely bound up with the nature of this research. In well-defined research areas, where problems can be clearly formulated, it is possible and at the same time useful to pre-define the type of user of a tool to be developed. In a pioneer research field such as integrated water assessment, it is more difficult and at the same time less useful to pre-define a particular user group. Such user groups are not readily available and if one succeeded in identifying them, it would probably be impossible to distinguish a clear-cut demand. For this reason, it has been chosen to use the general purposes of the tool rather than some kind of user-defined demand as the guideline for development.

Main characteristics of the tool
The above-mentioned requirements have resulted in a tool consisting of three main components: a simulation model, a set of indicators linked to this model and a framework of perspectives to analyse uncertainties. The simulation model forms the heart of the AQUA tool and describes a variety of 'pressures' on the water system, changes in the state of the water system, impacts on various socio-economic functions of the water system and societal responses to these impacts. Corresponding to this schematization, the simulation model consists of a pressure, state, impact and response sub-model. To enable the user of the model to analyse the effects of specific policy measures, the model contains a set of policy variables (in the form of 'manageable' parameters). By changing the values of policy variables, the user of the model can calculate the effects of individual water policy measures or - more advanced - integrated strategies. The model is relatively simple in all its individual parts. For most of the sub-models, especially for the state sub-model, far more detailed models are available. However, these generally require more data, have longer running times and are unsuitable for exploring highly uncertain long-term developments. Among the outcomes of the simulation model of AQUA are time series for: total and stable runoff, groundwater depletion, groundwater-level decline, the distribution of freshwater stocks over four functional water quality classes, potential water supply, specific and total water demands, actual supplies and supply costs, consumptive water use, wastewater production and treatment, and the percentages of the population with clean water supply and sanitation. If applied to the world as a whole, the model also provides information on sea-level rise and coastal impacts. The simulation model will be described in detail in the remaining sections of this chapter.

A tree of indicators is linked to the simulation model, as schematically represented in Figure 3.1. In this way, rough model output can be communicated to the user of the model in a more comprehensive form. In correspondence with the schematization of the simulation model, the tool distinguishes between pressure, state, impact and response indicators, thus linking up with an approach which is currently used in several international institutions. By distinguishing between different levels of aggregation, the user of the model can be provided with relevant information at different levels of comprehensiveness. A further discussion of water indicators can be found in Chapter 4.

The simulation model has been built in such a way that the user can apply different perspectives on how the system behaves (world-views) and on how it is or should be managed (management styles). For that purpose, different perspective-based model formulations have been included. If the user of the model chooses one

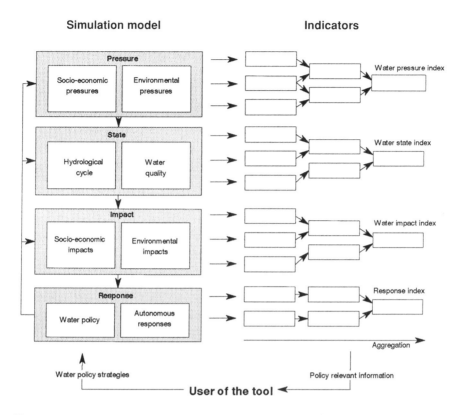

Figure 3.1. The link between the simulation model and the tree of indicators within the framework of AQUA.

particular perspective, one consistent set of model equations and parameter values will be employed. The user of the model can get an insight into the risks of a particular management style by varying the world-view while keeping the management style constant. An elaborate discussion on the implementation of perspectives in the AQUA model is in Chapter 5.

The tool has a user-friendly interface, so people can easily work with the model and experiment with different assumptions and policy strategies themselves. The tool has been implemented in the simulation and visualization environment 'M' and runs through Windows on a personal computer. In this study, however, I will not go further into details of model implementation, as this study is not intended to discuss and evaluate the technical aspects of implementation or the user-friendliness of the model.

AQUA has been designed as a generic tool, which means that it can be applied at different spatial levels. The tool has in practice been applied to the world as a whole and to the Zambezi basin. These two particular applications of the generic tool are referred to as the 'AQUA World Model' and the 'AQUA Zambezi Model'. A further application, for the Ganges-Brahmaputra basin, is not discussed here, but has served as a test case (Hoekstra, 1995; Van Rijswijk, 1995). The term generic implies that the different applications have the same model structure and use the same set of equations. However, each application requires its own particular input data and needs to be calibrated and tested separately. The application of AQUA to the world as a whole is reported in Chapters 6 and 7 and its application to the Zambezi basin in Chapters 8 and 9. However, I also refer to the two specific applications a number of times in this chapter, in order to illustrate how the generic framework of AQUA can be put into operation in a particular case.

3.4 Concepts and model schematization

The development of the simulation model of AQUA has been guided by some general concepts. Integration has been such a concept, referring to the need to describe the main interrelations between human activities and water-related processes. Systems theory has been another, being a way to look at the world in terms of entities which interact with each other. The concept of 'system dynamics' has been used as a specific way of studying the change of a system. The 'pressure-state-impact-response' schematization is the main ordering mechanism, so as to be able to position different types of processes in relation to one another. Finally, 'generic' has been used as a concept referring to the search for a model formulation which is applicable on different spatial levels. Below, I will show how the various concepts have been used to formulate the structure and elements of the AQUA simulation model.

Integration
The demand for an integrative approach has been decisive in choices with respect to the type of processes to be considered. From the integrated point of view, the main interest is in describing the interaction between various water-related processes, not in describing these processes themselves in detail. The following topics were identified as connected with each other: water demand and actual supply, water-use efficiency, 'water literacy', technological development, consumptive water use, actual and potential water re-use, supply costs per litre, water prices, expenditure in the water sector, economic growth, population growth, water export and import, relative dependence on different kinds of water sources, public water supply and sanitation coverage, hydroelectric power generation, artificial reservoirs, artificial

groundwater recharge, wastewater production and treatment, land-use changes (such as deforestation, decrease of wetlands, increase of irrigated areas, urbanization), the water balance on land, groundwater-level decline, total and stable runoff, potential water supply, water scarcity, water pollution, water policy, climate change, ice sheet dynamics, melting of glaciers, sea-level rise, and socio-economic impacts of sea-level rise. In actual fact, each topic mentioned might be regarded as a research item in itself. For instance, the dynamics of the ice sheets of Antarctica and Greenland are complex physical processes which can be best understood through a detailed analysis of accumulation, calving and melting rates, local instabilities, temperature sensitivities, response times, historical sea-level records, etc. To give another example, water prices are in fact the result of a complex societal process, which can be understood through considering the economics of water supply, the development of water supply technology, the rationale behind subsidies, the history of water as a free commodity, the cultural perception of water, etc. As stated above, the purpose of the AQUA model is not to describe each topic in all its complexity, but to represent the knowledge available on each topic in a simplified form and particularly to show the coherence of the different topics. A theory which has proved useful in analysing coherence and inter-linkages between different subjects is systems theory.

Systems theory
According to systems theory, a system is a coherent whole of entities which interact with each other. A system can often be subdivided into a number of sub-systems, each of which could be regarded as a system in itself. Interactions between entities or sub-systems can have different forms. Often, interactions are explained in terms of cause and effect. A particular type of interaction is feedback, where an entity or sub-system modifies as a result of its own activity. Negative feedback is a cyclic process which suppresses the signal that originally started the process. Positive feedback is a cyclic process reinforcing the original signal. A system is separated from its environment by the boundaries of the system. An open system is a system which is connected to and interacts with its environment. In a closed system, there is no interaction between the system and its environment.

A model is often understood as a representation of a system, in either a formal or an informal way. The simulation model of AQUA can be characterized as a formal model of an open system, which means that the modelled system is not isolated but subject to interaction with its surroundings. Most of the topics mentioned above are regarded as an endogenous part of the system we want to consider, which means that they actually form the system to be modelled. However, some of the topics are seen as exogenous developments, to be taken into account as determinants of the system under consideration, but not to be regarded as part of the system or to be analysed by themselves. The following variables have been classified as exogenous determinants: population, gross national product, value added in the industrial sector, demand for hydropower generation, land cover, irrigation demand, livestock, temperature and thermal expansion of the ocean. In order to account for the influence of these variables on the system considered, the model requires time series for them as input (Figure 3.2).

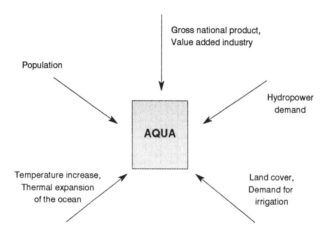

Figure 3.2. Required input with respect to exogenous developments.

System dynamics

The mathematical structure of system dynamics has been used as a formal method of describing systems' changes over time. The central question from a system dynamics point of view is: what makes a system change? The focus has been on changes over the long term. In the AQUA model, the most important external driving forces are population and economic growth, and the demand for irrigation. The most important internal driving force is technological development (improvement of water-use efficiency). An important negative feedback loop within the model is that increasing water demand will result in increasing water scarcity and supply costs, which in turn will lead to increasing water-use efficiency and decreasing water demand (see Figure 3.3). Changes in water demand are modelled as a function of changes in water prices. The expenditure needed to meet water demand is calculated on the basis of water demand in the current year and water costs in the previous year. The actual expenditure is equal to or lower than the demanded expenditure, depending on the availability of capital. Actual water supply

AQUA: a tool for integrated water assessment

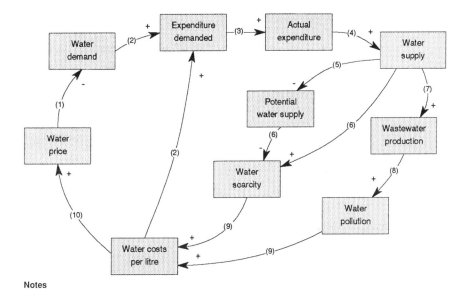

Notes

(1) An increase in prices results in a decrease in demand (Equation 3.7).
(2) The expenditure demanded is calculated as the product of total water demand and water costs per litre.
(3) The actual expenditure is equal to or lower than the expenditure demanded (Equation 3.55).
(4) Water supply increases if the expenditure for water supply increases (Equations 3.40-3.43).
(5) Potential water supply can (indirectly) be affected if actual water supply increases (see text).
(6) An increase in supply and a decrease in potential supply result in an increase in scarcity (Equation 3.44).
(7) An increase in water supply leads to an increase in wastewater production.
(8) Wastewater leads to pollution (Equations 3.36-3.37).
(9) An increase in scarcity and pollution results in an increase in costs (according to cost curves).
(10) An increase in costs leads to an increase in prices (Equation 3.50).

Figure 3.3. The mechanism of water demand and supply as represented in the model.

follows directly from the actual expenditure made. Figure 3.3 shows that, if water supply increases, water costs per litre will grow through two mechanisms. First, costs of purification will grow due to increased pollution, which has been modelled in a direct way, without delay. Second, costs of transport and storage of water will grow due to an increase in water scarcity. Primarily, the increase of water scarcity is a direct result of the increase in water supply. An indirect cause can be a decrease in potential water supply, which may occur if a part of the total volume of water supply leaves the region considered (through evaporation and export from the regional atmosphere). An increase in water costs directly translates into an increase in water prices. Negative feedback occurs as a result of the fact that an increase in price, caused by an increase in demand, will lower demand.

Because the model aims to describe changes in the long term, it is sufficient in many cases to consider changes in yearly totals and averages. For this reason, a large part of the model has a temporal resolution (time step) of one year. However, in the case of hydrological processes on land it is particularly interesting to consider long-term changes in seasonal dynamics. In this part of the system changes in annual totals and averages do not provide sufficient information, so it is necessary to simulate seasonal dynamics explicitly. Therefore, it has been chosen to use a time step of one month for the part of the model which describes terrestrial hydrology.

The purpose of the model is to analyse changes on a temporal scale of between a few decades and one or two centuries. The AQUA World Model has been made operational for the period 1900-2100. The period 1900-1990 has been used to calibrate and test the model. In the AQUA Zambezi Model a shorter simulation period is used, namely from 1990 to 2050. The main reason for not using a longer historical period was the lack of sufficient data for proper calibration and testing of the model.

The pressure-state-impact-response loop
In order to structure the various elements within the overall system considered, it has been chosen to distinguish a pressure, state, impact and response sub-system. If put in relation to each other, the four sub-systems form a closed causal loop. Corresponding to this schematization, the simulation model of AQUA has been designed as four interacting sub-models: the pressure, state, impact and response sub-model (Figure 3.4). The *pressure* sub-model calculates water demand from determinants such as population size, gross national product, value added in the industrial sector and demand for irrigated cropland. This sub-model also calculates consumptive water use and wastewater production and treatment. The *state* sub-model describes hydrological processes and freshwater quality. The hydrological cycle is modelled by distinguishing different water stores and by simulating the flows between these stores. This yields estimates of for example net precipitation, river runoff, groundwater-level decline, fossil groundwater depletion and sea-level rise. Water quality is described in terms of various water quality variables and quality classes. The *impact* sub-model calculates actual water supply to households, irrigated lands, livestock and industry; generation of hydroelectric power; and socio-economic impacts on coastal areas of sea-level rise. The *response* sub-model represents the societal response to water system changes and socio-economic impacts. Responses are regarded as partly manageable. Policy variables - in the form of 'manageable' parameters - can be interactively changed by the user of the model, and include investment in infrastructure, water pricing, and legislative and managerial measures.

Generic

The structure of the AQUA simulation model has been designed to be as generic as possible, to make the model easily applicable in different settings. For this reason, most of the schematizations are not fixed, but can be changed according to the needs of a particular study. In this way, it is possible to change for instance the temporal or spatial schematization, the number of water-demanding sectors, the number of water quality variables, etc. Table 3.1 gives an overview of the most important schematizations of the generic AQUA model and specifies the particular choices

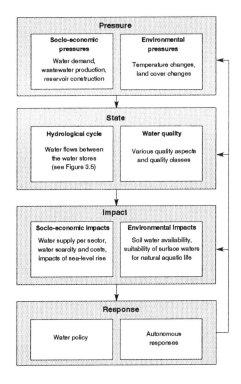

Figure 3.4. Main structure of the AQUA simulation model.

made in the AQUA World Model and the AQUA Zambezi Model. In a few cases, the generic model also provides alternative equations, to be chosen according to the preference of the user of the model. This applies to the calculation of specific water demands, the definition of potential water supply and water scarcity, and the computation of public water supply, sanitation and wastewater treatment coverage. The aim of distinguishing alternative model formulations is to be able to explore possible futures according to different perspectives (see also Chapter 5).

Table 3.1. Schematizations in the simulation model of AQUA.

Schematization	Symbol	AQUA World Model (Chapters 6-7)	AQUA Zambezi Model (Chapters 8-9)
Temporal schematization			
Simulation period	T	1900-2100	1990-2050
Time step	dt	Year or month[1]	Year or month[1]
Spatial schematization			
Spatial compartments	n_{comp}	Land, polar ice sheets, oceans and atmosphere	Land
Socio-economic regions	n_s	Land as one large region	Eight countries
Hydrological areas on land	n_h	Land as one large river basin	Eight sub-basins
Land cover types	n_{lct}	Sixteen types: see text	Five types: see text
Coastal clusters	n_{clust}	Sixty coastal clusters	Not applicable
Polar ice sheets	n_{ice}	Antarctica and Greenland	Not applicable
Atmospheric subdivision	n_{atm}	Terrestrial, polar and oceanic atmosphere	Terrestrial atmosphere
Other schematizations			
Water-demanding sectors	n_{sect}	Domestic, irrigation, livestock and industrial	Domestic, irrigation, livestock, industrial, export
Population categories	n_{pop}	No separate categories	Urban and rural population
Domestic demand categories	n_{dom}	Public and private demand	Public and private demand
Livestock demand categories	n_{livst}	No separate categories	Cattle and sheep/goats/pigs
Industrial demand categories	n_{ind}	No separate categories	No separate categories
Types of water source	n_{src}	Surface, renewable ground, fossil ground, and sea water	Surface and ground water
Types of water store	n_{store}	Ten types (see Figure 3.5)	Four types (see text)
Water quality variables	n_{wqv}	Four quality variables: NO_3^-, NH_4^+, DON, PO_4^{3-}	Four quality variables: NO_3^-, NH_4^+, DON, PO_4^{3-}
Water quality classes	n_{wqc}	Four classes: A, B, C and D	Four classes: A, B, C and D
Sea-level change components	n_{slr}	Eight components (see text)	Not applicable
Types of coastal protection	n_{prot}	Stone dykes, clay-covered dykes, sand dunes	Not applicable
Items of expenditure	n_{exp}	Nine items (see text)	Eight items (see text)

[1] The time step is one year in the pressure, impact and response sub-models and in that part of the state sub-model which describes oceanic and ice sheet processes. The time step is one month in the part of the state sub-model which describes terrestrial hydrology.

Spatial schematization

The generic model distinguishes four spatial compartments: land, ice sheets, oceans and the atmosphere. The parts of the model which refer to the last three compartments are only relevant for the application of AQUA at global level. If AQUA is applied to a specific land area, as in the Zambezi study, it is possible to omit the equations which refer to these compartments and only take account of the equations which refer to the land compartment. In this case, climate is no longer an endogenous part of the model, but has become a necessary input to the model. The generic model recognizes n_{ice} ice sheets (or ice sheet regions). In the AQUA World Model, it has been chosen to explicitly distinguish between the Antarctica and the Greenland ice sheet, but not to subdivide them into smaller regions. In the present version of the generic model, the oceans cannot be subdivided. However, the atmosphere has been schematized into three parts: above land, above the ice sheets and above the oceans.

In the generic model, the land area is subdivided into n_s socio-economic regions (e.g. countries or country clusters) and n_h hydrological areas (e.g. catchment areas). Within each hydrological area, n_{lct} land cover types are distinguished. An area of a particular land cover type in a certain hydrological area is called a physical basic area. The processes of precipitation, soil moisture change, evaporation, percolation and direct runoff are simulated at the level of physical basic areas. Other processes in the state sub-model, such as groundwater and river runoff and water quality changes, are simulated at the level of the catchment areas. The processes in the pressure, impact and response sub-models are simulated per socio-economic region. In the AQUA Zambezi model, the Zambezi basin has been schematized into eight sub-basins and eight socio-economic regions. The latter correspond to the territories of the eight African countries which make up the Zambezi basin. In the water balance calculations in the state sub-model, five land cover types have been distinguished: forest, grassland, rain-fed and irrigated cropland, and open water. The present version of the World Model does not explicitly differentiate between different socio-economic or hydrological areas, but regards the global land area as one large 'river basin'. In the state sub-model of the World Model, spatial variations are taken into account by distinguishing sixteen land cover types: forest, grassland, desert, wetland, rain-fed and irrigated cropland, urban area, and open water in two different climate zones. The original intention was to distinguish between a tropical and a temperate zone, but the form in which data were available forced me to use a division into developing and developed world, which however roughly corresponds to the climatic subdivision mentioned.

The generic model defines n_{clust} coastal clusters, spatial units used for the calculation of the socio-economic impacts of sea-level rise on the world's coasts. In the AQUA Zambezi Model, the coastal impact module has not been made operational, but in the AQUA World Model, this module has been expanded to quite a detailed level. With five hydraulic, three demographic and four economic classes,

> **Transformation of variables between sub-models**
>
> In the pressure sub-model, all variables are arrays of n_s elements, each element referring to one specific socio-economic region (e.g. country area). In the state sub-model, however, all variables are arrays of n_h elements, each element referring to one specific hydrological area (e.g. catchment area). Because the socio-economic and hydrological schematizations generally do not coincide, pressures on the water system per country are translated into pressures per catchment area[1]. Each country-specific pressure variable p[s] is translated into a catchment-specific pressure variable p[h] by the following operation:
>
> $$p[h](t) = \sum_{s=1}^{n_s} \frac{D[s,h] \times p[s](t)}{\sum_{h=1}^{n_h} D[s,h]} \qquad (3.1a)$$
>
> in which D[s,h] is a distribution matrix for which the elements add to one. For translating catchment-specific information from the state sub-model into country-specific information in the impact sub-model, a similar problem is encountered. The transformation of a catchment-specific state variable s[h] into a country-specific state variable s[s] is done in the opposite way to the above:
>
> $$s[s](t) = \sum_{h=1}^{n_h} \frac{D[s,h] \times s[h](t)}{\sum_{s=1}^{n_s} D[s,h]} \qquad (3.1b)$$
>
> The distribution matrix used in the above equations can differ per type of variable transformed. It may for example reflect the population distribution or the area distribution. A variable such as water withdrawal may be supposed to be spatially distributed in a similar way to the population, so that the population distribution matrix can be used, but a variable such as stable runoff is more likely to be distributed according to the area distribution. If information is available on the precise distribution of a specific variable, a specific distribution function might be used (e.g. for irrigation).
>
> ---
>
> [1] The present version of the AQUA World Model does not need such a translation, because it does not have subdivisions into different countries or catchment areas.

the world's coasts have been subdivided into sixty clusters, each one characterized by a combination of hydraulic, demographic and economic properties.[1]

---

[1] The calculation of coastal impacts for sixty clusters is rather detailed if compared to the high aggregation levels applied in other parts of the model. This is a result of the fact that the coastal impact module has received particular attention in a separate sub-project during the study, carried out by RA (1994).

Other schematizations

In the AQUA World Model, four water-demanding sectors are distinguished: the domestic, livestock, irrigation and industrial sectors. In the Zambezi Model, a fifth demand sector is taken into account: water export from the basin. Both models make a distinction between public and private domestic water demand. The Zambezi Model additionally distinguishes between urban and rural demand. The World Model calculates livestock water demand on the basis of one aggregated livestock category, including cattle, sheep, goats, pigs and poultry (assuming that one 'ox equivalent' represents one cow, eight sheep, goats or pigs, or 150 hens). The Zambezi Model explicitly recognizes two livestock categories: cattle as one and sheep, goats and pigs as another (neglecting water demand from hens). In the World Model, four types of water source are distinguished: fresh surface water, renewable fresh ground water, fossil fresh ground water and sea water. The Zambezi Model only takes into account fresh surface water and renewable fresh ground water. A kind of water source present in the generic version of AQUA, but not in the two applications, is water import from a neighbouring area, as a balance to water export.

The generic model makes a distinction between ten types of water store: glaciers and snow, soil moisture, biological water, fresh surface water (including rivers and lakes), renewable fresh ground water, fossil fresh ground water, salt ground water, water stored in ice sheets, ocean water, and atmospheric water. Ground ice in permafrost regions (see Table 2.1) is assumed to be isolated and is therefore not taken into account. In the World Model, all types of water store are considered. The AQUA Zambezi Model only distinguishes four types of store: fresh ground water, fresh surface water, soil moisture and biological water. Both the World and the Zambezi Models take four water quality variables into account: nitrate (NO_3^-), ammonium (NH_4^+), dissolved organic nitrogen (DON) and phosphate (PO_4^{3-}). To characterize the overall quality of a body of water, four water quality classes are distinguished:

- class A - good quality, suitable for the maintenance of natural aquatic ecosystems,
- class B - adequate quality, does not meet natural conditions but is suitable for most human purposes,
- class C - inadequate quality, unsuitable for both natural aquatic ecosystems and drinking, and
- class D - poor quality, unsuitable also for agricultural and industrial purposes.

Each water quality class is characterized by certain quality standards, i.e. maximum concentrations of the substances considered.

The AQUA World Model considers eight components of sea-level change: thermal expansion, glaciers, Greenland, Antarctica, ground water, deforestation, loss of wetlands and artificial reservoirs. The model calculates the separate contributions from each of these components, except for the contribution from thermal expansion of the ocean, which is considered an external input scenario. As stated above, impacts of sea-level rise on the world's coasts are calculated by distinguishing between sixty coastal clusters. For the calculation of the costs of dyke building, three types of coastal protection are distinguished: stone dykes, clay-covered dykes and sand dunes.

The generic model makes a distinction between nine items of water-related expenditure, viz. for public water supply, sanitation, irrigation, livestock water supply, industrial water supply, hydropower generation, domestic and industrial wastewater treatment and coastal defence. The World Model includes all nine components; the Zambezi Model excludes the last item.

3.5 The pressure sub-model

Environmental pressures on the water system, such as changes in land cover and temperature, are input scenarios for AQUA. Socio-economic pressures on the water system, such as water demand, wastewater production and reservoir construction, are driven by demographic, economic and technological developments and are discussed below.

Water demand
Water demand is simulated separately for the domestic, irrigation, livestock and industrial sectors. Demands are considered to be the driving force behind supplies, but actual supplies (simulated in the impact sub-model) may be lower than demands due to allocation constraints (simulated in the response sub-model). Demands are thus conceived as latent demands which are not necessarily met. Total water demand WD_{tot} is calculated as the sum of the above mentioned sector demands plus a possible demand for water export out of the area under consideration:

$$WD_{tot}(t) = WD_{dom}(t) + WD_{irr}(t) + WD_{liv}(t) + WD_{ind}(t) + WD_{exp}(t) \quad [kg/yr] \quad (3.2)$$

Domestic water demand WD_{dom} includes demand from households, municipalities, commercial establishments and public services. The model discriminates between public and private supply. One reason for this distinction is that the number of people receiving public supply is an indication of the number of people with *proper* water supply, which is a determinant of human health. Another reason is that public supply generally implies a larger demand per capita. Public water supply systems include public hand pumps, standpipes and house taps. People relying on self-supply obtain their water via dug-wells, tube-wells or yard taps, or directly from rivers, canals, lakes or ponds. The total domestic water demand is calculated as the sum of public and private demand:

$$WD_{dom}(t) = \left(rpp + (1-rpp) \times COV_{pws,dem}(t)\right) \times POP(t) \times WD_{dom,pc}(t) \quad [\text{kg/yr}] \quad (3.3)$$

in which POP represents the population in a certain area, $WD_{dom,pc}$ the domestic water demand per capita (in the case of public supply), $COV_{pws,dem}$ the public water supply coverage demanded (see Equation 3.51) and rpp the ratio of private to public water demand per capita. Optionally, the equation can be applied to urban and rural areas separately. Irrigation water demand WD_{irr} is calculated by multiplying the irrigated cropland area (in km^2) by the irrigation water demand per hectare (in kg/yr/ha):

$$WD_{irr}(t) = 100 \times A_{irr}(t) \times WD_{irr,pha}(t) \quad [\text{kg/yr}] \quad (3.4)$$

Livestock water demand results from the number of heads and the water demand per head:

$$WD_{liv}(t) = LIV(t) \times WD_{liv,ph} \quad [\text{kg/yr}] \quad (3.5)$$

This equation is applied to each of the n_{livst} livestock categories distinguished. Total livestock water demand is equal to the sum of the demands per category. Each livestock category has its own specific water demand per head, which has been assumed to be constant. Industrial water demand consists of demands from various industrial sectors, including demand for manufacturing, cooling water demand from thermoelectric power plants and groundwater withdrawal requirements of mining industries. The aggregate demand is calculated as the product of the total value added in the industrial sector (in US$/yr) and the average water demand per dollar value added (in kg/US$):

$$WD_{ind}(t) = VA_{ind}(t) \times WD_{ind,p\$}(t) \quad [\text{kg/yr}] \quad (3.6)$$

An alternative could have been to calculate industrial water demand on the basis of a specific industrial demand *per capita* instead of per dollar value added (e.g. Margat, 1994). However, this approach is not suitable if one wants to account for the effect of industrialization, where gross national product and value added in the industrial sector grow much faster than the population (see Figure 2.1). There are also authors who calculate industrial water demand separately for each industrial sector and use specific variables such as water demand per ton of manufactured goods or water demand per joule of thermoelectric power generation (e.g. Raskin *et al.*, 1995, 1996). This is certainly more elaborate, but it also needs more data and it could be questioned whether - for long-term forecasts - a more accurate model formulation will improve the model outcome if the extra input data needed for this more accurate formulation are not available. Working with an average value as in the equation above implicitly assumes that there will be no major changes from water-intensive industrial sectors (e.g. thermoelectric power generation) towards less intensive sectors (e.g. manufacturing) or the reverse.

Specific water demands

The variables $WD_{dom,pc}$, $WD_{irr,pha}$, $WD_{liv,ph}$ and $WD_{ind,p\$}$ in the above equations are so-called specific water demands or water-use intensities. For livestock, specific demand has been assumed to be constant. For each of the three remaining sectors, specific water demand WD_{spec} is calculated as:

$$\frac{dWD_{spec}(t)}{dt} = r(t) \times WD_{spec}(t) \quad [\text{kg/yr/yr}] \tag{3.7}$$

in which the factor r is calculated according to one of the following three alternatives:

$$r(t) = El_G(t) \times \frac{dGNP_{pc}(t)/dt}{GNP_{pc}(t)} - \frac{dEff_{act}(t)/dt}{Eff_{act}(t)} \quad \text{Alternative I}$$

$$r(t) = El_G(t) \times \frac{dGNP_{pc}(t)/dt}{GNP_{pc}(t)} + El_P \times \frac{dWP(t)/dt}{WP(t)} \quad \text{Alternative II}$$

$$r(t) = El_G(t) \times \frac{dGNP_{pc}(t)/dt}{GNP_{pc}(t)} + El_P \times \frac{dWP(t)/dt}{WP(t)} - \frac{dEff_{act}(t)/dt}{Eff_{act}(t)} \quad \text{Alternative III}$$

The variable GNP_{pc} represents gross national product per capita, WP the water price and Eff_{act} the actual water-use efficiency. The model distinguishes a different water price and water-use efficiency for each sector. The parameters El_G and El_P are sector-specific growth and price elasticities respectively. The growth elasticity is defined as a function of GNP_{pc} and is thus time-dependent (it is assumed that the response of demand to economic growth will decrease if a certain stage of development has been reached). The first alternative takes economic growth and efficiency improvements as the most important determinants of specific water demand. A similar approach can be found in for example Raskin *et al.* (1995, 1996, 1997) and Shiklomanov (1997), although they use different formulations. The second alternative takes economic growth and water price as the most important determinants (see e.g. Kindler and Russell, 1984). In this alternative, efficiency improvements are included in the price mechanism. The third alternative is a combination of alternatives I and II and supposes that prices are one factor in influencing efficiency, but that there are forces other than prices alone which may lead to efficiency improvements. In alternative III, $dEff_{act}$ represents efficiency improvements which are typically non-price-driven, but driven by for instance independent technological innovation or increased public awareness of the environmental impacts of excessive water use. In Chapter 5 I will return to the question of which alternative to use under which conditions.

For each sector, actual water-use efficiency is supposed to depend on the water-conserving technology available, i.e. the maximum possible efficiency Eff_{max}, and the extent to which this technology is in fact being used. Actual efficiency is supposed to move towards the maximum possible, as an autonomous process but possibly accelerated by policy measures. Although improving water-use efficiencies through education (increasing public awareness) is often mentioned as an important policy instrument (Postel, 1992), knowledge of the effectiveness of this instrument is poor. A simple logistic curve with a diffusion rate d has been assumed, to simulate the diffusion of water-conserving technology:

$$\frac{dEff_{act}(t)}{dt} = d \times \left(Eff_{act}(t) - Eff_{min}\right) \times \left(Eff_{max}(t) - Eff_{act}(t)\right) \quad [1/\text{yr}] \quad (3.8)$$

The minimum and maximum efficiency values determine the bottom and ceiling of the logistic curve. In the current applications of AQUA, the minimum efficiency value has been assumed to be zero, which means that, if the actual exceeds half of the maximum efficiency, one is in the second half of the logistic curve (the concave part). In the case of the domestic and industrial sectors, water-use efficiency is a relative concept, which means that an efficiency value has meaning only if compared to a previous efficiency value. Irrigation efficiency, however, has an absolute physical meaning, defined as the fraction of the total water withdrawal which actually benefits the crop (i.e. the part taken up and transpired by the plant). The remainder consists of water losses through evaporation and groundwater recharge and this is often more than the water actually used by the plant. The maximum possible efficiency in the case of irrigation has a natural upper limit of

100 per cent. For the domestic and industrial sectors, the development of Eff_{max} is considered to be an input scenario:

$$\frac{dEff_{max}(t)}{dt} = TD \quad [1/\text{yr}] \quad (3.9)$$

in which TD is a measure of (non-price-driven) technical development with a value greater than or equal to zero.

The effect of water use on the hydrological cycle
The actual pressure of water use on the hydrological cycle is a result of the overall balance of the water used: from where it is derived (the sources) and where it is left (the sinks). At present, in most parts of the world, the main part of the water for human purposes is derived from fresh surface water and shallow renewable fresh ground water, although other sources used are deep fossil ground water and saline water. Water supply from source *src* is calculated as:

$$WS[src](t) = wsf[src](t) \times WS_{tot}(t) \quad [\text{kg/yr}] \quad (3.10)$$

where WS_{tot} represents the total water supply and $wsf[src]$ the water source fraction of source *src*, indicating the ratio of the supply from source *src* to the total water supply. Because the relative use of the different sources is partly subject to national water policies, this is regarded as a policy variable (see Section 3.8). The sinks for used water are the fresh groundwater and surface water stores, and the atmosphere. The amount of *potential water re-use* is defined as that part of the total freshwater withdrawal which can possibly be re-used. It complements the volume of *consumptive water use*, defined as the part of the water withdrawal which does not return to one of the freshwater stores but is lost through evaporation. Consumptive water use WS_{cons} is calculated as:

$$WS_{cons}(t) = \sum_{sect=1}^{n_{sect}} fcwu[sect](t) \times WS[sect](t) \quad [\text{kg/yr}] \quad (3.11)$$

The fractions consumptive water use *fcwu* for the domestic, livestock and industrial sectors are assumed as exogenous scenarios, based on historical data and expected trends. For export water, the fraction consumptive water use is equal to one. The fraction consumptive water use for irrigation $fcwu_{irr}$ is calculated as a function of the irrigation efficiency $Eff_{act.irr}$:

$$fcwu_{irr}(t) = 1 - (1 - loss_{evap}) \times (1 - Eff_{act,irr}(t)) \quad [-] \quad (3.12)$$

where $loss_{evap}$ is the fraction of the water which is not taken up and transpired by the crop but nevertheless is lost through evaporation. This factor is assumed to be constant.

A part of the wastewater from the domestic, livestock and industrial sectors is treated before it drains into rivers and lakes. The volume of treated wastewater is calculated for each sector as a function of the wastewater treatment coverage:

$$WW_{tr}(t) = COV_{wwt,act}(t) \times WW_{tot}(t) \quad [kg/yr] \quad (3.13)$$

WW_{tot} represents the total wastewater disposal for a specific sector and $COV_{wwt,act}$ the wastewater treatment coverage for this sector (see Equation 3.53).

The effect of artificial reservoirs
Artificial freshwater reservoirs are planned for a variety of reasons: hydropower generation, water supply, flood control, navigation, recreation and aquaculture. However, in terms of total artificial reservoir volume and area, one can conclude that the bulk of reservoirs is - if not primarily at least secondarily - related to hydropower (Gleick, 1993a). The model therefore takes the demand for hydropower generation as the driving force of dam construction. The water storage in artificial reservoirs S_{res} is supposed to be proportional to the hydropower generation capacity HGC and the total reservoir area A_{res} to the reservoir storage:

$$S_{res}(t) = HGC(t) / \theta \quad [kg] \quad (3.14)$$
$$A_{res}(t) = 10^{-9} \times S_{res}(t) / d_{res} \quad [km^2] \quad (3.15)$$

where θ represents the average hydropower generation capacity per unit of reservoir storage (in MW/kg) and d_{res} the average reservoir depth (in m). The increased evaporation as a result artificial reservoirs is calculated in the state sub-model.

3.6 The state sub-model

The model distinguishes the water stores and flows as shown in Figure 3.5. For each water store, the model contains a mass balance in the form of:

$$\frac{dS(t)}{dt} = \sum F_{in}(t) - \sum F_{out}(t) \quad [kg/yr] \quad (3.16)$$

in which S is the storage, ΣF_{in} the sum of the inflows and ΣF_{out} the sum of the outflows. The storage will change if inflows and outflows are not equal. Three of the water stores distinguished are assumed to be in dynamic equilibrium, which means that the storage remains constant. The saline groundwater storage remains constant because the outflow into the oceans is supposed to equal the inflow from fresh ground water. This probably reflects and is certainly close to reality on a time scale of a few decades (which is a brief period from a geological point of view). The

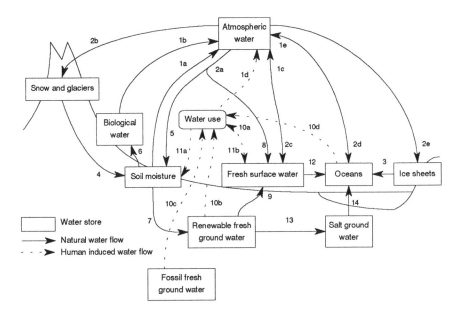

Water flows

1 Evaporation from
  a soil moisture (Eqation 3.24)
  b terrestrial biota (Equation 3.24)
  c fresh surface water (Equation 3.23a)
  d water withdrawals (Equation 3.11)
  e oceans (Equation 3.20)
2 Precipitation on
  a land surface
  b snow and glaciers
  c fresh surface water
  d oceans
  e ice sheets
3 Calving/melting of ice sheets (Equation 3.17)
4 Melting of snow and glaciers (Equation 3.18)
5 Infiltration (=2a - 8)

6 Plant uptake (Equation 3.24)
7 Percolation (Equation 3.28b)
8 Direct runoff (Equation 3.28a)
9 Delayed runoff (Eqation 3.31a)
10 Water withdrawal from
  a fresh surface water (Equation 3.10)
  b renewable fresh ground water (Equation 3.10)
  c fossil fresh ground water (Equation 3.10)
  d oceans (Equation 3.10)
11 Return flow to
  a land surface
  b fresh surface water
12 River runoff (Equation 3.33)
13 Subsurface runoff (Equation 3.31b)
14 Salt groundwater seepage

Figure 3.5. Schematization of the hydrological cycle.

atmospheric water content does not change because global precipitation is supposed to equal global evaporation. With a time step of one month, this assumption is allowed, because atmospheric water vapour has a renewal time of only eight to ten days (Korzun et al., 1978; L'vovich, 1979). As a result of this approach, the model is unsuitable for simulating possible long-term changes in atmospheric water storage. However, this cannot have a significant influence on the model results, because atmospheric water amounts to only 0.001 per cent of the total water stock on earth (Shiklomanov, 1993). The biological water storage remains constant in the model by assuming that water uptake and transpiration by plants are equal. Although seasonal and long-term biomass changes are realistic (Houghton et al., 1983) and important for the global distribution of for instance *carbon*, one may neglect them in a calculation of the global distribution of *water* because biological water forms only 0.0001 per cent of the global water stock, equivalent to a 3 mm sea-level rise.

A change in the oceanic water storage corresponds to a change in the average sea level. The separate contributions to sea-level change from ice sheets, glaciers, fresh ground water, fresh surface water and soil moisture (Figure 2.6) are calculated by dividing changes in water storage by the global ocean surface (361×10^6 km^2). Total sea-level rise is the outcome of the net increase in oceanic water storage plus the volume increase as a result of thermal expansion. The average decline in groundwater level is calculated by dividing the decrease of the renewable fresh groundwater storage by the total land area (133×10^6 km^2, excluding Antarctica and Greenland).

Ice sheet dynamics
The model distinguishes n_{ice} ice sheets or ice sheet regions. An ice sheet grows as a result of precipitation and shrinks through calving and melting. Due to the cold, evaporation from ice sheets is negligible and can therefore be ignored. In Antarctica melting is also negligible because of the extremely cold climate. The complex dynamics of ice sheets are simplified in the model by applying the static sensitivity approach as described in Warrick et al. (1996). In conformity with the IPCC, storage changes are expressed in sea-level equivalents (1 mm sea-level rise = 361×10^{12} kg ice storage decrease). The water storage S_{ice} in an ice sheet is calculated as follows:

$$\frac{dS_{ice}(t)}{dt} = -10^6 A_{oc} \times (\alpha \times T_{incr}(t) + \beta) \qquad \text{[kg/yr]} \qquad (3.17)$$

in which α represents the static ice sheet sensitivity in mm/yr/°C, β the initial imbalance in mm/yr, A_{oc} the ocean surface and T_{incr} the increase in global mean surface air temperature in °C since the first year of simulation.

Melting of glaciers and snow
The storage of water as glaciers and snow in mountainous regions is subject to short-term and long-term dynamics. The short-term dynamics are driven by temperature variations within a year: addition by snowfall during the cold period of the year and depletion by melting in the warm period. The current version of AQUA does not account for these short-term dynamics. The long-term change in the average storage in glaciers is simulated as a function of the global mean temperature increase, according to the equation proposed by Wigley and Raper (1995) and used in the latest climate assessment by the IPCC (Warrick et al., 1996):

$$S_{glac}(t) = S_{glac,i} - S_{glac,loss}(t) \qquad [kg] \qquad (3.18a)$$

$$\frac{dS_{glac,loss}(t)}{dt} = \frac{-S_{glac,loss}(t) + S_{glac,i} \times T_{incr}(t)/T_{glac}}{\tau} \qquad [kg/yr] \qquad (3.18b)$$

where S_{glac} represents the actual glacier storage, $S_{glac,i}$ the initial storage, $S_{glac,loss}$ the storage loss since the initial year, T_{incr} the global mean temperature increase since the initial year in °C, T_{glac} the temperature increase required to raise the equilibrium line altitude of a glacier to the top of the glacier (hereafter called the critical temperature increase) and τ the glacier response time in years. To account for regional variations in the altitudinal ranges and response times of glaciers, ranges have been assumed for the glacier properties: $[T_{glac,min}, T_{glac,max}]$ and $[\tau_{min}, \tau_{max}]$, for the latter taking a range of ±30 per cent around a middle value (following Wigley and Raper). It has been assumed that T_{glac} is uniformly distributed with respect to $S_{glac,i}$ and that τ is related to T_{glac} (smaller τ for smaller T_{glac}). This approach describes the actual glacier dynamics more accurately than the equation used in an earlier version of AQUA (Hoekstra, 1995, 1997), which was taken from Oerlemans (1989). The new approach accounts for the fact that smaller glaciers with a small critical temperature increase and response time will disappear first, which will increase the global mean response time.

Oceanic evaporation and precipitation
The water balance of the oceanic atmosphere reads:

$$\frac{dS_{atm,oc}(t)}{dt} = E_{oc}(t) - P_{oc}(t) - AMT(t) \qquad [kg/yr] \qquad (3.19)$$

where E_{oc} represents oceanic evaporation, P_{oc} oceanic precipitation and AMT the net advective moisture transport from the oceanic to the terrestrial atmosphere. The change in water storage in the oceanic atmosphere is supposed to be nil on the time scale of simulation, so that oceanic precipitation equals oceanic evaporation minus advective moisture loss to land. Oceanic evaporation is simulated on an annual basis with a simple equation which relates the increase in oceanic evaporation to the global mean temperature increase:

$$E_{oc}(t) = (\lambda \times T_{incr}(t) + 1) \times E_{oc,i} \quad \text{[kg/yr]} \quad (3.20)$$

where λ is a temperature sensitivity factor in $^{\circ}C^{-1}$. The advective moisture transport from oceans to land is supposed to change as a function of the global mean temperature increase and the global consumptive water use:

$$AMT(t) = (\mu \times T_{incr}(t) + 1) \times AMT_i - v \times WS_{cons}(t) \quad \text{[kg/yr]} \quad (3.21)$$

where μ represents a temperature sensitivity factor in $^{\circ}C^{-1}$ and v the fraction of consumptive water use which is lost to the oceanic atmosphere. The latter parameter is supposed to have a value between zero and one.

Terrestrial evaporation and precipitation
The water balance of the terrestrial atmosphere reads similarly to the balance of the oceanic atmosphere:

$$\frac{dS_{atm,land}(t)}{dt} = E_{land}(t) - P_{land}(t) + AMT(t) \quad \text{[kg/yr]} \quad (3.22)$$

where E_{land} represents land evaporation (including consumptive water use) and P_{land} land precipitation. It has been assumed that water storage in the terrestrial atmosphere does not change significantly on the time scale of simulation, so that land precipitation is equal to land evaporation plus advective moisture transport to land. The distribution of land precipitation over the physical basic areas distinguished is derived from IIASA's climate database (Leemans and Cramer, 1991). Evaporation from land is calculated per physical basic area on a monthly basis, using Thornthwaite's empirical equations. Potential evaporation E_p is calculated according to Thornthwaite (1948) on the basis of monthly temperatures T_m in $^{\circ}C$:

$$E_p(t) = \begin{cases} lcf \times (h/12) \times (N/30) \times 16(10T_m(t)/I(t))^{a(t)} & T_m(t) \geq 0 \\ 0 & T_m(t) < 0 \end{cases} \quad \text{[mm/month]} \quad (3.23a)$$

$$I(t) = \sum_{j=1}^{12} I_m(t - \frac{j-1}{12}) \quad \text{with} \quad I_m(t) = \begin{cases} (T_m(t)/5)^{1.514} & T_m(t) \geq 0 \\ 0 & T_m(t) < 0 \end{cases} \quad [-] \quad (3.23b)$$

$$a(t) = 0.49 + 1.79 \times 10^{-7} I(t) - 7.71 \times 10^{-5} I^2(t) + 6.75 \times 10^{-7} I^3(t) \quad [-] \quad (3.23c)$$

In these equations, I is an annual heat index, I_m a monthly heat index, a a time-dependent exponent, h the mean number of daylight hours per day and N the number of days in a month. In the current applications of AQUA, average values have been assumed for the last two parameters: $h=12$ and $N=30$. The land cover factor lcf is not included in the equation proposed by Thornthwaite, but used here to change the value of potential evaporation for full vegetation cover into a value for the actual cover. In addition, the factor is used as a calibration parameter to equate the simulated actual evaporation with observed data (cf. Kwadijk, 1993). The above equation for potential evaporation has frequently been criticized for its simplicity (e.g. Van Wijk and De Vries, 1954). However, Penman (1956) concluded that, considering its inherent simplicity and obvious limitations, Thornthwaite's equation works surprisingly well. Calder *et al.* (1983) show that more detailed equations for potential evaporation, such as those of Penman (1948), Priestley and Taylor (1972) or Thom and Oliver (1977), do not necessarily give better results than a simple approach. They find that optimization of the parameters involved has a greater effect on the model fit than the choice of equation. Yates and Strzepek (1994) add that physical methods are more soundly based than an empirical method such as Thornthwaite's, but are more data intensive as well as data sensitive. Pereira and Paes de Camargo (1989) argue that there is no reason to avoid Thornthwaite's equation, except in the case of strong advection of dry air from surrounding areas, an oasis situation.

Actual evaporation E_a is calculated as proposed by Thornthwaite and Mather (1955, 1957) and results from the potential evaporation E_p, the precipitation P and the soil moisture content S_{soil} as follows:

$$E_a(t) = \begin{cases} P(t) - dS_{soil}(t)/dt & P(t) < E_p(t) \\ E_p(t) & P(t) \geq E_p(t) \end{cases} \quad \text{[mm/month]} \quad (3.24)$$

$$S_{soil}(t) = \begin{cases} S_{cap} \times \exp(-APWL(t)/S_{cap}) & P(t) < E_p(t) \\ \min(S_{cap}, S_{soil}(t-dt) + P(t) - E_p(t)) & P(t) \geq E_p(t) \end{cases} \quad \text{[mm]} \quad (3.25)$$

$$APWL(t) = \begin{cases} APWL(t-dt) + E_p(t) - P(t) & P(t) < E_p(t) \\ 0 & P(t) \geq E_p(t) \end{cases} \quad \text{[mm]} \quad (3.26)$$

The variable *APWL* is the so-called accumulated potential water loss, a measure of the potential deficiency of water in the soil. The parameter S_{cap} represents the water-holding capacity of the soil, i.e. the water available for plant uptake if the soil is at field capacity, and has a specific value per land cover type. The equations imply that if precipitation exceeds potential evaporation, a possible soil moisture deficit is replenished to the water-holding capacity and the potential evaporation becomes actual. The rest of the precipitation is then available for runoff. However, if the precipitation is less than the potential evaporation, there is nothing available for runoff. Actual evaporation from *fresh surface water* is assumed to equal the potential evaporation. Thornthwaite's equations have been proven to be adequate in several earlier attempts at river-basin modelling, for instance by Gleick (1987a, 1987b) for the Sacramento basin in California, Vörösmarty et al. (1989) for South America, Vörösmarty and Moore (1991) for the Zambezi basin, Thompson (1992) for Kansas and Missouri, Kwadijk (1993) for the Rhine basin, Van Deursen and Kwadijk (1994) for the Ganges-Brahmaputra basin and Conway et al. (1996) for the Nile basin.

The largest part of the evaporation from land consists of transpiration by plants. The transpiration loss of plants is balanced by the uptake of soil moisture. The remaining part of the land evaporation consists of direct evaporation from the soil. The transpiration fraction of total evaporation differs per land cover type.

Runoff

The precipitation available for runoff, the net precipitation P_{net}, is calculated per physical basic area as the precipitation minus the evaporation and the soil moisture change:[1]

$$P_{net}(t) = P(t) - E_a(t) - \frac{dS_{soil}(t)}{dt} \qquad [\text{kg/yr}]^2 \qquad (3.27)$$

---

[1] The model does not account for 'storm runoff' in the case of unsaturated soil. The Thornthwaite equations recharge the soil up to field capacity before water is available for runoff. Accounting for storm runoff would require a small adaptation of the model (cf. Gleick, 1987a).

[2] From here on, water flows are expressed in kg/yr, but the actual calculations are first carried out in kg/month, using a time step of one month. The volume of water in kg is calculated from the volume in mm (as used in the previous equations for precipitation, evaporation and soil moisture change) on the basis of area data.

The net precipitation partly flows into the fresh surface water store as *direct runoff*; the remaining part *percolates* into the fresh groundwater store. The model calculates direct runoff R_{dir} and percolation R_{perc} per land cover type as follows:

$$R_{dir}(t) = \varphi \times P_{net}(t) \quad \text{[kg/yr]} \tag{3.28a}$$

$$R_{perc}(t) = (1-\varphi) \times P_{net}(t) \quad \text{[kg/yr]} \tag{3.28b}$$

where φ represents a land-cover-specific ratio. The outflow from the fresh groundwater store is assumed to relate to the storage as:

$$R_{fgw}(t) = \left(\frac{S_{fgw}(t)}{\kappa_{fgw}}\right)^p \quad \text{[kg/yr]} \tag{3.29}$$

where κ_{fgw} represents a response factor and p the linearity of the relationship. The lag time of the groundwater store, k_{fgw}, defined as the ratio of groundwater storage to outflow, is then a function of the groundwater storage:

$$k_{fgw}(t) = \kappa_{fgw}^p \times S_{fgw}(t)^{1-p} \quad \text{[yr]} \tag{3.30}$$

If exponent p is equal to one (groundwater outflow linearly relates to the storage), the lag time is constant and equal to κ_{fgw}. If p has a value greater than one (which is more plausible than a value smaller than one), the lag time increases if the storage decreases. This means that the dynamic behaviour of the groundwater store changes if the groundwater level declines. Exponent p is an input parameter of the model. The response factor κ_{fgw} is calculated by the model on the basis of the initial values for the groundwater storage and outflow, assuming that the groundwater store is initially in balance. Using a constant partition factor χ, the total groundwater outflow is divided into two components: *delayed surface runoff* and *subsurface runoff*:

$$R_{del}(t) = \chi \times R_{fgw}(t) \quad \text{[kg/yr]} \tag{3.31a}$$

$$R_{subs}(t) = (1-\chi) \times R_{fgw}(t) \quad \text{[kg/yr]} \tag{3.31b}$$

The *undelayed river runoff* R_{riv0} is calculated as follows:

$$R_{riv0}(t) = R_{dir}(t) + R_{del}(t) + P_{fsw}(t) - E_{fsw}(t) - WS_{fsw}(t) + WW_{tot}(t) + R_{riv\,up}(t) \quad \text{[kg/yr]} \tag{3.32}$$

in which R_{dir} and R_{del} are the direct and delayed runoff, P_{fsw} the precipitation onto surface water, E_{fsw} the evaporation from surface water, WS_{fsw} the surface water withdrawal and WW_{tot} the total domestic, livestock and industrial wastewater disposal into surface waters. The term $R_{riv,up}$ represents the river runoff from the upstream basin(s) where applicable. The *delayed river runoff* R_{riv} is calculated as:

$$R_{riv}(t) = \begin{cases} \dfrac{S_{fsw}(t-T_{fsw}) - S_{fsw,min}}{k_{fsw}} & k_{fsw} > 0 \\ R_{riv0}(t-T_{fsw}) & k_{fsw} = 0 \end{cases} \quad [\text{kg/yr}] \quad (3.33)$$

where $S_{fsw,min}$ represents a certain minimum water storage below which no river runoff will occur. Above this minimum level, the outflow from the surface water system is a linear function of the storage, determined by a lag time k_{fsw}. In addition, it has been assumed that there is a delay T_{fsw} before the river runoff generated within a watershed actually leaves the basin. If k_{fsw} is taken as zero, river runoff behaves according to the running time principle: it takes a certain amount of time before a water particle which enters a river reaches the ocean. If k_{fsw} has a given value and T_{fsw} is zero, river runoff behaves according to the storage principle: river outflow depends on the water storage in the river, not directly on the river inflow. Whereas the running time principle postpones a runoff peak, but leaves it intact, the storage principle smoothes a peak. The storage principle particularly applies to rivers with large lakes which buffer runoff. The *total runoff* from land to oceans consists of the river runoff plus the subsurface runoff:

$$R_{tot}(t) = R_{riv}(t) + R_{subs}(t) \quad [\text{kg/yr}] \quad (3.34)$$

In an equilibrium situation, where there is no long-term change in water on land, the total runoff in a year is equal to the net precipitation P_{net} in that year. *Stable runoff* is calculated over a year and defined as that part of the total runoff on which one can rely throughout the year. The largest stable runoff component is the natural one, consisting of the natural groundwater recharge (percolation of net precipitation). Stable runoff is reduced by groundwater withdrawals WS_{fgw}, but increased by drainage of irrigation water DR and artificial groundwater recharge $AGWR$. The contribution of artificial reservoirs to stable runoff is accounted for by taking a fixed percentage f of the artificial reservoir volume S_{res}:

$$R_{stable}(t) = R_{perc}(t) - WS_{fgw}(t) + DR(t) + AGWR(t) + f \times S_{res}(t) \quad [\text{kg/yr}] \quad (3.35)$$

This definition does not include a possible contribution by natural lakes to stable runoff, but this component is probably small in most river basins. On a global level, L'vovich (1973a, 1973b) estimates that the stable runoff from regulation by natural lakes is 2.4 per cent of the stable runoff from subsurface origin. Only in some basins, such as the St. Lawrence basin (which includes the Great Lakes), would the above definition of stable runoff clearly not satisfy, and in that case it can be replaced by a definition which takes the minimum monthly river runoff in a year as a measure of stable runoff. In other cases I prefer the above definition, because it makes it easy to trace the different components of stable runoff, which is interesting if one wants to get an insight into the causes of changes in stable runoff.

Freshwater quality

To characterize the quality of fresh surface waters, the model distinguishes n_{wqv} water quality variables, each reflecting the average concentration of a specific chemical substance. It has been chosen to include a very simple water quality module in AQUA and make the model flexible, so that it is easy to link a more detailed water quality model to the AQUA simulation model. In both the global study and the Zambezi study, links with other models have been made (Rotmans and De Vries, 1997; UNEP, 1995b, 1996), but in this study I confine myself to the simple approach applied within AQUA. The average concentration of a substance in the river at the outflow of a basin is simulated as follows:

$$c_{avg}(t) = c_{avg,nat} + tmc \times \frac{c_{ww} \times WW_{untr}(t)}{R_{riv}(t)} \quad [\text{mg/l}] \quad (3.36)$$

where $c_{avg,nat}$ represents the natural concentration, c_{ww} the concentration in untreated wastewater, WW_{untr} the annual disposal of untreated wastewater, tmc a transmission coefficient and R_{riv} the river runoff from the basin studied. The transmission coefficient is a factor between zero and 100 per cent, indicating the fraction of the emitted substance which leaves the basin through river runoff. This approach does certainly not do justice to the complex water quality processes which actually take place, but it may be satisfactory as a method to arrive at a rough indication of the effect of wastewater disposal on water quality, provided the transmission coefficient is properly calibrated. A limitation of this approach is that it does not take into account the exchange of chemicals between water and sediments, so that the effect of belated delivery of chemicals from sediments to the water cannot be considered. However, this is only relevant for particular substances such as heavy metals. Depending on the purpose of a particular application of AQUA, it might be necessary to add a more advanced water quality module.

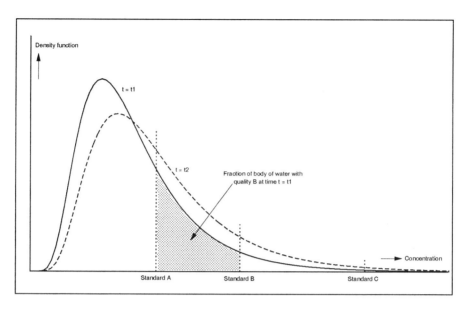

Figure 3.6. Log-normal distribution of the concentration of a substance in a specific body of water at two points in time. The fraction of the body of water falling within a particular quality class is determined on the basis of standards per quality class. In the example given, the fraction of the body of water in quality class A decreases in the period between t=t_1 and t=t_2, while the fractions in classes B, C and D increase.

Data on average concentrations are processed in three steps, to arrive at a distribution of the fresh surface water stock over the n_{wqc} water quality classes (see Section 3.4). First, a log-normal distribution is projected on the average concentrations. Second, it is calculated - for each water quality variable - how the water stock is distributed over the different water quality classes. For this purpose each class is characterized by standards (maximum concentrations) for all water quality variables (Figure 3.6). The fraction Q_A of a body of water which falls within water quality class A, the best quality, is calculated as follows:

$$Q_A(t) = F\left\{ \frac{\ln(c_{st,A}) - \ln(c_{avg})}{\sqrt{\ln\left((c_{dev}/c_{avg})^2 + 1\right)}} \right\} \quad [-] \quad (3.37a)$$

in which c_{avg} represents the average concentration in the body of water, c_{dev} the deviation of the log-normal distribution, $c_{st,A}$ the standard for water quality A and F

the standard normal distribution function. Supposing that three further water quality classes are distinguished, the fractions of the body of water which fall within the other quality classes are calculated by:

$$Q_B(t) = F\left\{\frac{\ln(c_{st,B}) - \ln(c_{avg})}{\sqrt{\ln\left((c_{dev}/c_{avg})^2 + 1\right)}}\right\} - Q_A(t) \qquad [-] \qquad (3.37b)$$

$$Q_C(t) = F\left\{\frac{\ln(c_{st,C}) - \ln(c_{avg})}{\sqrt{\ln\left((c_{dev}/c_{avg})^2 + 1\right)}}\right\} - Q_A(t) - Q_B(t) \qquad [-] \qquad (3.37c)$$

$$Q_D(t) = 1 - Q_A(t) - Q_B(t) - Q_C(t) \qquad [-] \qquad (3.37d)$$

In the third step, the variable-specific distributions over the different water quality classes are translated into one general water quality distribution. The water quality variable showing the worst specific distribution is taken as decisive for the general distribution.

3.7 The impact sub-model

The impact sub-model simulates potential water supply, actual supply to the domestic, irrigation, livestock and industrial sectors, freshwater scarcity, water supply costs and hydropower generation. It further computes population and capital at risk as a result of sea-level rise.

Potential water supply
Potential water supply WS_{pot} is interpreted as the maximum possible annual amount of fresh water which can be supplied from a long-term point of view. Because there are different perceptions of how potential water supply should be defined (see Chapter 2), three alternatives have been implemented:

$$WS_{pot}(t) = \begin{cases} R_{stable}(t) + WS_{fgw}(t) - DR(t) & \text{Alternative I} \\ R_{stable}(t) - iaf \times R_{perc}(t) + WS_{fgw}(t) - DR(t) & \text{Alternative II} \\ R_{tot}(t) + WS_{cons}(t) & \text{Alternative III} \end{cases} \quad [\text{kg/yr}] \quad (3.38)$$

where R_{stable} represents stable runoff, WS_{fgw} total groundwater withdrawal, DR drainage of irrigation water, *iaf* the inaccessible fraction of the natural stable runoff

R_{perc}, R_{tot} total runoff and WS_{cons} consumptive water use. Substitution of the stable runoff component in the first alternative by Equation 3.35 shows that potential water supply in this alternative is considered equal to natural and artificial groundwater recharge plus a fraction of the artificial reservoir volume. The second alternative is similar, but a reduction is made to account for inaccessible water flows. Water flows in remote areas are regarded as (partly) inaccessible only if a river basin is uninhabited or thinly populated throughout, as is the case in the Amazon basin. Water flows in remote mountainous upstream parts of a river basin are not regarded as inaccessible if people can use the water downstream. In the third alternative, potential water supply is defined as the natural total runoff, which is equal to the actual total runoff plus the amount of consumptive water use.[1]

Actual water supply
Total water supply is the sum of the supplies to the domestic, livestock, irrigation and industrial sectors, plus the supply for export out of the area under consideration:

$$WS_{tot}(t) = WS_{dom}(t) + WS_{irr}(t) + WS_{liv}(t) + WS_{ind}(t) + WS_{exp}(t) \quad [kg/yr] \quad (3.39)$$

From an economic point of view, water supply will generally be equal to or approach demand, because prices continuously drive demand and supply in the direction of equilibrium. However, water demand is often not or only partially determined by the market mechanism. In that case, if demand is not an *economic* but a *planned* demand, demand might exceed actual supply (see Section 2.2). Water demand can be managed by, for instance, setting policy targets with respect to the cropland area to be irrigated or the percentage of people to be connected to public water supply systems. If that is the case, expenditure required might exceed available budgets, so that the demand cannot be met due to lack of finance. Therefore the model applies so-called allocation factors to the demands, which have a value between zero and one (a value of one means that supply equals demand). As will be discussed in the next section, the allocation factors are calculated in the response sub-model, on the

---

[1] Each of the three alternatives takes into account the possible effect of consumptive water use on potential water supply as discussed in Section 2.3. In the first two alternatives, consumptive water use (in so far as it returns to land as precipitation) will enlarge net precipitation and percolation, and thus stable runoff and potential water supply. In the third alternative, consumptive water use is explicitly added, so that it will increase potential water supply (again provided it is not lost to the oceanic atmosphere). In Chapter 5 I will show how the three alternatives are applied as a function of the perspective chosen.

basis of maximum allowable expenditure. Domestic, irrigation, livestock and industrial water supplies are calculated as:

$$WS_{dom}(t) = \left(rpp + (1-rpp) \times AF_{pws} \times COV_{pws,dem}(t)\right) \times POP(t) \times WD_{dom,pc}(t) \quad \text{[kg/yr]} \quad (3.40)$$
$$WS_{irr}(t) = AF_{irr}(t) \times WD_{irr}(t) \quad \text{[kg/yr]} \quad (3.41)$$
$$WS_{liv}(t) = AF_{liv}(t) \times WD_{liv}(t) \quad \text{[kg/yr]} \quad (3.42)$$
$$WS_{ind}(t) = AF_{ind}(t) \times WD_{ind}(t) \quad \text{[kg/yr]} \quad (3.43)$$

where AF_{dom}, AF_{irr}, AF_{liv} and AF_{ind} are the respective allocation factors for the four sectors. The amount of water withdrawn for export is assumed to equal the demand.

Freshwater scarcity
As discussed in Chapter 2, not only the concept of potential water supply but also that of water scarcity can be interpreted in different ways. I have implemented two alternative physical definitions of freshwater scarcity, which actually include others if the term 'potential water supply' is substituted by one of its alternative definitions:

$$SCARC(t) = \begin{cases} WS_{tot}(t) / WS_{pot}(t) & \text{Alternative I} \\ WS_{cons}(t) / WS_{pot}(t) & \text{Alternative II} \end{cases} \quad \text{[-]} \quad (3.44)$$

The former definition employs total water supply and the latter consumptive water use for comparison with potential water supply. In Chapter 5 I will discuss how these alternative definitions of scarcity are applied in combination with the different definitions of potential water supply.

Water supply costs
Water supply costs per litre are the outcome of cost curves, which differ for each type of water source, sector and quality of intake water. The cost curves are an input into the model and can be different per application (see Chapters 6 and 8). For withdrawals from fresh surface water and renewable ground water, supply costs are assumed to be a function of freshwater scarcity: the scarcer the water, the higher the cost per litre. The supply costs also depend on the sector: water for drinking is generally more expensive than water for irrigation or industrial use. Finally, the supply costs depend on the quality of the intake water: the worse the quality, the higher the costs, due to additional treatment requirements. It is assumed that the quality of the intake water is distributed over the different water quality classes in accordance with the quality distribution simulated in the state sub-model. The costs of water supply from fossil ground water and sea water are a time-dependent input

into the model. In all cases, calculated costs include both depreciation and investment costs to renew or extend supply infrastructure (fixed costs) and operational and maintenance costs (variable costs).

Hydropower generation

Annual hydropower generation HG is calculated as the product of the hydropower generation capacity HGC and the utilization fraction uf:

$$HG(t) = 31536 \times uf \times HGC(t) \quad [\text{GJ/yr}] \quad (3.45)$$

where 31536 is the conversion factor between MW and GJ/yr. The hydropower generation capacity (in MW) is a result of the annual expenditure on hydropower and the average costs per MW. The model calculates total discounted annual costs, including both depreciation and investment costs and operational and maintenance costs. The annual costs of hydropower are supposed to increase if the hydropower generation capacity increases:

$$\frac{dCOST_{hydr}(t)}{dt} = COST_{hydr}(t) \times \frac{1}{El_C} \times \frac{dHGC(t)/dt}{HGC(t)} \quad [\text{US\$/MW/yr/yr}] \quad (3.46)$$

where El_C is the cost elasticity of hydropower supply. Because the increase in costs per MW will be small in an early phase of hydropower development, but larger if a larger fraction of the hydropower potential is already being used, El_C is supposed to be a function of the ratio of actual to potential hydropower generation capacity, getting smaller if the ratio increases.

Impacts of sea-level rise

The impacts of sea-level rise are calculated according to the method developed by Hoozemans *et al.* (1993) and RA (1994). The world's coasts have been classified into n_{clust} clusters, based on hydraulic characteristics, population density and economic situation. The impacts are calculated per coastal cluster and expressed in terms of people and capital at risk. Each cluster is characterized by a critical water level H_{crit} at which flooding of the coastal lowlands will occur. The critical flooding probability, i.e. the probability that the critical water level will be exceeded, is calculated by:

$$Pr_{crit}(t) = 10^{(\gamma_1 - H_{crit}(t) + slr(t))/\gamma_2} \quad [1/\text{yr}] \quad (3.47)$$

in which *slr* represents the sea-level rise since the initial year. A possible additional increase in the critical flooding probability as a consequence of extra storm surge or local subsidence are ignored, as is a possible natural response by coastlines to sea-level rise. The parameters γ_1 and γ_2 are cluster-specific constants in the sea level-probability relationship. For each coastal cluster, the model calculates how many people (POP_{coast}) and how much capital (CAP_{coast}) will be flooded at probabilities Pr in the range $[0, Pr_{crit}]$, distinguishing a limited number of distinct probability steps. The number of people and the amount of capital to be flooded is larger at the lower probabilities, because higher grounds also will be subject to flooding. To calculate the Pr-dependent variables POP_{coast} and CAP_{coast}, each coastal cluster has been subdivided into n_{elev} elevation classes, each having its own specific area, population density and capital density. The number of people and the amount of capital at risk are calculated as:

$$POP_{rsk}(t) = \sum_{Pr=0}^{Pr_{crit}(t)} \left(\frac{dPOP_{coast}[Pr](t)}{dPr} \times Pr \right) \qquad \text{[people/yr]} \qquad (3.48)$$

$$CAP_{rsk}(t) = \sum_{Pr=0}^{Pr_{crit}(t)} \left(\frac{dCAP_{coast}[Pr](t)}{dPr} \times Pr \right) \qquad \text{[US\$/yr]} \qquad (3.49)$$

Population and capital at risk can be interpreted as the expected number of people and amount of capital subject to annual flooding events.

3.8 The response sub-model

The response sub-model represents societal responses to the developments as simulated in the pressure, state and impact sub-models. Most of the responses are regarded as manageable to some extent and are therefore represented as manageable parameters. An overview of such parameters in the AQUA model is given in Table 3.2. Below I discuss the various response options in the order presented in the table.

Inter-basin water trade
Water abundance in one area and water scarcity in a neighbouring area may lead to a mutual wish to trade water. Actual water trade always depends on a number of factors, not least those of a political nature. Whether water trade takes the form of water trade agreements or of free trade depends on the political situation as well. At present, inter-basin water trade is unusual, but as an option for the future it cannot be neglected.

Table 3.2. An overview of manageable parameters in AQUA.

Manageable parameter	Symbol	Unit	Equation	Manageable through ...
Water demand for water export	WD_{exp}	kg/yr	3.2	Water trade or contracts with neighbouring countries
Technological diffusion rate	d	1/yr	3.8	Information and educational programmes, subsidies
Technological development rate	TD	1/yr	3.9	Research and development programmes
Ratios water price / actual cost	rpc	-	3.50	Water pricing, water taxes
Water source fractions	wsf	-	3.10	Water resources management, subsidy or tax per source
Minimum river runoff after water withdrawal	-	kg/yr	-	Water resources management, legislation
Artificial groundwater recharge	$AGWR$	kg/yr	3.35	Water resources management
Public water supply coverage	$COV_{pws,pol}$	-	3.51	Construction of infrastructure, investments, subsidies
Sanitation coverage	$COV_{san,pol}$	-	3.52	Construction of infrastructure, investments, subsidies
Wastewater treatment coverage	$COV_{wwt,pol}$	-	3.53	Construction of infrastructure, investments, legislation
Maximum acceptable critical flooding probability	$P_{crit,max}$	1/yr	3.54	Dyke heightening in accordance with a policy target
Maximum water expenditure as fraction of GNP	q	-	3.55	Budgeting of public spending

The increase of water-use efficiency: water demand policy

The model distinguishes three mechanisms of response to inefficient water use: the development of water-conserving technology, the actual introduction of new technology and water pricing. The key parameters in the first two mechanisms are the technological development rate and the technological diffusion rate (see Section 3.5). The third mechanism depends on developments regarding the price of water, i.e. the tariff charged to the consumer. Per water-demanding sector, water prices are related to actual water supply costs by:

$$WP[sect](t) = rpc[sect](t) \times COST_{ws}[sect](t) \quad [\text{US\$/kg}] \quad (3.50)$$

where the time-dependent parameter *rpc[sect]* is the ratio of water price to actual costs for water-demanding sector *sect*. Water pricing is understood as a government policy to change the fraction of actual supply costs charged to the consumer. Throughout the world, water prices have traditionally been lower than actual water supply costs, so that today water pricing is generally taken to mean increasing water prices. Water pricing and the other two mechanisms for improving water-use efficiency are partly autonomous processes, but there is in each case also room for policy intervention. The parameters *TD*, *d* and *rpc* are therefore regarded as manageable parameters. These parameters are handles which can be played with by the user of the model, to analyse where policy efforts might be effective.

Water resources management
In the model, the relative use of different water sources is determined by the so-called water source fractions (Equation 3.10). A water source fraction is the ratio of the supply from a certain source to total water supply. The different types of water source distinguished are fresh surface water, renewable ground water, fossil ground water, salt water and imported water. It is difficult to find universal mechanisms which change water source fractions. Factors such as the extent of pollution (mostly of surface waters), the unreliability of surface water availability or energy costs (pumping of ground water costs relatively more energy than diverting surface water) are for example relevant indicators for the choice between surface water and ground water, but it is difficult to quantify the importance of each of these factors. Which source is chosen depends very much on regional circumstances, but - within a given region - the choice appears to differ per water resources development project. Because there are no clear rules which determine water source fractions and government policy can significantly affect them, I have chosen not to model the fractions, but to consider them as manageable parameters. The choice of a certain water source fraction for surface water can be overruled by choosing a minimum monthly river runoff below which no further surface water withdrawals are allowed. If that happens, ground water will be taken. In this way a minimum dry season river runoff is guaranteed for maintaining ecosystems. Another manageable parameter in the model with regard to water resources management is artificial groundwater recharge through deep well injection, meant to increase stable runoff and purify the water injected.

Public water supply and sanitation
In analysing a large number of national data on public water supply coverage and gross national product per capita, I found a weak statistical correlation between the two, but not enough to postulate a causal relationship. Nevertheless, the water supply coverage of a nation seems to be mainly a socio-economic issue. Environmental factors such as natural water availability or pollution, for example, do not correlate at all. The same is true for the number of people with access to

sanitation facilities. Two schools of thought can be distinguished with respect to the development of public water supply coverage and sanitation. One states that economic development helps to improve water supply and sanitation conditions, while the other maintains that it is improved water supply and sanitation which creates the conditions for economic development. In accordance with the latter school, Young *et al.* (1994) assert that an increase in the provision of basic water and sanitation services is a *prerequisite* for improved health and for sustainable social and economic advancement. To conform to the two schools, I have implemented two alternative formulations of the public water supply and sanitation coverage:

$$COV_{pws\,dem}(t) = \begin{cases} \text{Function}(GNP_{pc}(t)) & \text{Alternative I} \\ COV_{pws.pol}(t) & \text{Alternative II} \end{cases} \quad [-] \quad (3.51a)$$

$$COV_{pws\,act}(t) = AF_{pws}(t) \times COV_{pws\,dem}(t) \quad [-] \quad (3.51b)$$

$$COV_{san\,dem}(t) = \begin{cases} \text{Function}(GNP_{pc}(t)) & \text{Alternative I} \\ COV_{san.pol}(t) & \text{Alternative II} \end{cases} \quad [-] \quad (3.52a)$$

$$COV_{san\,act}(t) = AF_{san}(t) \times COV_{san\,dem}(t) \quad [-] \quad (3.52b)$$

In alternative I, the coverage demand is a function of gross national product per capita. In alternative II, the demand depends on the policy target set. In both cases, the actual coverage depends on the demand and the allocation of means. AF_{pws} and AF_{san} are the allocation factors for public water supply and sanitation respectively (see Equation 3.56).

Wastewater treatment
Domestic and industrial wastewater treatment has been modelled in a similar way to public water supply and sanitation. The fraction of wastewater which is treated before disposal is supposed to be a function of either economic development or policy:

$$COV_{wwt\,dem}(t) = \begin{cases} \text{Function}(GNP_{pc}(t)) & \text{Alternative I} \\ COV_{wwt.pol}(t) & \text{Alternative II} \end{cases} \quad [-] \quad (3.53a)$$

$$COV_{wwt\,act}(t) = AF_{wwt}(t) \times COV_{wwt\,dem}(t) \quad [-] \quad (3.53b)$$

Coastal defence
Actual or expected impacts of a sea-level rise will result in a demand for coastal defence measures such as heightening of dykes and dune reinforcements. It is assumed that the basic demand for additional coastal protection is to keep, per

coastal cluster, the critical flooding probability Pr_{crit} below a certain maximum level $Pr_{crit,max}$. The maximum level used in the model is the critical flooding probability in 1990. However, as a policy option, the basic demand for coastal protection can be adjusted by changing the maximum level. To meet this demand, the actual critical water level H_{crit} should be raised as follows:

$$H_{incr\_dem}(t) = \gamma_1 - \gamma_2 \times log(Pr_{crit,max}(t)) + slr(t) - H_{crit}(t) \quad [m] \quad (3.54)$$

The investment needed for this increase in the critical water level is calculated in three steps. First, the length of the coast to be strengthened is calculated, followed by the average rise in height and, finally, the costs of protection, which are a result of the protection length, the average rise, and the cost per unit length and unit rise. The model takes three types of coastal protection into account (stone-protected sea dykes, clay-covered sea dykes and sand dunes), for each of which the cost of increasing the critical water level is different.

Expenditure
The annual expenditure required to meet a given water demand is calculated as the product of the demand and the costs of water supply. Expenditure needs are calculated separately for the domestic, irrigation, livestock and industrial sectors. In a similar way, the model calculates required expenditure for sanitation, hydropower, domestic and industrial wastewater treatment and coastal defence. The calculated expenditure includes both investment and operational and maintenance costs. The costs of water supply from fresh surface or ground water are calculated in the impact sub-model, as are the costs of hydropower generation (Equation 3.46). Costs of desalination (in US$/litre), fossil groundwater use (in US$/litre), sanitation (in US$/cap/yr), domestic and industrial wastewater treatment (in US$/litre) and dyke heightening (US$/m/km) are regarded as exogenous scenarios. For each item, the required expenditure is compared to the maximum allowable expenditure. The actual expenditure EXP_{act} is the lesser of the demanded expenditure EXP_{dem} and the available budget EXP_{max}:

$$EXP_{act}(t) = min(EXP_{dem}(t), EXP_{max}(t)) \quad [US\$/yr] \quad (3.55a)$$
$$EXP_{max}(t) = q(t) \times GNP(t) \quad [US\$/yr] \quad (3.55b)$$

where q represents the maximum allowable expenditure expressed as a fraction of gross national product. Parameter q may differ by item and vary in time, and is regarded as a manageable parameter. Per item, the allocation factor is calculated as:

$$AF(t) = EXP_{act}(t) / EXP_{dem}(t) \quad [-] \tag{3.56}$$

Allocation factors naturally have a value between zero and one.

3.9 Discussion

After presenting the principal equations and assumptions of the model, let me briefly summarize some of the main characteristics of the model in terms of possibilities and limitations. The model is typically designed to explore changes in the long term, in the order of several decades to one or two centuries. This means that the model can be useful in formulating long-term priorities, but not in supporting five- or ten-year planning studies. A major limitation is that depreciation and investment costs on the one hand and operational and maintenance costs on the other are expressed in one formula, which means that the costs calculated represent *long-term average* costs and not necessarily the actual costs in a specific year. The model assumes that depreciation costs of the existing infrastructure are taken into account. If this is in fact not done, it will result in relatively low total costs after investment has been made and relatively high total costs if new investment is needed. For planning on a scale of between five years and one or two decades it would be important to distinguish between fixed and variable costs, but for exploring long-term futures this is probably less important.

Another characteristic of the model is its design as a policy support tool. For this reason, several manageable parameters have been distinguished. I would like to point out that this choice strongly influences the possibilities and limitations of the model. By 'parameterization' of policy responses to a large extent, the model does typically not *explain* policy response or human behaviour as it will evolve. On the contrary, the model considers policy response open to question. Whereas a purely descriptive approach is intended to find and lay down the rules which determine human behaviour, the approach followed here is meant to analyse the effect of alternative policy strategies. Policy-making is often nothing less than an effort 'to set the rules', an effort which is better supported by a model where different policy response strategies can be fed in in the form of input parameters, than by a model where responses are defined endogenously according to certain fixed rules. However, the other side of the coin is that one should be careful in explaining model outcomes in terms of plausibility. Through the possibilities in the model to 'steer' the future, one could get the idea that it is possible to obtain any particular future 'as desired'. But the plausibility of a given model outcome depends heavily on the plausibility of the parameter values fed into the model. The fact that some parameters have been labelled 'manageable' does not imply that all parameter values are possible or equally plausible. This should be kept in mind when using the model. To obtain some perception of what might be regarded as possible and what not, Chapter 5 will discuss a number of perspectives on the 'manageability' of different processes. These perspectives will be applied in the chapters where the model is actually used.

4 Water indicators

The simulation model of AQUA has too many variables and the output of one run is too extensive to easily get an impression of what is the most essential information produced. This is in fact the case with most simulation models and it often challenges the usefulness of these models for policy analysis. To bridge the gap between multiple model results and the demand for comprehensive information, indicators have been linked to the simulation model, aggregating output data into meaningful information. To be able to provide information of different degrees of detail, a tree of information has been designed, consisting of three different levels of comprehensiveness. The lowest, most detailed level consists of the relevant model outcome. On the second level, the model outcome has been processed into composite indicators, often a weighting of different pieces of information or a comparison of a certain variable with some reference variable. The third, most comprehensive level provides the most aggregated information, in the form of overall pressure, state, impact and response indices.

4.1 The link between indicators and simulation model

In the world of policy makers and analysts, an indicator is generally understood to be an instrument for communicating key information about a system in a simplified form to policy makers and the general public. As Ott (1978) puts it, an indicator is a means devised to reduce a large quantity of data down to the simplest form, while retaining their essential meaning for the questions asked. Many authors have addressed the question of what properties a useful indicator should have. Some of the characteristics often mentioned are: policy-relevant, scientifically-based, recognizable, appealing, observable, reproducible, sensitive to change in time, comprehensive, linking up with available data and relatively easily applicable (Liverman *et al.*, 1988; Bakkes *et al.*, 1994). A property I would like to stress here is the possibility of being linked to the output of model studies, so that model results can be made easily accessible. Particularly in discussions on sustainable development, where the long-term future is so much part of the debate and where forecasting studies necessarily play an important role, there is a great need for indicators which translate model results into useful information.

Within the AQUA framework indicators have been linked to the simulation model, as schematically shown in Figure 3.1. The tree of indicators is represented in enlarged and elaborated form in Figure 4.1. Three levels of information are distinguished, to be able to provide information of different comprehensiveness. The first, most detailed level consists of relevant model outcomes, such as water demand per sector, wastewater production, potential water supply, sea-level rise, groundwater-level decline, actual water supply and water supply costs per sector. On the second level, the model outcome has been processed into composite indicators, mostly a weighting of different pieces of information or a comparison of a certain variable with a reference variable. The third, most comprehensive level provides aggregate information in the form of overall pressure, state, impact and response indices.

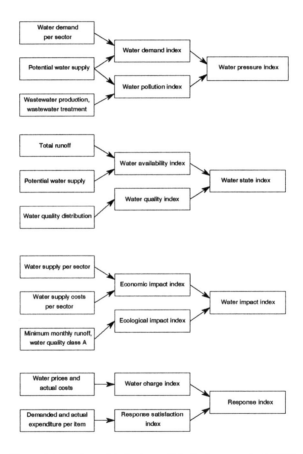

Figure 4.1. The tree of indicators within the framework of AQUA.

The water pressure index includes information on water demand and pollution compared to the amount of water available. The water state index contains information on freshwater availability and freshwater quality. The water impact index includes information on both socio-economic and ecological impacts of changes in the water system. The response index gives information on the extent to which required expenditure is actually spent and the degree to which water charges cover actual costs. Normative elements, included in the aggregation procedures, can be set according to the user's preference. The distinction between pressure, state, impact and response indicators links up with an approach which is currently used in several international institutions, such as the Organisation for Economic Co-operation and Development, the United Nations Environment Programme, the United Nations Commission on Sustainable Development, and the World Bank (see e.g. OECD, 1993; Bakkes et al., 1994; Swart and Bakkes, 1995; Gouzee et al., 1995; Bakkes and Van Woerden, 1997). Several authors hold the notion that indices refer to a higher level of aggregation than indicators (e.g. Rotmans, 1997). However, because it is difficult to make a clear distinction, the terms are often interchangeable. At each level of aggregation, pieces of information have a particular range. Indicators on a relatively detailed level tell something about one particular thing and not about the whole system. Aggregate indices say something about a collection of things (possibly a whole system or subsystem), but they refer to only a few system properties and are sensitive for a limited number of system changes.

4.2 Pressure indicators

The purpose of a water pressure indicator is to provide information on the pressure of human activities on the water system. Two of the most important of these pressures are offstream water use and disposal of untreated wastewater. As a measure of the first kind of pressure, a water demand index WDI is calculated as:

$$WDI(t) = \frac{WD_{tot}(t)}{WS_{pot}(t)} \qquad [-] \qquad (4.1)$$

in which WD_{tot} represents total water demand and WS_{pot} potential water supply. If the demand index is one, the annual volume of water demand equals the upper limit of available water resources. The pressure of untreated wastewater on the water system can be expressed by multiplying the total untreated wastewater flow by a dilution factor and dividing the product by the potential water supply. In this way, a water pollution index WPI is calculated:

$$WPI(t) = \frac{df \times WW_{untr}(t)}{WS_{pot}(t)} \qquad [-] \qquad (4.2)$$

where WW_{untr} stands for the total annual disposal of untreated wastewater. The dilution factor df represents the extent to which concentrations of pollutants in untreated wastewater have to be diluted in order to attain an 'acceptable' level, say ten to twenty times (see Section 6.3). An overall pressure index PI has been defined as the sum of the water demand and the water pollution index:

$$PI(t) = WDI(t) + WPI(t) \qquad [-] \qquad (4.3)$$

This index shows the combined pressure from water demand and pollution and illustrates the possible trade-off between the two. Demand and pollution draw upon the same amount of water, and therefore increased demand requires decreased pollution if one does not want to add to the pressure on the water system.

4.3 State indicators

Whereas water pressure indicators refer to human activities which will change the conditions of the water system, water state indicators relate directly to the conditions of the water system. In the past few decades a large number of state indicators has been proposed and applied, from various hydrological coefficients and moisture and aridity indices (e.g. L'vovich, 1973a, 1973b; Korzun et al., 1978; Shiklomanov, 1989) to several water quality indices (e.g. Couillard and Lefebvre, 1985; House, 1990). It has become clear that different objectives require different indicators, so that a single 'best indicator' does not exist. The indicators discussed below are considered useful in the context of long-term explorative water studies and have been incorporated in the AQUA framework. The first such indicator is the runoff coefficient, defined as the ratio of total runoff to total land precipitation:

$$C_R(t) = \frac{R_{tot}(t)}{P_{land}(t)} \qquad [-] \qquad (4.4)$$

If the water storage on land remains constant, this is equal to one minus the ratio of evaporation to precipitation. The runoff coefficient is used so often by hydrologists that I felt it should be calculated within the AQUA framework, but one could question what this coefficient actually shows. It is often assumed that evaporation is a 'loss' of water, so that a large runoff coefficient, indicating relatively large runoff, would be preferable. However, as Savenije (1996) argues, this is not necessarily true. Let us consider for instance what happens if afforestation or consumptive water use leads to increased evaporation ΔE. A part of ΔE will be lost from the area under consideration ($v \times \Delta E$) and the remaining part will return to the land as precipitation

($\Delta P = (1-v) \times \Delta E$). As a consequence, total runoff will decrease to some extent ($\Delta R = -v \times \Delta E$). The runoff coefficient will become smaller for all values of v, but the new situation is not necessarily worse. If v approaches one, the new situation is indeed worse (from a hydrological point of view!), because the additionally evaporated water is lost to a neighbouring area (or the ocean) and total runoff decreases. But if v is near zero, water is being recycled (precipitation increases) and total runoff remains constant, which is not undesirable at all. One can conclude from this example that a change in the value of the runoff coefficient alone does not say much about the worsening or improvement of the hydrological situation. In fact, if the coefficient decreases as a result of increased precipitation it is a positive development, but if it decreases as a result of decreased runoff it is a negative development.

Another hydrological indicator often used and calculated in AQUA is the ratio of actual to potential evaporation, sometimes called the evaporation efficiency (Rind *et al.*, 1990). Following Leemans and Van den Born (1994), I use the term moisture index:[1]

$$MI(t) = E_a(t) / E_p(t) \quad [-] \tag{4.5}$$

The index is calculated per physical basic area and represents the extent to which atmospheric demand for water is met. It gives information on the moisture conditions in a specific area. The index is calculated per month, but the annual average is also computed. According to Korzun *et al.* (1978), the productivity of the vegetation cover and the overall reserves of phytomass decrease abruptly when the moisture index over a year falls below 0.5. A further moisture indicator calculated per physical basic area is the soil moisture index *SMI*, the ratio of actual soil water content to the soil water-holding capacity:

$$SMI(t) = \frac{S_{soil}(t)}{S_{cap}} \quad [-] \tag{4.6}$$

In addition to the runoff coefficient and the moisture indices, a water availability index *WAI* is calculated, as the fraction of total runoff which is considered available for water use:

$$WAI(t) = \frac{WS_{pot}(t)}{R_{tot}(t)} \quad [-] \tag{4.7}$$

---

[1] A reciprocal *aridity* index can be obtained by taking one minus the moisture index (Van Deursen, 1995).

An increasing value in the water availability index may indicate, for instance, the beneficial effect of artificial surface reservoirs or artificial groundwater recharge. A decreasing value on the other hand might indicate the negative effect of for example deforestation, loss of wetlands or urbanization. Climate change could have both a positive and a negative effect, depending on the assumptions made and the region considered. The interpretation of the water availability index depends on the perception of the concept of potential water supply (see Sections 5.3 and 5.4). Other relevant indicators with respect to the state of the water system are sea-level rise and groundwater-level decline, both of which are readily available in the output of the model.

A simple water quality index WQI is defined as the fraction of fresh surface water falling in water quality class A (good quality, suitable for the maintenance of natural aquatic ecosystems, see Equation 3.37a):

$$WQI(t) = Q_A(t) \qquad [-] \qquad (4.8)$$

An overall water state index is computed as the product of the water availability index and the water quality index:

$$SI(t) = WAI(t) \times WQI(t) \qquad [-] \qquad (4.9)$$

Defined in this way, the overall state index is a measure of the fraction of total runoff which is available and of good quality.

4.4 Impact indicators

The aim of impact indicators is to show what environmental changes mean for the functioning of a society or ecosystem. Some relevant impact indicators are readily available from the model output. Water supply costs per sector for instance are useful measures for the possible effect of increasing water scarcity on the several water-demanding sectors. Population and capital at risk in coastal areas are useful indicators of the possible socio-economic impacts of sea-level rise. The distribution of the total body of fresh surface water over different water quality classes tells something about the suitability of surface waters to maintain natural aquatic ecosystems. Below, I discuss how some model outputs are further processed into more aggregate indices. *Average* costs of water supply are calculated on the basis of the costs per sector and the relative water use per sector:

$$COST_{ws}(t) = \frac{\sum_{sect=1}^{n_{sect}} WS[sect](t) \times COST_{ws}[sect](t)}{\sum_{sect=1}^{n_{sect}} WS[sect]} \qquad [\text{US\$/kg}] \qquad (4.10)$$

An economic impact index II_{econ} is computed as the relative increase in water supply costs since a specific reference year:

$$II_{econ}(t) = \frac{COST_{ws}(t) - COST_{ws,i}}{COST_{ws,i}} \qquad [-] \qquad (4.11)$$

A value of zero means that the economic value of water has not changed since the reference year. A value above zero means that water has become scarcer in an economic sense. A value below zero means that water has become cheaper, in which case there are no negative impacts of water on the economy.

Apart from these economic impact indicators, some ecological impact indicators are calculated. For each physical basic area, the length of the growing period *LGP* is calculated as the number of months within one year in which precipitation exceeds 50 per cent of the potential evaporation and temperature exceeds five degrees centigrade (Oldeman and Van Velthuyzen, 1991). Changes in the length of the growing period are regarded as useful information on the water supply to terrestrial ecosystems.

Another measure calculated is the 'rain erosivity index' *REI*. 'Rain erosivity' is one of the factors determining surface erosion and may change if rainfall patterns change, for instance as a result of global warming, land use changes or consumptive water use. Many different indices of rain erosivity have been developed (Bergsma, 1981). Because of its simplicity, it has been chosen to use the rain erosivity index devised by Arnoldus (1978), which depends on monthly precipitation data only. This index is calculated over a year, per physical basic area, as:

$$REI(t) = \frac{\sum_{j=1}^{12}\left(P(t-(j-1)/12)\right)^2}{\sum_{j=1}^{12} P(t-(j-1)/12)} \qquad [-] \qquad (4.12)$$

The functioning of aquatic ecosystems depends on both hydrological and water quality variables. As a measure of the hydrological situation, the minimum monthly river runoff in a year is taken. Especially consumptive water use can significantly reduce river runoff in the driest month. As a measure of the water quality conditions, it has been chosen to take the fraction of the surface-water stock classified as quality class A. This is the 'good quality' class, representing water suitable for the maintenance of natural aquatic ecosystems. An ecological impact index II_{ecol} is calculated as the fraction of water which does not meet quality A, multiplied by the relative change in the minimum monthly river runoff:

$$II_{ecol}(t) = \frac{R_{riv,min,i}}{R_{riv,min}(t)} \times (1 - Q_A(t)) \qquad [-] \qquad (4.13)$$

where $R_{riv,min}$ represent the river runoff in the driest month in a given year (in kg/month) and $R_{riv,min,i}$ the same but in the reference year. The ecological impact index gives information about the loss of natural, pristine bodies of water and the loss of water in the driest month of the year. The value is equal to or greater than zero. A value of zero can be interpreted as an absence of ecological impacts. An overall impact index II is calculated as the weighted sum of the economic and the ecological impact index:

$$II(t) = \frac{II_{econ}(t) + II_{ecol}(t)}{2} \qquad [-] \qquad (4.14)$$

In this equation, the two components have the same weight, but the user of the model can apply alternative weighting factors.

4.5 Response indicators

The aim of a response indicator is to provide policy-relevant information on the societal response to water system changes or socio-economic or ecological impacts. The water expenditure index WEI has been defined as the total expenditure in the water sector as a fraction of gross national product:

$$WEI(t) = \frac{\sum EXP_{act}(t)}{GNP(t)} \qquad [-] \qquad (4.15)$$

A water charge index WCI represents the average ratio of water price to actual water supply costs and is calculated as the weighted average of the sector-specific ratios:

$$WCI(t) = \frac{\sum (rpc[sect] \times WS[sect])}{\sum WS[sect]} \quad [-] \quad (4.16)$$

Parameter rpc_{sect} represents the ratio of water price to actual water supply costs for sector *sect* (see Section 3.8) and WS_{sect} is the water supply to this sector. It is interesting to consider the relationship between the water charge index and the water expenditure index. An increase in water prices, to cover the actual costs, will increase the water charge index but may reduce the water expenditure index (because of increased water-use efficiency). This illustrates that water policy does not necessarily just cost money, it can also save money.

A response satisfaction index *RSI* gives information about the extent to which the response demanded is satisfied and this is defined as the ratio of actual to demanded expenditure in the water sector:[1]

$$RSI(t) = \frac{\sum EXP_{act}(t)}{\sum EXP_{dem}(t)} \quad [-] \quad (4.17)$$

An overall response index *RI* is calculated as the weighted sum of the response satisfaction index and the water charge index:

$$RI(t) = \frac{RSI(t) + WCI(t)}{2} \quad [-] \quad (4.18)$$

A value of one means that demanded expenditure is met and water charges cover full costs.

---

[1] The response satisfaction index can be regarded as the 'allocation factor' for the water sector as whole (cf. the allocation factors per item in Equation 3.56).

5 Perspectives on water

In the previous two chapters the generic AQUA tool has been described in detail, but one topic has been left for this chapter: how to handle uncertainties. This chapter focuses on one particular method of uncertainty analysis, the perspective approach, which plays a central role in the rest of this book. It has been chosen to use four perspectives from the cultural theory as a rationale in explaining historical developments and exploring possible futures: the hierarchist, egalitarian, individualist and fatalist perspectives. Each includes a specific perception of how the world functions (world-view) and how people act (management style). A short introduction to the cultural perspectives in general is followed by a more elaborate discussion of perspectives on water. The various perceptions of different water issues which have been discussed in Chapter 2 are now placed within the context of four coherent points of view. The last section of this chapter discusses how the perspectives have been implemented in the simulation model of AQUA by applying different model formulations.

5.1 Uncertainties: a perspective approach

People have always been interested in the future, not only in what might happen, but also in what their role could be in shaping it. As great uncertainties are inherent in any exploration of long-term future developments, one needs an approach which somehow takes these uncertainties into account. In the first chapter of this book, it was mentioned that there are many different types and sources of uncertainty and various methods to deal with unpredictability were discussed. In policy research two approaches in particular have become widespread, sometimes applied separately and sometimes in combination. One of these methods is probability theory, in which the *uncertainty range* around a central projection is calculated. This approach is very useful if, except for the values of a number of parameters involved, the functioning of a system is relatively well known. Because this is not the case in most real world systems, especially not if humans are involved, this approach has often not been entirely satisfactory. The second approach is to introduce *scenarios*, defined here as alternative future developments which could evolve under different assumptions. The main advantage of this approach is that it makes it possible to vary the basic assumptions behind projections of the future, thus taking not only uncertainties in the values of some system parameters into account, but also uncertainties in the rules which determine the system's behaviour. In practice, however, researchers often develop scenarios by varying just a few parameter values, so that the scenarios are relatively close to each other. The advantage of presenting different scenarios is then turned into the risk of suggesting a relatively great certainty about the future, by showing the diversity within *one type* of future only. Other types of future, which differ much more fundamentally, remain unexplored. The recent water supply scenarios developed by the Stockholm Environment Institute (Raskin *et al.*, 1997), which were discussed in Section 2.2, are an example of this approach. It seems that the desire to know the future makes it difficult to accept that the range of possible or plausible futures is both very large and divergent. Another reason might be that few people - least of all policy makers - will be merciful to researchers who conclude that in fact *anything* can happen. Policy makers generally expect conclusions from researchers on what will *probably* happen, preferably as accurate as possible. However, it is unlikely that policy makers are effectively helped if researchers suggest a greater certainty about future developments than can be justified. It might make decision-making easier, but such decisions will be injudicious, based on a limited view of possible futures, which might result in undesirable surprises.

Occasionally, the scenario approach is used in combination with uncertainty ranges. This has been done for example by the Intergovernmental Panel on Climate Change (IPCC) in its projections of climate change. Various *scenarios* for greenhouse gas emissions have been developed and, for each scenario, *uncertainty ranges* have been introduced to show the climatic response (Houghton *et al.*, 1996). Although this combined approach is relatively refined, it still carries a problem which is inherent in both the scenario approach and the probability method. This problem is that none of the approaches has an answer to the question of how alternative basic assumptions or numerical uncertainties relate to each other.[1] Because most real world systems include many parts where alternative assumptions can be made and where unknown parameter values play a role, a large number of possible combinations of assumptions and parameter values exists, which however are not all necessarily equally plausible. Some combinations might for example be internally contradictory. If one does not know how different assumptions might interdepend or how uncertainties are related, one cannot do other than compose a few scenarios on an *ad hoc* basis. Therefore, as part of the scenario approach, one needs a rationale for choosing which sets of basic assumptions are more plausible than others, i.e. a rationale for choosing which scenarios to develop. Similarly, for the probability method one needs a rationale for the interdependence of different numerical uncertainties. Although any theory about such interdependency might

---

[1] 'Relate' can be understood here in both a physical sense - how do uncertainties relate from a *material* point of view - and a spiritual sense - how do uncertainties relate according to people's *perception*. See discussion in Section 10.2.

itself be uncertain and questionable, one should be aware that assuming independence is equally a kind of rationale, by no means a priori more plausible than any other.

There have been numerous attempts to use a rationale for composing scenarios. *Storytelling* is probably the most common approach: making scenarios plausible by explaining how they come into being. This approach has been followed by for instance Meadows *et al.* (1972, 1991), in their exploration of possible future worlds on the basis of the World3 model. In this approach, the plausibility of a scenario depends on the cogency of the story. Within the tradition of storytelling it has become the practice to give scenarios befitting names, to explain what kind of future one is exploring. *Business as Usual* and *Conventional Development* for example refer to futures which will evolve if people 'stick to their present way of life'. The term *Sustainable Development* is used for a future in which people 'care for their children'. The problem, however, is that the choice of story is rather arbitrary. Concepts such as 'our present way of life' or 'caring for our children' can be understood in several different ways, hence for instance the controversy between environmentalists and economists in the debate on sustainable development. Ecological sustainability is clearly not the same concept as sustained economic growth, and a future based on the former concept will differ considerably from a future based on the latter (Lélé, 1991). More systematic approaches than storytelling use coherent views of how the world functions and how people behave. In their report on possible sustainable pathways for the Netherlands for example, the Dutch Scientific Council for Governmental Policy uses the four management perspectives utilizing, saving, controlling and protecting (WRR, 1994). Each of these perspectives includes a given coherent view on scarcity of resources and risks to nature and mankind. Nevertheless, one can still ask why it is these perspectives which are considered and not others. A few researchers have noticed this problem and have started to look for an *empirical* basis for perspectives, which should be found in the social sciences.

The social sciences offer a large variety of more or less empirically-founded theories on people's perceptions of life and nature. Max Weber for instance formulated a number of 'styles of life', each having a particular world-view and certain moral values. He distinguished three pure styles: traditional, charismatic and legal domination. In the first type people put a high value on what had always existed, in the second on leadership, and in the third on rationality (Bendix, 1977). Hofstede's theory (1991) introduced a four-dimensional cultural typology, which categorized cultures along four dimensions: power distance (from small to large), collectivism versus individualism, femininity versus masculinity and risk avoidance (from weak to strong). A fifth dimension was identified later: directed by long-term or short-term interests. In a comprehensive analysis of cultures by Thompson *et al.* (1990), cultures are characterized along two dimensions: degree of incorporation of individuals into groups and extent to which the life of individuals is determined by others. In this theory, known as the 'cultural theory', five 'viable ways of life' are distinguished, each with its own distinctive myth of nature: the hierarchist, egalitarian, individualist, fatalist and hermit. Van Amstel *et al.* (1988) formulated five basic 'views on nature': a classical, development, functional, ecosophical and sustainable technology view. Colby (1991) described five fundamental 'paradigms' of environmental management: frontier economics, environmental protection, resource management, eco-development and deep ecology, each characterized by a distinctive human-nature relationship and management strategy, a dominant imperative and particular threats and themes. Philosophers also contribute to the discussion, by proposing various classifications of attitudes towards nature, using opposites such as anthropocentrism and ecocentrism or gradual classifications from despotism, enlightened rule, stewardship, partnership, and participation in nature to oneness with nature (Passmore, 1980; Barbour, 1980; Achterberg, 1986; Zweers and De Groot, 1987). As noted, a few researchers have started to use such life-style and world-view classifications as a rationale for developing scenarios. Bossel (1996) for example has developed two global change scenarios using the perspectives competition and partnership. In this research, however, I experiment with using the cultural theory of Thompson *et al.* (1990), following Rotmans and De Vries (1997), Van Asselt *et al.* (1996) and Van Asselt and Rotmans (1996).

The reason for using the cultural theory as a rationale in exploring possible futures is that - in my opinion - this theory is most appropriate if one uses the following criteria: empirical basis, universality (in time and space), comprehensiveness (does the theory cover both life-styles and world-views), elaborateness, and applicability to the subject (in this case 'management of natural resources'). Most of the social classifications mentioned above lack a sound empirical basis. Especially the philosophical classifications have a highly theoretical nature, reflecting how people can, could or should think and act rather than how people actually *do* think and act. Of the theories mentioned, the theories of Weber, Hofstede and Thompson are probably those relatively well-founded in empiricism. Although these theories are also criticized for their empirical basis, one should consider this within the current context, in which *all* theories in the social sciences are criticized for not being empirically sound. Cultural theory is not immune to this kind of criticism either, but with roots in anthropological field studies it does not have a particularly weak empirical basis if compared to the other theories. The second criterion, universality, can lead to the same kind of problems as that of empirical soundness. In fact, none of the theories really deserves the designation universal, because there are too many points of ambiguity in each. The theories can at best *explain* something. Their predictive value is low, but again, *all* theories in the social sciences suffer from these defects. Nevertheless, the authors of the cultural theory claim that their theory is universally applicable. References are made to different cultures throughout the world, including primitive cultures. The five ways of life are supposed to evolve from one into another, a certain way of life may dominate in a given period, but each way of life is considered to be essential at all times. By contrast, some of the other theories include a notion of development or evolution in one direction only. Colby (1991) for instance suggests a paradigmatic evolution from the primordial opposites of frontier economic and deep ecology towards environmental protection, resource management and finally eco-development. A similar notion of evolution can often be found in the philosophical classifications of the relationship between man and nature. This makes these theories less useful for my purpose, because the different perspectives are not equal, but one (the most modern one) is considered to be preferable to the others. Most of the theories mentioned are quite comprehensive, in the sense that they treat a large variety of subjects, from social organizations, values and beliefs to perceptions of the functioning of the world. However, some, such as the theories of Hofstede, Van Amstel and Colby, are as yet less elaborate than the others. Finally, the theories differ in their applicability to the field of natural resources management. Several have been developed with a strong emphasis on the relationship between man and environment and thus naturally address the issue of environmental management (Van Amstel, Colby and the philosophical classifications), but others lay more emphasis on social organization (Weber and Hofstede). The precursors of the cultural theory are anthropological descriptions of ways of life, but the current theory as formulated in Thompson *et al.* (1990) provides for a strong relationship between ways of life and ideas on nature and management. It is not my intention to evaluate the cultural theory further. I prefer to leave this to social scientists and limit myself to taking the theory as the starting point for my research.

I will now explain the approach followed. As shown in Figure 5.1, six steps are distinguished in the perspective-based scenario approach. The first step is to identify major uncertainties and controversies which have to be handled. Which type of uncertainties have to be dealt with depends largely on the type of policy questions to be answered. As shown in Chapter 2, questions on the sustainability of freshwater use bring with them a number of uncertainties on fundamental assumptions, for instance on the definition of concepts such as water demand, potential water supply and water scarcity. The second step is to detect the areas in the model which represent weak knowledge. These areas are the places where assumptions will have to be varied in order to explore alternative futures. In the third step, coherent perspectives on water are formulated, on the basis of the uncertainties and controversies identified

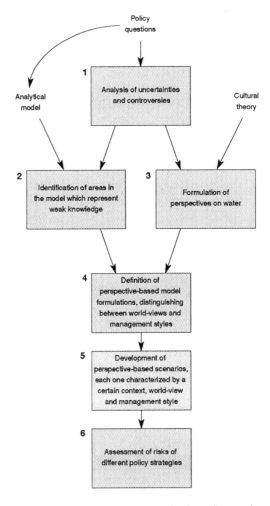

Figure 5.1. Distinctive steps in the perspective-based scenario approach.

and with the help of the cultural theory. In the fourth step, it is analysed how the uncertain quantities and relationships in the model will be interpreted according to the different perspectives. In this step, a number of perspective-based model formulations is defined. A distinction is made between model formulations which refer to the autonomous behaviour of the system (world-views) and model formulations which refer to how the system is or should be managed (management styles). Applying a specific perspective means using a particular set of model formulations. The model is calibrated for each of the perspectives, on the basis of historical data. In the fifth step, perspective-based scenarios are developed. Utopias are scenarios in which world-view and management style correspond to one particular perspective. Additionally, exogenous, contextual developments are assumed to conform to this particular perspective. Dystopias are scenarios where world-view, management style and contextual developments do not all correspond to the same perspective. In the sixth step, the risks of different policy strategies are analysed by analysing the dystopian futures: what happens if a certain management style, which corresponds to one particular perspective, is applied in a world which appears to behave according to another perspective. Such risk analysis can result in recommendations for policy priorities to safeguard a sustainable supply of fresh water. By analysing the risks of policy strategies at a level beyond that of one particular perspective, it is possible to formulate priorities which exceed the preferences of the individual perspectives.

5.2 Cultural theory

Thompson *et al.* (1990) describe cultural theory as a theory about socio-cultural viability which explains how ways of life maintain (or fail to maintain) themselves. They argue that the viability of a way of life depends upon a mutually supportive relationship between a particular cultural bias and a particular pattern of social relations. It is claimed that there are five (and only five) viable ways of life. As a result of societal and environmental changes, people can be dislodged from their way of life into a different way of life. A persistent pattern of surprises, understood as discrepancies between the expected and the actual, forces individuals to look for alternative ways of life which can provide a more satisfying fit with the world as it appears to be. It is claimed that each way of life depends upon each of the four rival ways of life, so that the plurality is not merely fortuitous but a prerequisite. As Thompson *et al.* (1990) formulate it: to destroy the other is to murder the self. As main references for the cultural theory, I have used Thompson (1988), Thompson *et al.* (1990) and Schwarz and Thompson (1990). However, one of the roots of the cultural theory lies in the anthropological research of Mary Douglas.

In her book *Natural Symbols* (1970), Douglas introduced a group-grid typology of cultures, based on a comparison of social structures and corresponding ideas about ritual, sin and self. She argues that the character of social relations can be described along two axes: group and grid. The group axis represents the degree to which an individual is incorporated into confined units. A positive score on the group axis means that an individual strongly feels that he or she belongs to a group. The grid axis denotes the extent to which an individual's life is circumscribed by externally imposed prescriptions, or in other words, the extent to which external rules determine someone's behaviour. A positive score on the grid axis indicates a high role definition, strong regulation of interactions between people and little room for individual choice. On the basis of the two dimensions, Douglas proposes to distinguish four types of social relations. The combination of high group and high grid refers to social groups where individuals are involved with other people, but separated from them by numerous limits and boundaries. In the case of high group but low grid, all status is insignificant apart from one kind, the status involved in belonging to a defined group. Low group plus low grid means that individuals are free from social constraints: group organization barely exists and fixed rules for behaviour are lacking. In the last combination, low group and high grid, individuals do not belong to a circumscribed group, but they are nevertheless constrained in their relations with other people. After a detailed account of social relations, Douglas describes how four different types of cosmological perception overlay the four types of social structure. A high group/grid society is accompanied by a perception of the cosmos as rational and regulative. In the case of high group but low grid, the cosmos is irrational, divided into good and evil, inside and outside, dominated by witchcraft and sorcery. In a low group/grid society, the cosmos is experienced as benign, unstructured and unmagical. Finally, the combination of low group and high grid is accompanied by a perception of the cosmos as irrational, not regulative.

The group-grid typology reappears and has been further developed in the cultural theory, where the four types of social structure and cosmology are called ways of life and myths of nature. The four ways of life are described as the hierarchist, egalitarian, individualist and fatalist. To these, a fifth way of life has been added: the hermit, autonomous or ineffectual way of life, where the individual withdraws from coercive or manipulative social involvement altogether. The hermit escapes social control by refusing to be controlled or to control others. In this research only the first four ways of life will be considered, following Schwarz and Thompson (1990), because the hermit is not supposed to really influence the processes which are described in this study. Although they can and do exist, hermits are not involved in social transactions within the fourfold system of hierarchists, egalitarians, individualists and fatalists and as such can be ignored. It has also been said that hermits 'leave the social map' in their ambition to attain total enlightenment.

Although here it is *me* who has chosen to exclude the hermits from this study, it is what hermits would have done themselves given the choice.

Table 5.1 shows how hierarchists, egalitarians, individualists and fatalists fit into the group-grid schematization by Douglas, and gives some further characteristics of the four different ways of life. Thompson *et al.* (1990) and Schwarz and Thompson (1990) point out that in particular the archetypes of the hierarchists and individualists have also been described in various social theories other than the cultural theory (Williamson, 1975; Lindblom, 1977). The hierarchist represents the bureaucrat, the technocrat, the manager-engineer. Hierarchies consist of confined social groups which are neatly ranked and ordered in relation to each other. By contrast, the individualist represents the entrepreneur, pioneer, adventurer, liberal, capitalist. A central element in market cultures is the autonomy of individuals and their freedom to bid against and bargain with each other. According to the authors of the cultural theory, this dichotomy cannot adequately describe the diversity of cultures. Many people reject both the individualism of the market and the inequalities of the hierarchy: egalitarians for instance, who stress the importance of co-operation and strive for social equality and voluntary relationships. The egalitarian represents the communard, the sectarian, the original socialist, the green movement. Finally, there are the fatalists, the marginal members of society, who experience life as a lottery, in which they are not able to influence events. Fatalists feel their behaviour is prescribed to such an extent that freedom of choice is minimal.

In the context of this study, it is most interesting to consider how each way of life faces issues such as scarcity, growth, technology, and the management of needs and resources. The typical response of hierarchists to resource scarcity is to allocate physical quantities by direct, bureaucratic means. Needs are regarded as given and unmanageable, so that the only strategy available to prevent shortages is to increase resources. By contrast, egalitarians believe that resources are given and finite, so that needs have to be reduced to ensure a reasonable supply of resources. Scarcity is perceived as a depletion of resources, due to over-exploitation of nature. The solution is to change people's life-style and to opt for small-scale technology. Egalitarians have little interest in economic growth, because it will make it more difficult to attain equality. Individualists reject the idea that natural resources are limited, arguing that human skill and knowledge are the ultimate resource. Scarcity is perceived as the driving force behind a further refinement of this skill and knowledge. Human ingenuity is multiple: both needs and resources can be managed. Economic growth is considered a prerequisite for the continued development of science and technology and thus for survival. Fatalists, finally, regard both their

Table 5.1. Characteristics of the hierarchist, egalitarian, individualist and fatalist.

	Hierarchist	Egalitarian	Individualist	Fatalist
Social structure	High group, high grid	High group, low grid	Low group, low grid	Low group, high grid
Myth of nature	Perverse, tolerant	Ephemeral	Benign	Capricious
Rationality	Procedural	Critical	Substantive	Fatalistic
Knowledge	Almost complete, organized	Imperfect, holistic	Sufficient, timely	Irrelevant
Needs	Given, unmanageable	Social, manageable	Individual, manageable	Unmanageable
Resources	Scarce, manageable	Depleting, unmanageable	Abundant, manageable	Lottery, unmanageable
Management style	Control, regulatory	Preventive	Laissez-faire, adaptive	Passive
Learning style	Anticipation	Trial without error	Trial and error	Luck
Desired systems' properties	Controllability	Sustainability	Exploitability	Capability
Ideal scale	Large	Small	Appropriate	No preference
Economic growth	Desirable, with conditions	Undesirable	Desirable, unconditionally	Desirable, good fortune
Desired technology	High-technology	Small-scale technology	Cheap technology	No preference
Salient risks	Loss of control	Catastrophic developments	Threats to the free market	Surprises
Risk-handling style	Institutionalization	Reduction	Taking the opportunities	Acceptance

Sources: Thompson *et al.* (1990), Schwarz and Thompson (1990).

needs and their resources as unmanageable. Their management strategy involves coping with an environment over which they have no control. Nature is perceived as a lottery-like cornucopia, where resources might be abundant, but it has to be seen whether and when they will be available. Fatalists regard economic growth as good fortune for some and do not believe they can actively increase their own wealth. Chance may bring it their way.

Another issue of interest to this study is risk. Douglas and Wildavsky (1982) argue that the traditional distinction between objectively calculated *physical risks* and subjectively biased individual *perceptions of risks* is inappropriate to understand current controversies over risks. They show that these controversies can only be understood if the concept of risk is regarded as a social construct. Both private, subjective perception and public, physical science are closely connected to culture, shared beliefs and values. The apparatus of scientific investigation is as unique to a specific culture as are its results. As a consequence, risk assessment cannot be disconnected from cultural bias. Schwarz and Thompson (1990) and Thompson *et al.* (1990) elaborate on this argument and include it in their cultural theory. Each way of life has a particular attitude towards risks. The largest fear of hierarchists is that they might lose control. In response they strive hard to manage the entire risk system, which explains their readiness to set acceptable levels of risks. The largest threat to egalitarians is unbridled growth, resulting in catastrophic, irreversible and inequitable developments. Egalitarians tend to do everything they can to avert these risks. The largest threat to individualists is an improper functioning of markets. However, individualists are generally optimistic: risk is opportunity. If there was no uncertainty or danger of loss, there would be no prospect of personal reward and hence no scope for entrepreneurs. Lastly, the primary concern of fatalists is to cope with the surprising events the future will bring to them. Fatalists do not knowingly take risks, but through their passivity, they in fact accept ambient risks, whatever these risks may be.

5.3 Perspectives on water

In this section, the hierarchist, egalitarian, individualist and fatalist perspectives are elaborated for issues which in some way relate to water. Two lines of thinking have been followed and brought together. First, reasoning along the line of cultural theory, it was asked: what perspectives on water can be deduced from the cultural theory? Second, taking current controversies on water policy issues as a starting point, it was asked: what coherent perspectives may underpin the different points of view? As material for the latter, I used of the analysis in Chapter 2, where different points of view are discussed with respect to issues such as water demand, water availability and water scarcity. It proved possible in the end to bring both lines together reasonably well, without deviating from the main theses in the cultural theory and without distorting the prevalent views in the world of water policy researchers and analysts. The main characteristics of the four perspectives on water which have resulted from the two-way approach are presented in Table 5.2.

Table 5.2. The four perspectives on water.

	Hierarchist	Egalitarian	Individualist	Fatalist
Water demand	A given need	A manageable desire	Price-driven	An unmanageable desire
Water trade	Controlled trade	No water trade	Free trade	Trade is for the rich
Water-conserving technology	Large-scale technology push	Small-scale technology push	Price-driven	No policy
Water price policy	Incremental price increase	Water taxing	Market pricing	No policy
Potential water supply	Stable runoff	Stable runoff in inhabited areas	Total runoff or no limits	Irrelevant to individuals
Water scarcity	Supply problem	Demand problem	Market problem	Problem of individuals
Groundwater use	Inevitable	Below sustainable level	Profitable	Profitable to a few
Artificial groundw. recharge	Solution to water scarcity	Should not be necessary	Desirable if cost-effective	No policy
Artificial surface reservoirs	Solution to water scarcity	Undesirable	Desirable if cost-effective	No policy
Hydrological cycle	Robust within limits	Vulnerable to perturbations	Robust	Unpredictable
Sensitivity of sea level	Moderately sensitive	Highly sensitive	Insensitive	Unknown
Public water supply	Incremental improvements	Basic supply to everyone	Driven by economic growth	Given to the rich
Water quality evaluation	Functional quality standards	Pristine quality as reference	Economic value	No reference
Wastewater policy	Treatment to meet standards	Treatment, decrease production	'Polluters pay' principle	No policy
Flooding risks	Divergent risk levels	Equal risk principle	Economic trade-off	Risk acceptance

The hierarchist perspective

A typical characteristic of hierarchists is to regard scarcity as a supply problem. Their management strategy is to look how they can manage their resources. Water scarcity is translated into a problem of how to increase supply in order to meet demand. Water demand is regarded as a given need, emanating from facts such as population size, economic development and need for irrigation. Water resources are available within certain limits. Stable runoff may be regarded as an appropriate measure of potential water supply and can be enlarged through construction of surface reservoirs and artificial groundwater recharge. However, the ultimate limit to potential water supply is total runoff. Hierarchists do not reject further building of large dams to increase stable runoff, although they recognize that the negative aspects of dam construction - often intangible and difficult to compare to the benefits - should be mitigated as far as possible. Groundwater use is regarded as inevitable, but the danger of over-exploitation of aquifers is recognized. Artificial groundwater recharge might be a good solution, having the additional advantage of natural purification. Hierarchists are willing to strive for more efficient water use, but regard efficiency improvements as seriously hampered by all kinds of social and economic constraints. As a result, one should not have great expectations of programmes aimed at the development or introduction of water-conserving technology, particularly not if one expects changes to come from collective efforts to introduce small-scale water-conserving technology. People should rather aim to develop and introduce of high technology on a large scale (on the supply side, not the consumer side). Hierarchists are not inclined to push water prices strongly in the direction of real costs (market pricing), because a rapid increase in water prices would disturb socio-economic stability to an unacceptable extent. As their ultimate goal, hierarchists aim for a situation where water charges fully cover operational and maintenance costs. It is considered fair that water consumers should have to repay only part of the investment costs.[1] Inter-basin or international trade of water is regarded as a possible way of improving the allocation of water, but it is seen as an issue to be regulated by governments rather than by free enterprise, due to the public character of water. Public water supply and proper sanitation facilities for everyone are desirable goals, but policy targets should be realistic, as was shown during the International Drinking Water Supply and Sanitation Decade in the 1980s, when targets appeared to be unattainable despite successful efforts to include the Decade within the various international aid programmes.

Hierarchists, who perceive nature as tolerant, consider that disturbances such as global warming, land use changes and consumptive water use will alter the hydrological cycle to some extent, but not uncontrollably. It is assumed that disturbances can be assimilated as long as they do not reach critical levels. An issue such as wastewater treatment becomes important if water quality standards are not

---

[1] As an argument for this position, one could for instance mention the importance of investments in public water supply for improving public health. In the case of irrigation, one could say that irrigation investments stimulate the general economy and that hidden taxes are often already imposed on farmers through price controls for agricultural products (Peterson, 1987).

reached. Standards can differ by type of water source and type of intended use. Hierarchists typically advocate the diversification of water use: clean ground water for drinking, slightly polluted surface or ground water for manufacturing or irrigation, more severely polluted surface water for cooling, etc. Risks of flooding are if possible regulated by formulating maximum acceptable risk levels and improving dykes or other defences to conform to these levels. Acceptable risk levels vary for different areas, from relatively high in undeveloped areas to comparatively low in highly developed areas.

The egalitarian perspective
Egalitarians, who perceive nature as fragile, are prudent in assessing water resources and take account of temporal and spatial variability. Stable runoff in inhabited areas may be an appropriate measure of potential water supply, but in addition indicators of excessive water use, such as for example the actual decline of groundwater tables and the remaining amount of high-quality water, are essential. Water scarcity is regarded as a problem caused by growing water demand and pollution. The solution is supposed to be the management of human needs. Water demand is seen as a manageable desire which can be changed by policy incentives and shifts in social customs and preferences. Applying small-scale water-conserving and re-use technology can lower water-use intensities in all sectors. The egalitarian is more sensitive to communal programmes to introduce new technology than to an increase in the water price. As Gibbons (1986) observes, the risk-averse farmer facing water cost increases will be the last to switch to new irrigation techniques or to different crops which use less water. Nevertheless, to accommodate the environmental consequences of excessive water use, egalitarians advocate that such impacts are included in the price of water as a tax. However, everyone should have access to water to fulfil basic needs, which means that water should be free to people who otherwise would not have it. Because egalitarians attach great importance to equity, access to safe drinking water and sanitation facilities for everyone is a principal policy goal. It would be typically egalitarian to promote a second International Drinking Water Supply and Sanitation Decade.[1] Water trade in any form is considered undesirable, because water is seen as public property. Transfer of water

---

[1] The (first) International Drinking Water Supply and Sanitation Decade (1981-1990) was also an egalitarian initiative. The subsequent institutionalization of the Decade was in contrast hierarchist achievement. In the egalitarian view, too much bureaucracy has meant that the goals of the Decade have been far from achieved. From a hierarchist point of view, this was instead a consequence of several kinds of inevitable social and political constraints, such as for instance insufficient involvement by women and political resistance to cost-sharing (see Christmas and De Rooy, 1991).

between different river basins is rejected from an ecological point of view. According to the egalitarian, wastewater should as a matter of principle be treated before disposal. Even better than wastewater treatment is a reduction in wastewater production. Bodies of water should if possible return to their pristine quality. According to the egalitarian, the fragile dynamic equilibrium of the water balance is easily disturbed by human activities. Intensive water use, human-induced temperature change and deforestation may considerably affect stable runoff and the sea level. Fertilizers and household and industrial waste will not only remain in some hot spots, but will spread throughout the world (witness the fact that several manmade chemicals have already been found in Antarctica). Egalitarians are strongly opposed to further building of large dams, arguing that the social and ecological costs of dam construction by far outweigh possible benefits. Groundwater use should be reduced to stop over-exploitation of aquifers. As a measure of an acceptable level of groundwater withdrawal, one should not only look at natural recharge (which might give an overestimate), but also at the actual effect of withdrawals on water tables. Artificial groundwater recharge is not regarded as a real solution, because the water would have to be taken from surface waters which are vulnerable to over-exploitation as well. In the egalitarian view, risks of flooding should be first reduced in areas where risks are highest (equal risk principle). Egalitarians are most concerned with the protection of less developed regions, where poor but densely populated areas are exposed to regular flooding. In the case of increased flooding frequency as a result of land cover changes, erosion or climate change, preventive strategies are preferable to defensive strategies.

The individualist perspective
The perspective of individualists largely coincides with what has been described in Chapter 2 as the economic point of view: water is an economic good and should be managed as such. Individualists regard all options to improve water supply conditions as realistic, provided they are cost-effective. Efficiency improvements which reduce demand are often profitable, as they save not only water but also money. However, extending the resource base - for instance through exploitation of untouched aquifers or increasing desalination capacity - can be profitable as well. The right mixture of water supply and demand management will be a function of circumstances which are different in time, per region and per user. Individualists consider total runoff the proper measure of potential water supply. Remote or flood flows can be made available if demand is large enough (which means if people are willing to pay). If water recycling and desalination techniques become more efficient and economically feasible on a large scale, water might even become an

unlimited resource, so that the problem will no longer be one of availability but one of the efficient exploitation of water. The hydrological variability of water in time and space is not a real limiting factor to individualists, who regard free trade as the ultimate solution to carry water and water-intensive products (virtual water) to the demand areas. Water demand is determined by the price mechanism: higher prices as a result of increased scarcity will lower demand and stimulate the development of more efficient technology. If new water-conserving techniques become cost-effective, they will replace older techniques, prices will drop and demand will rise again. Individualists strongly discourage subsidies on water, at present common practice all over the world. Water prices should be established by market mechanisms. In cases of high water scarcity, high-tech options for water supply (e.g. desalination) could be stimulated by government institutions, but always on pay-back basis. Individualists do not pursue an active policy in public water supply and sanitation, because they believe that economic development will increase public water supply and sanitation coverage adequately. Economic development is even regarded as a prerequisite for water supply and sanitation improvements. According to the individualist point of view, wastewater treatment is an economic trade-off. Application of the 'polluters pay' principle will force polluters to treat wastewater if this is preferable to paying for the damage caused by pollution (which has to be expressed in financial terms in some way). The value of a body of water of a certain quality depends on its economic value.[1] In the individualist view, reducing or accepting risks of flooding is an economic trade-off, which means that acceptable risk levels are a function of economic development. In line with their perception of nature as robust, individualists tend to regard possible disturbances of the hydrological cycle as of minor importance. If intensive water use, land use changes or global warming have some effect on the hydrological cycle, the resulting changes will occur slowly enough for people to adapt.

The fatalist perspective
According to fatalists, there are so many uncertainties that changes to the hydrological cycle can in practice be regarded as unpredictable. If even scientists disagree on global carrying capacity and possibilities of growth, and if policy makers propose contradictory types of measures, there is little reason to believe that people can knowingly improve their own future. Whether people are provided with enough clean water or not is seen rather as a matter of individual luck than as a

---

[1] There are several methods to determine the economic value of a body of water (Gibbons, 1986). The effect of pollution on the value of water can for instance be calculated as the damage with regard to different types of use (Keilani *et al.*, 1974).

matter of regional water shortage or abundance. Why else are people dying from waterborne diseases in many places in the world where water is said to be abundant? Questions such as 'is water a finite or infinite resource' or 'should potential water supply be measured as total or stable runoff' are considered academic questions which are irrelevant to individual people. Water scarcity is seen as a problem of

> **Perspectives on water in the literature**
> One can easily recognize the different perspectives on water in the scientific literature. However, an attempt to classify each individual research paper or book into one of the four perspectives would fail, because various elements of different perspectives are often mixed in the message from one author. Nevertheless, in several cases one perspective clearly dominates. Below, I cite three authors who might be considered representative of the hierarchist, egalitarian and individualist perspective respectively. I have to admit that I could not find a proper representative of the fatalist perspective in the scientific literature, probably due to the fact that people who are predominantly fatalist do not publish scientific papers (the fatalist considers knowledge irrelevant).
>
> Shaw (1994) formulates an important element of the hierarchist perspective as follows: "Once the needs of an area have been established, be it for a single town's domestic supply, the irrigation water for a commercial plantation, or the total demands of a whole country, and some continuing requirements for the future made, then the engineer must investigate the availability of the resources." Postel (1996) responds according to the egalitarian perspective: "For all its impressive engineering, modern water development has adhered to a fairly simple formula: estimate the demand for water and then build new supply projects to meet it. It is an approach that ignores concerns about human equity, the health of ecosystems, other species, and the welfare of future generations. In a world of resource abundance, it may have served humanity adequately. But in the new world of scarcity, it is fueling conflict and degradation. (...) Three principal forces conspire to create scarcity and its potential to incite conflict or dispute: the depletion or degradation of the resource, which shrinks the "resource pie"; population growth, which forces the pie to be divided into smaller slices; and unequal distribution or access, which means that some get larger slices than others. Although all three often play a part, it appears that unequal distribution often has the most important role (...) A top priority is to ensure that both people and ecosystems get at least the minimum amount of good-quality water they need to remain healthy and to function productively." Anderson (1995), as a typical exponent of the individualist perspective, believes above all in regulation through the market mechanism: "Experience around the world has demonstrated over and over again that the only successful way to avoid shortages is to rely on free market pricing and allocation. The same is true for water. (...) Eliminating laws against water marketing and establishing private water rights would give consumers an incentive to use water more efficiently. (...) If markets in water were permitted, demand would be reduced, supply would be increased, water would be reallocated, and the specter of water crisis would vanish."

individuals. Water is given to the rich, both in water-poor and water-rich parts of the world. The poor seem to lack adequate water supply and sanitation under all circumstances. Water demand is regarded as an unmanageable desire, which is or is not satisfied. Water trade is only possible for the rich people in power and does not benefit the poor. Risks of flooding are accepted and have to be handled, because fatalists do not feel they can reduce them. Fatalists are not in favour of any particular management strategy, which means that their management strategy essentially comes down to doing nothing, merely coping with whatever situation evolves. In the fatalist view, people are unable to control the future, and even if they could, interests diverge and strategies would counteract each other to such extent, that the net result would be a lottery. An increase in water prices or the introduction of water taxes would not make sense, because one could question who would profit and who would suffer. Several studies show that it is the poor who pay most for their water. The chances are that increasing prices would make life even worse for the poor, without affecting the rich who can pay easily.[1] Improving 'water literacy' among people through education, in order to conserve water, is not regarded as a very useful policy either. According to the fatalist point of view, the people who might learn most from these kinds of programmes are often not those members of society in a position to conserve much water, while the people who are in such a position are already well educated, but probably unwilling to give up their privileges (shower, garden, private swimming pool).

5.4 Perspective-based model formulations

In this section, I show how the four perspectives on water have been implemented in AQUA in the form of different model formulations. In the implementation of perspective-based model formulations, a distinction is made between a world-view and a management style. Each world-view consists of a particular set of equations and initial and parameter values, representing a specific perspective on how the world works. Each management style consists of a particular set of parameter values, representing a specific policy strategy. The model has two 'handles', one which can switch between three world-views (hierarchist, egalitarian and

---

[1] At present, it is probably the urban poor who spend the highest proportion of their income on water. For example, in Port-au-Prince, Haiti, the poorest households sometimes spend 20 per cent of their income on water. In Onitsha, Nigeria, the poor pay about the same percentage of their income on water during the dry season, while upper-income households pay 2-3 per cent. In several cities in the world households purchasing water from street vendors pay as much as twenty-five to fifty times more per litre than households connected to the municipal water supply system (World Bank, 1993).

individualist) and one which can switch between four management styles (hierarchist, egalitarian, individualist and fatalist). The world-view handle sets the equations, initials and parameters as indicated in Table 5.3. The management-style handle sets the parameters as presented in Table 5.4. The tables only show relative values, because absolute values can differ per application, as discussed in Chapter 6 (AQUA World Model) and Chapter 8 (AQUA Zambezi Model). Because the fatalist perceives the world as a lottery, no specific fatalist world-view has been implemented. The fatalist's world is supposed to randomly follow either the hierarchist, the egalitarian or the individualist world-view.

The position of the hierarchist has been made operational in the model by assuming that specific water demands primarily depend on economic growth and efficiency improvements (Figure 5.2). Medium estimates are taken from the literature for growth elasticities; price elasticities are assumed to be nil. Relatively high estimates are used for the technological development and diffusion rates. Inter-basin or international water trade may increase, depending on the particular circumstances in the area under consideration. Stable runoff is considered a measure of potential water supply. The ratio of consumptive water use to stable runoff is taken as a measure of water scarcity. Water-supply costs are supposed to increase as a function of this ratio (Figure 5.3). A weak water-pricing policy is applied: prices are assumed to increase incrementally to 75 per cent of the total actual costs. At this level, operational and maintenance costs are fully covered, but investment costs only partly. Water will thus remain subsidized, but less so than today. It has been assumed that there will be no change in the relative use of the different types of water source. Artificial groundwater recharge is stimulated in places where ground water would otherwise be depleted. Public water supply, sanitation and wastewater treatment are assumed to be primarily subject to public policy. 'Realistic' policy targets are assumed for the improvement of public water supply and sanitation and for the extension of wastewater treatment capacity. Medium estimates from the literature are used for most of the hydrological parameters. As a result, the hydrological balance will respond moderately to disturbances such as global warming, land cover changes, groundwater withdrawals and consumptive water use. Medium estimates have also been assumed for standard deviations of water quality distributions. As a rule, one can say that the hierarchist view has been made operational by taking the most traditional approaches and central values ('best estimates') from the literature.

The egalitarian view is represented in AQUA by assuming that specific water demands are influenced by economic growth, water prices and (non-price-driven) efficiency improvements (Figure 5.2). Growth and price elasticities and technological development and diffusion are relatively low. There will be no significant inter-basin water transfers. A prudent estimate of available water resources has been made by taking potential water supply as equal to stable runoff in inhabited areas. Water supply costs greatly increase if water becomes scarcer (Figure 5.3). Water scarcity is calculated as the ratio of total water supply to stable runoff in inhabited areas. A strong water-pricing policy (water-taxing) is applied, with prices increasing to more than 100 per cent of the total actual costs. To prevent groundwater

Table 5.3. Implementation of three world-views in AQUA.

	Equation	Hierarchist world-view	Egalitarian world-view	Individualist world-view
Equations [1]				
Specific water demands (see also Figure 5.2)	3.7	Alternative I	Alternative III	Alternative II
Potential water supply	3.38	Alternative I	Alternative II	Alternative III
Water scarcity (see also Figure 5.3)	3.44	Alternative II	Alternative I	Alternative II
Public water supply coverage	3.51	Alternative II	Alternative II	Alternative I
Sanitation coverage	3.52	Alternative II	Alternative II	Alternative I
Wastewater treatment coverage	3.53	Alternative II	Alternative II	Alternative I
Initials				
Initial imbalances of ice sheets	3.17	Medium	High	Low
Initial stock of renewable fresh ground water	-	Medium	High	Low
Parameters				
Growth elasticities water demand	3.7	Medium	Low	High
Price elasticities water demand	3.7	Zero	Low	High
Fractions consumptive water use	3.11	Medium	High	Low
Evaporation fraction of irrigation loss	3.12	Medium	High	Low
Ice sheet sensitivities	3.17	Medium	High	Low
Critical temperature increase glaciers	3.18	Medium	Low	High
Response time glaciers	3.18	Medium	Low	High
Sensitivity oceanic evaporation	3.20	Medium	High	Low
Loss of consumptive water use	3.22	Medium	High	Low
Exponent groundwater runoff	3.29	Medium	Low	High
Ratio delayed runoff / total groundwater runoff	3.31	Medium	Low	High
Reservoir storage contributing to stable runoff	3.35	High	Low	High
Standard deviations water quality	3.37	Medium	High	Low
Dilution factor	4.2	Low	High	Low
Water supply costs (see also Figure 5.3)	-	Increase moderately	Increase rapidly	Increase slowly
Desalination costs	-	No change	No change	Decrease
Potential hydropower generation capacity	-	Medium	Low	High

[1] For the alternative formulations of equations: see Chapter 3.

Perspectives on water

depletion (and possible sea-level rise), the egalitarian promotes the use of more surface instead of ground water. Ambitious policy targets are set to bring wastewater treatment coverage to 100 per cent, so that the quality of surface water will return to a more or less natural state. There are also ambitious policy targets with respect to public water supply and sanitation. Relatively high estimates from the literature are used for hydrological response parameters, so that disturbances will have a relatively large effect.

Table 5.4. Implementation of four management styles in AQUA.

Manageable parameter[1]	Hierarchist management style	Egalitarian management style	Individualist management style	Fatalist management style
Inter-basin water import / export	Medium increase	Zero	Large increase	Zero
Technological diffusion rate	High	Low	Zero[2]	Zero
Technological development rate	High	Low	Zero[2]	Zero
Ratios water price / actual cost	Increase to 75%	Increase to above 100%	Increase to 100%	Constant
Water source fractions	No change	Less groundwater use	More desalination	No change
Minimum river runoff after withdrawal	Medium	High	Low	Zero
Artificial groundwater recharge	Increase	Zero	Increase	Zero
Public water supply coverage	'Realistic' policy targets	Ambitious policy targets	Function of GNP_{pc} [3]	No policy targets
Sanitation coverage	'Realistic' policy targets	Ambitious policy targets	Function of GNP_{pc} [3]	No policy targets
Wastewater treatment coverage	'Realistic' policy targets	Ambitious policy targets	Function of GNP_{pc} [3]	No policy targets
Maximum acceptable critical flooding probability	Reduction of risks	Equal risk principle	Function of GNP_{pc} [3]	No measures
Maximum water expenditure as fraction of GNP	Medium	High	Not applicable	No limit

[1] For an overview of manageable parameters in AQUA: see Table 3.2.

[2] In the individualist world-view, water prices are the driving force behind technological innovation. Zero-values have therefore been assumed for the parameters which represent *non-price-driven* technological development and diffusion.

[3] In the individualist world-view, public water supply, sanitation and wastewater treatment coverage are not a function of policy targets but of economic growth (alternative I in the Equations 3.51-3.53). The same applies to the maximum acceptable critical flooding probability.

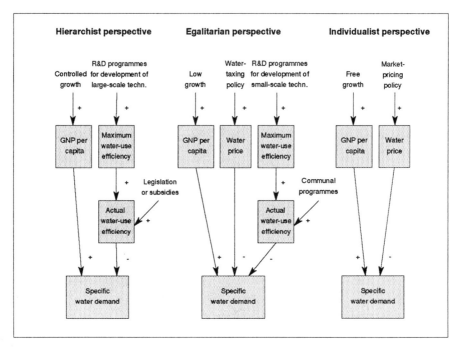

Figure 5.2. Perspective-based model formulations with respect to water demand.

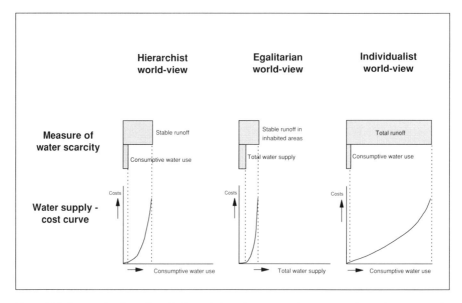

Figure 5.3. Perspective-based model formulations in respect of water scarcity and water supply - cost curves. Consumptive water use is defined as the part of the total water supply which is lost through evaporation and can therefore not be re-used.

The individualist point of view has been made operational by taking specific water demands as a function of growth and prices (Figure 5.2). Relatively high estimates are made for growth and price elasticities. Efficiency improvements which are not price-driven have been assumed to be nil. Total runoff has been taken as a measure of potential water supply. Water-supply costs increase as a function of the ratio of consumptive water use to total runoff (Figure 5.3). Water prices will increase until full costs are covered, both operational and maintenance costs and investment costs. The increase in public water supply and sanitation coverage is supposed to be subject to economic development, as is the fraction of wastewater treated before disposal. In the future, global desalination capacity will grow, because supply from freshwater sources will become more expensive in an increasing number of regions, while desalination of salt or brackish water will become cheaper, due to more efficient techniques. Desalination will be used especially for preparing drinking water and for industrial purposes. With regard to the response of the water system to disturbances such as global warming, land use changes and water use, relatively low response values are assumed. By assuming low standard deviations for water quality distributions, water pollution is supposed to be concentrated in some areas.

The fatalist management style has been implemented by assuming 'no response': technological development and diffusion are nil, there is no water-pricing policy, no inter-basin water trade, the relative use of different water sources remains the same and there are few public attempts to improve public water supply, sanitation, wastewater treatment or coastal protection. Application of the fatalist management style can be useful to obtain a reference for 'if we do nothing'.

6 The AQUA World Model

This chapter describes a specific application of the generic AQUA tool: the AQUA World Model. The aim of this model is to improve understanding of the long-term interaction between water and development on a global scale. The model has been calibrated for the period 1900-1990, for each of the hierarchist, egalitarian and individualist world-views. In this way, historical developments are explained according to three alternative points of view. A sensitivity analysis, to study the response of model outcomes to small variations in parameter values, shows that model behaviour depends greatly on the perspective chosen. The chapter concludes with a discussion of some analytical results for the 20th century.

6.1 Introduction

The purpose of the AQUA World Model is to get an insight into the long-term interaction between water and human beings on a global scale. Questions to be addressed have been formulated in Chapter 2. They are for example: to what extent do different kinds of human activity and environmental change alter the hydrological cycle, how does a limited water availability influence the development of water supply costs and water demand, and how can people respond to increasing water scarcity? The model should be useful to researchers or policy analysts in international organizations who participate in discussions on water and sustainable development or are concerned with negotiations on the agreement of policy priorities. In this chapter, the model itself is discussed, putting emphasis on historical simulation, calibration and validation. In the next chapter, the model will be used in exploring possible futures, putting emphasis on future simulation, interpretation of model results and translation of model output into policy-relevant information.

An overview of the schematization of the AQUA World Model has already been given in Section 3.4 (Table 3.1). Let me briefly summarize the most important characteristics. The starting year for simulation is 1900. The world has been schematized into four spatial compartments: land, polar ice sheets, oceans and the atmosphere. The land compartment includes the entire global land area, excluding the ice sheets of Antarctica and Greenland. The oceans are regarded as one large water store and the global land area as one large river basin. Sixteen land cover types are distinguished to account for spatial varieties. Four aggregated water-demanding sectors are distinguished: the domestic, irrigation, livestock and industrial sectors. Domestic water demand comprises public and private demand. The model uses one

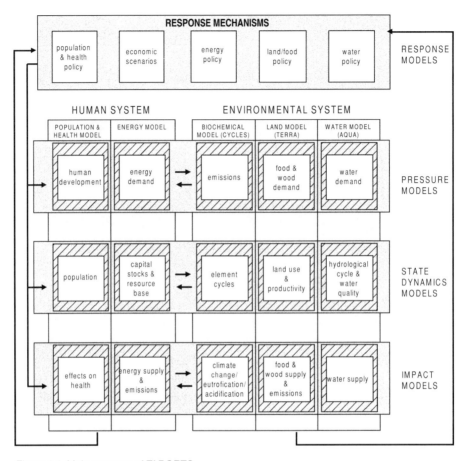

Figure 6.1. Main structure of TARGETS.

aggregated livestock category, which includes cattle, sheep, goats, pigs and hens. Ten types of water store are distinguished, as shown in Figure 3.5. The model recognizes four water quality variables (nitrate, ammonium, dissolved organic nitrogen and phosphate) and four water quality classes (good quality class A to bad quality class D). Finally, there are nine items of 'water expenditure': public water supply, sanitation, irrigation, livestock water supply, industrial water supply, hydropower generation, domestic and industrial wastewater treatment and coastal defence.

The AQUA World Model can be run 'stand-alone' or as integral part of the model framework TARGETS, which is an acronym for *T*ool to *A*ssess *R*egional and *G*lobal *E*nvironmental and Health *T*argets for *S*ustainability. Beside water, TARGETS covers four other fields: population and health, energy, food, and global cycles of the basic elements carbon, nitrogen, phosphorus and sulphur. If AQUA is run as part of TARGETS, it can support a policy analyst in developing and evaluating water policy as part of an overall policy for sustainable development. Although I limit myself in this study to using AQUA as a stand-alone tool, I would like to show briefly how the AQUA World Model is embedded within TARGETS. For further information and experiments with the TARGETS model as a whole, the reader is referred to a recently published book by Rotmans and De Vries (1997). A schematic diagram of the framework of TARGETS is given in Figure 6.1. There are a number of inter-linkages between the several sub-models of TARGETS. The inputs of exogenous developments into AQUA, as summarized in Figure 3.2, are provided by other sub-models within TARGETS. Equally, AQUA provides inputs into the other sub-models. The models have been inter-linked dynamically, which means that feedback between models is taken into account. The Population and Health Model calculates the population size, one of the variables used in AQUA to calculate water demands. Demographic and health dynamics simulated in the Population and Health Model, however, depend among other factors on the coverage of public water supply and sanitation, data which are provided by AQUA. The percentage of the population with public water supply and proper sanitation facilities is a rough indication of the number of people with access to safe drinking water and living under hygienic conditions. The remainder of the population runs a higher risk of being exposed to bacteriologically contaminated water. The demand for expansion of the hydropower generation capacity from the Energy Model determines the demand for new water reservoirs in AQUA. The CYCLES model, describing the global cycles of carbon, nitrogen, phosphorus and sulphur, inter-links with AQUA in many ways. The first link concerns the effects of global warming, simulated in CYCLES, on the hydrological processes in AQUA. Second, CYCLES calculates warming and thermal expansion of the ocean, one of the contributors to total sea-level rise as calculated in AQUA. Third, the hydrological cycle influences the elements cycles: the outflow of elements from ground and surface water is driven by the water outflow from these stores. Fourth, AQUA simulates domestic and industrial wastewater treatment, which determines the fate of nutrients in CYCLES. Finally, if AQUA runs as an integral part of TARGETS, the calculation of concentrations of nitrate, ammonium, dissolved organic nitrogen and phosphate in AQUA is replaced by inputs of concentrations from CYCLES. The set of water quality variables taken into account now corresponds to the substances distinguished in CYCLES. Average surface-water concentrations of the following substances are provided: dissolved organic and inorganic nitrogen, particulate organic nitrogen, dissolved organic and inorganic phosphorus, and particulate organic phosphorus.

For ground water, CYCLES provides the concentrations of dissolved organic and inorganic nitrogen. Land cover changes simulated in the TERRA model affect the processes of evaporation, groundwater recharge and river runoff simulated in AQUA. Livestock calculated in TERRA determines livestock water demand in AQUA and the demand for irrigated cropland influences the water demand for irrigation. Conversely, actual irrigation and natural water availability as calculated in AQUA influence the productivity of cropland in TERRA. Finally, one of the factors determining erosion, the rain erosivity index, is calculated in AQUA on the basis of rainfall distribution throughout the year. Gross world product and value added in the industrial sector from the 'economic scenario generator' within TARGETS are used within AQUA to calculate water demand. Conversely, increasing supply costs if water becomes scarcer require more investment and may hinder economic development. Water models alone never account for such feedback because they start from exogenous population and economic scenarios. Because this may lead to inconsistency between starting point and model outcome in the long run, we have begun to explore such feedback mechanisms through the comprehensive framework of TARGETS (Rotmans and De Vries, 1997). However, in this study the AQUA World Model is run stand-alone, so that time series for the exogenous developments have to be taken as input.

6.2 Exogenous developments and model initialization

The model uses three types of input data: time series for exogenous developments, initial values and parameter values. In this section, I discuss the first two types of input data; the parameters will be discussed in the next section. For population, livestock and economic development in the period 1900-1990, data have been taken from Klein Goldewijk and Battjes (1997): see Table 6.1. The different types of livestock have been aggregated into one category, expressing the number of heads in 'ox equivalents'. One ox equivalent is either one cow, eight sheep, goats or pigs, or 150 hens. This means that one sheep, goat or pig on average is supposed to use eight times less water than one cow and one hen 150 times less, which falls within the ranges given by Euroconsult (1989) and Van der Leeden et al. (1990). For historical data on global hydropower generation capacity, I have used estimates by De Vries and Janssen (1997) for 1900-1940 and UN data from Brown et al. (1993) for 1950-1990.

According to the IPCC, the global mean surface air temperature has increased by about 0.3 to 0.6 ^{0}C since the late 19th century, and by about 0.2 to 0.3 ^{0}C over the last forty years, the period with the most credible data (Nicholls et al., 1996). As raw

Table 6.1. Population, livestock, economic production, hydropower and temperature during the period 1900-1990.

Input variable	1900	1910	1920	1930	1940	1950	1960	1970	1980	1990
Population[1] (10^9)	1.64	1.77	1.91	2.08	2.28	2.51	3.01	3.69	4.43	5.27
Cattle[1] (10^9)	0.37	0.39	0.47	0.51	0.55	0.64	0.87	1.02	1.22	1.29
Sheep, goats and pigs[1] (10^9)	0.54	0.62	0.72	0.91	0.90	1.12	1.66	1.96	2.35	2.66
Poultry[1] (10^9)	0.83	0.83	0.87	1.20	1.20	1.37	1.79	2.73	7.13	10.8
Gross world product[1] (10^{12} 1990-US\$/yr)	1.45	1.87	2.26	2.68	3.30	4.20	6.73	11.4	16.2	21.0
Value added industry[1] (fraction of GWP)	0.42	0.41	0.41	0.41	0.40	0.39	0.38	0.37	0.36	0.35
Hydropower generation capacity[2] (GW)	5	7	10	15	24	45	157	282	467	628
Temperature increase since 1900[3] (°C)	0.00	-0.10	-0.01	0.11	0.23	0.21	0.26	0.23	0.31	0.50

[1] Klein Goldewijk and Battjes (1997).
[2] De Vries and Janssen (1997) for 1900-1940 and Brown et al. (1993) for 1950-1990.
[3] Based on Jones et al. (1994), who give annual temperature anomalies relative to the period 1950-1979. To smooth the temperature record, decadal anomalies have been calculated: the decadal anomaly for 1900 as the average of the annual anomalies for 1895-1904, etc. Temperature increase since 1900 has been defined as the decadal anomaly in a given year minus the decadal anomaly for 1900.

data for historical temperature variations, the data from Jones *et al.* (1994) have been used. Figure 6.2 shows that global warming occurred mainly during two periods, between 1910 and 1940 and since the mid-1970s. For the spatial distribution of monthly temperatures in the defined physical basic areas, IIASA's climate database has been used (Leemans and Cramer, 1991).

There are only a few estimates of historical land cover changes on a global scale and the sources available are difficult to compare. Each source covers its own specific period and uses its own classification of land use types and regional breakdown. A comprehensive analysis is provided by Richards (1990), who gives data for the period 1700-1980 for ten world regions, distinguishing between 1. forests and woodlands, 2. grasslands and pastures and 3. croplands. Another source is the FAOSTAT database of the Food and Agriculture Organization, containing data for the period 1961-1994, which has a country resolution and differentiates between 1. forests and woodland, 2. permanent meadows and pastures, 3. cropland and 4. other land, including urban areas, wasteland and barren land (FAO, 1996). Cropland is subdivided into rain-fed cropland and irrigated cropland, and arable land and land under permanent crops. The HYDE database of Klein Goldewijk and Battjes (1997) gives a useful overview for the period 1700-1980 for ten world regions, based on FAO data and Richards (1990), and provides data for four categories: forests, grasslands, pastures and croplands. Because my particular interest is the effects of

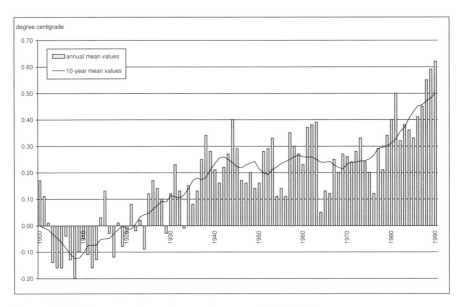

Figure 6.2. Global mean surface air temperature record 1900-1990.

irrigation, deforestation, urbanization and drainage of wetlands on the global hydrological cycle, none of these sources is directly useful. I have therefore used additional data on wetlands and urban areas from some other sources, as will be discussed below. Interpretation of the different sources gives the land transformations shown in Table 6.2.

The data for forests have been based on the HYDE database of Klein Goldewijk and Battjes (1997) for the period 1900-1960 and the FAOSTAT database of the FAO (1996) for 1961-1990. The FAO data have been accepted without modification, but the HYDE data have been scaled down in order to link up with the FAO data (for 1960, HYDE gives a global forest area which is 25 per cent larger than the area given by FAO). Deserts have been assumed as a constant, adding up to a global area of 20×10^6 km^2. My basic reference for this value is the land cover database of Olson et al. (1985), from which I derived a global desert area of 18.2×10^6 km^2 by adding the categories cool, sand and hot desert. The value also corresponds to the sum of cool semi-desert and hot desert in the IMAGE database (Leemans and Van den Born, 1994). For the period under consideration, the desert areas have been assumed to be constant, as the major land cover changes in the 20th century affected deserts to a limited extent only. It has been estimated that only 2 per cent of the global area converted to cropping in the period 1860-1978 was originally desert (Williams, 1990), which corresponds to a loss of desert area of less than 1 per cent (0.17×10^6 km^2 over 20×10^6 km^2). Reliable data for the reverse process, the increase of deserts through desertification, are not available, although there are some indicative estimates. According to Rozanov *et al.* (1990), desertification may have taken place in an area of about $0.57\text{-}0.95\times10^6$ km^2, which corresponds to an increase in desert area of about 3-5 per cent. Earlier estimates of desertification on a global scale, such as produced by UNEP in 1977 and updated in 1984 (UNEP, 1987), gave a more alarming picture, which was later criticized as 'the myth of the marching desert' (Forse, 1989; Binns, 1990). As Helldén (1991) suggests, a more thorough study of desertification based on scientific principles is probably needed before one can form a sounder picture of what has been and is in fact happening.

A problem in estimating historical trends in wetland areas is that most comprehensive analyses of land cover changes do not include wetlands as a separate category. Another problem is the interpretation of the concept of wetlands, which gives rise to a large variation in estimates. Mitsch and Gosselink (1993) estimate the area of wetlands in the world at 8.56×10^6 km^2, which is more than 6 per cent of the global land area, based on data from Bazilevich *et al.* (1971) and Maltby and Turner (1983). Matthews and Fung (1987) estimate the global wetland area at 5.3×10^6 km^2, about 4 per cent of the global land area. From Olson *et al.* (1985), a total wetland area of 3.71×10^6 km^2 can be derived. Korzun *et al.* (1978) and Shiklomanov (1993) give a swamp area of 2.68×10^6 km^2, an underestimate for wetlands as a whole. As input data for the model, a global wetland area of 8.56×10^6 in 1970 has been assumed (Mitsch and Gosselink, 1993), with a global average annual wetland loss linearly increasing from 0.4 per cent in 1900 to 1 per cent in 2000, an estimate based on data for Asia from Gleick (1993a) and data for the USA from Mitsch and Gosselink (1993). This means a loss of wetlands during the 20th century of about 50 per cent, which corresponds to an estimate by Öquist and Svensson (1996). The division of the global wetland area over developing and developed regions of the world has been assumed to correspond to the division over tropical and temperate zones as given by Mitsch and Gosselink (1993).

The data for rain-fed and irrigated cropland for the period 1961-1990 have been taken without modification from FAO (1996). The data for irrigated cropland for 1900-1960 have been taken from Gleick (1993a). Rain-fed cropland for this period has been based on the HYDE database (minor adjustments were needed in order to correlate with the FAO data in 1960). The urban area data in Table 6.2 have been obtained by dividing the global urban population (derived from Berry, 1990) by an estimated average population density in urban areas of 5000 people per km^2. According to data from the WRI (1994), this density is approximately the density found in 1980 in cities such as New York, Mexico City and Seoul, but half that of the densities in Tokyo, Jakarta and Delhi and three times higher than the densities in

Table 6.2. Land cover changes during the 20th century (10^6 km^2).

Land cover type	1900	1910	1920	1930	1940	1950	1960	1970	1980	1990
Developing world										
Forest	27.18	26.98	26.79	26.13	25.45	24.79	23.90	23.35	22.59	21.87
Grassland	23.84	24.15	24.48	24.89	25.32	25.75	25.94	26.37	26.97	27.60
Desert	14.00	14.00	14.00	14.00	14.00	14.00	14.00	14.00	14.00	14.00
Wetland	7.31	7.01	6.67	6.32	5.95	5.56	5.17	4.78	4.39	4.01
Rain-fed cropland	3.17	3.34	3.52	4.06	4.57	5.00	5.69	5.89	6.01	6.10
Irrigated cropland	0.36	0.37	0.39	0.44	0.54	0.70	1.01	1.23	1.51	1.75
Urban area	0.01	0.01	0.02	0.02	0.03	0.06	0.09	0.14	0.20	0.28
Fresh surface water	0.84	0.84	0.85	0.85	0.85	0.86	0.91	0.96	1.04	1.11
Total	76.37	76.37	76.37	76.37	76.37	76.37	76.37	76.37	76.37	76.37
Developed world										
Forest	19.04	18.90	18.77	18.65	18.54	18.42	18.40	18.71	19.14	19.54
Grassland	18.72	18.70	18.71	18.80	18.92	19.02	19.19	19.05	18.86	18.71
Desert	6.00	6.00	6.00	6.00	6.00	6.00	6.00	6.00	6.00	6.00
Wetland	5.77	5.53	5.27	4.99	4.69	4.39	4.08	3.78	3.47	3.17
Rain-fed cropland	4.87	5.24	5.60	5.88	6.12	6.36	6.34	6.33	6.17	6.06
Irrigated cropland	0.12	0.13	0.14	0.16	0.19	0.24	0.36	0.44	0.58	0.66
Urban area	0.03	0.04	0.05	0.06	0.07	0.09	0.11	0.14	0.16	0.18
Fresh surface water	0.40	0.40	0.40	0.40	0.40	0.41	0.45	0.49	0.56	0.61
Total	54.94	54.94	54.94	54.94	54.94	54.94	54.94	54.94	54.94	54.94
Total world	131.3	131.3	131.3	131.3	131.3	131.3	131.3	131.3	131.3	131.3

Sources: forests based on FAO (1996) and Klein Goldewijk and Battjes (1997); grassland own estimate, between FAO (1996) and Richards (1990); deserts based on Olson et al. (1985); wetlands based on Mitsch and Gosselink (1993), Gleick (1993a) and Öquist and Svensson (1996); rain-fed and irrigated cropland based on FAO (1996), Gleick (1993a) and Klein Goldewijk and Battjes (1997); urban areas own estimate; area of natural fresh surface water from Korzun et al. (1978); area of artificial fresh surface water based on L'vovich and White (1990). The division of the world into a developing and a developed region corresponds to the division made by FAO (1996), but Greenland is excluded from the developing region. The total land area refers to the area of the continents, excluding Antarctica and Greenland, and includes fresh surface water, but excludes inland salt water areas.

São Paulo and Shanghai. L'vovich and White (1990) estimate the global urban area in the mid-1980s at 1.2-1.4×10^6 km^2 and further assume that about 25 per cent of this area consists of impermeable surface. According to my own estimate, the urban area in the mid-1980s was about 0.4×10^6 km^2.

The areas of natural fresh water have been assumed constant, using data from Korzun et al. (1978), who give a global figure of 1.24×10^6 km^2. The growth in the expanse of artificial surface reservoirs has been based on L'vovich and White

(1990), assuming a division between the developing and developed world of 55 to 45 per cent.

The grassland areas have been used as the closing entry, calculated by subtracting the areas of all other land cover types from the total land area. The resulting estimates are much lower than those by Richards (1990), because his schematization into four land cover categories is so rough that his grassland category necessarily includes areas which may come under deserts, wetlands or urban areas in my schematization. My grassland estimates are however higher than those in FAO (1996), because the definition of grasslands in this study is broader than FAO's permanent meadows and pastures: they also include part of the land which is included in FAO's category 'other land'.

Model initialization
Initial values are required for the model variables which are computed by means of differential equations. For instance, the model needs initial values for specific water demands, water-use efficiencies and water stocks. An overview of the initial data used in the AQUA World Model is given in Table 6.3. Data on domestic water demand per capita, irrigation water demand per hectare and industrial water demand per dollar production in the initial year 1900 have been calculated from the total demands in that year given by Shiklomanov (1993). For this calculation, I used the data for population, irrigated cropland area and value added in the industrial sector as presented in Table 6.1. For the calculation of specific demands for public and private domestic supply, I used the ratio of private to public demand per capita from Table 6.5 and the public supply coverage as presented in Table 6.9. Water demand per livestock unit has been assumed at 12,000 kg/yr, a rough estimate of the average water demand of cattle based on various sources. L'vovich (1979) for example assumes that cattle use 14,600 kg/yr/head and Euroconsult (1989) gives a water demand of 7,300-14,600 kg/yr/head for cattle and 25,550-36,500 kg/yr/head for milk cows, but the World Bank (1987) gives an oxen demand of only 5,840-6,570 kg/yr/head. According to Van der Leeden *et al.* (1990), calves require 1,650-8,000, range cattle 5,800-11,600 and milk cows 10,000-32,000 kg/yr/head.

As it is difficult even to give reliable figures for present water-use efficiencies, it is certainly hard to make an assumption about efficiencies a hundred years ago. It is probably easiest for irrigation. According to Postel (1992, 1993), today's irrigation efficiency is estimated to average less than 40 per cent. Serageldin (1995) gives a figure of 45 per cent. Although there has been considerable progress in irrigation technology, it may be assumed that there has not yet been a major worldwide change in actual irrigation efficiencies, so that average efficiency in 1900 might for instance

Table 6.3. Initial values of the AQUA World Model (for the initial year 1900).

Variable	Symbol	Equation	Unit	Initial value
Domestic water demand per capita - in case of public supply	$WD_{dom,pc}$	3.7	kg/yr/cap	12,500
Domestic water demand per capita - in case of private supply	-	-	kg/yr/cap	9,400
Domestic water demand per capita - average	-	-	kg/yr/cap	9,800
Livestock water demand per head	$WD_{liv,ph}$	-	kg/yr/head	12,000
Irrigation water demand per hectare	$WD_{irr,pha}$	3.7	kg/yr/ha	11×10^6
Industrial water demand per dollar	$WD_{ind,p\$}$	3.7	kg/yr/US$	68
Domestic water-use efficiency	$Eff_{act,dom}$	3.8	-	0.50
Irrigation water-use efficiency	$Eff_{act,irr}$	3.8	-	0.35
Industrial water-use efficiency	$Eff_{act,ind}$	3.8	-	0.30
Water stocks (excluding renewable fresh ground water)	S	3.16	kg	Table 2.1
Renewable fresh ground water (see Section 6.3)	S_{fgw}	-	kg	$1 \times 10^{18} - 4 \times 10^{18}$
Oceanic evaporation	P_{oc}	3.20	kg/yr	505×10^{15}
Advective moisture transport to land	AMT	3.21	kg/yr	44.15×10^{15}
Advective moisture transport to Greenland	-	-	kg/yr	0.55×10^{15}
Advective moisture transport to Antarctica	-	-	kg/yr	1.66×10^{15}
Land precipitation per physical basic area	P	-	mm/month	See text

be estimated at about 35 per cent. According to Postel (1992), industries of many types can cut their water demand by between 40 and 90 per cent through the use of presently available technologies, without sacrificing economic output. The current industrial water-use efficiency is thus 10 to 60 per cent at most (possibly less if one also accounts for technologies which are available but not economically feasible). In their assessment of future water demands, Falkenmark and Lindh (1974, 1976) consider a case in which industries cut their demand by 72 per cent (thus assuming a current efficiency of only 28 per cent), and note that a super-effective policy is necessary to achieve such a reduction. Due to a lack of better data, industrial water-use efficiency in 1900 has been assumed at a value of 30 per cent. It seems that the potential reduction in water use in the domestic sector is smaller than in the industrial one. Postel (1992) gives a number of success stories of water conservation in cities and mentions savings between 10 and 30 per cent. She states that cities could cut their water demand by a third with technologies available today, without sacrificing quality of life. In this study, domestic water-use efficiency in 1900 has been estimated at 50 per cent.

The best estimates of global water stocks are probably those of Korzun et al. (1978), also used by Shiklomanov (1990, 1993, 1997) and presented in Table 2.1. However, the uncertainties are great and this may explain the wide ranges obtained if one compares estimates by different authors (Table 2.3). Changes in the global water stocks which may have occurred during the 20th century are small if compared to these uncertainty ranges in absolute stock volumes. As an example: the content of the oceans may have increased over the last 100 years by 10 to 25 cm (Warrick *et al.*, 1996), which corresponds to 36-90×10^{15} kg, but the total ocean storage is estimated at between 1320×10^{18} and 1370×10^{18} kg. The fresh surface-water stock may have increased by about 5×10^{15} kg (L'vovich and White, 1990; Shiklomanov, 1993), but estimates of the total stock vary between 30×10^{15} and 150×10^{15} kg. In the model, the estimates of Korzun *et al.* (1978) have been used as initial values for the year 1900, in the realization that we are interested in the simulation of *changes* to stocks rather than in the simulation of absolute stock values. As long as the uncertainties about the initial stocks do not hamper a proper simulation of stock changes, these uncertainties are not a real problem. In a sensitivity analysis, I looked at the effect of higher or lower initial stock values on some important model outcomes and found that the model outcomes are completely insensitive to most of the initial stock values taken. The only exceptions are the initial glacier stock and the initial renewable fresh groundwater stock, which both play a role in processes determining sea-level rise. This sensitivity of the model will be discussed further in the next section.

Oceanic evaporation in 1900 has been assumed at 505×10^{15} kg/yr, the value given by Korzun *et al.* (1977,1978). The initial values of atmospheric moisture transports from oceans to land and ice sheets have been estimated on the basis of Korzun *et al.* (1977,1978) and Warrick *et al.* (1996). For oceanic evaporation and atmospheric moisture transport, the same can be said as for the size of the water stocks: changes during the 20th century may have occurred, but they are relatively small if compared to uncertainty ranges for the total absolute values. Here also, a sensitivity analysis has shown that the simulated changes are not significantly influenced by the initial values, and therefore these uncertainties are not further analysed. The initial distribution of land precipitation over the physical basic areas under consideration has been assumed according to data from IIASA's climate database, which contains average data for the period 1931-1960 (Leemans and Cramer, 1991).

6.3 Model parameters: sensitivities and calibration

This section discusses the parameter values of the AQUA World Model. Some of these values have been taken directly from the literature as input for the model and others have been determined by calibration, which is understood here as the process of adjusting the parameter values within certain possibility ranges in order to correlate simulation results with observed data. The model has been calibrated three times: for the hierarchist, egalitarian and individualist world-views.[1] A sensitivity analysis has been carried out in order to get an insight into the effects of model parameters on model outcomes. The sensitivity of the simulation result to different parameter values gives information about the 'importance' of the various parameters. If the modelled system is 'underdetermined' (there are too many parameters to provide one analytical solution, as is generally the case in open systems), it is best to first *choose* the most plausible values for the less important parameters and then *determine* the more important ones by calibration. One could therefore suggest the carrying out of a sensitivity analysis before calibrating the model. However, the values of the parameters influence the outcome of the sensitivity analysis, so that any sensitivity analysis before calibration only yields provisional results. In fact, finding proper parameter values through sensitivity analysis and calibration is an iterative process. Here I do not present each step of this process, but only the final results. After a brief overview of the sensitivities of the calibrated model, I explain the calibration procedures followed and discuss the ultimate parameter values obtained.

A sensitivity analysis has been carried out for each of the three perspectives.[2] All model parameters have been considered, plus some initial values discussed in the previous section, namely the initial stocks of glaciers and renewable fresh ground water. For each perspective the parameters have been ranked according to their influence on some model results. Table 6.4 shows the first few parameters of each ranking list. Because the rankings sometimes change during the simulation period, the results have been presented for two points in time: the years 2000 and 2100.[3] An interesting result is that the rankings appear to differ for each perspective. This demonstrates that a perspective-based sensitivity analysis differs essentially from a traditional sensitivity analysis. Where a traditional analysis can be used to analyse the behaviour of *one particular model* of a system, the perspective approach considers

---

[1] Each time, the management style was chosen in correspondence with the world-view. However, where historical values of manageable parameters were available or could be estimated, I took the data of the realized management path instead of data corresponding to a particular management style. This means that, with respect to the past, no distinction was made between different management styles except for the manageable parameters for which no information was available. This was the case only for the technological diffusion and development rates.

[2] The sensitivity analysis was carried out with UNCSAM (UNCertainty analysis by Monte Carlo SAMpling techniques), a package designed for performing sensitivity and uncertainty analyses on a wide range of mathematical models (Janssen et al., 1994). Per perspective, I have varied each parameter within a small interval around the parameter value x for the perspective in question, according to a normal distribution with a mean value $\mu = x$ and a standard deviation $\sigma = 0.001 \times |\mu|$ (for $\mu \neq 0$) or $\sigma = 0.001$ (for $\mu = 0$).

[3] For the period 2000-2100 scenarios have been used which will be discussed in Section 7.2

the behaviour of *different representations* of the system. In this way, one can obtain an insight into the structural uncertainties in this behaviour. For example, if one considered only the hierarchist world-view, the conclusion (like that of the IPCC) would be that future sea-level rise is most sensitive to the ice sheet parameters of Antarctica and Greenland. This is not necessarily *the* behaviour of the system, which can be illustrated by the egalitarian world-view, where land evaporation and

Table 6.4. Sensitivity of model outcomes to a variation of parameter values, measured in the years 2000 and 2100.

Output variable	Ranking of parameters regarding their influence on the output variable[1]					
	Hierarchist		Egalitarian		Individualist	
	2000	2100	2000	2100	2000	2100
Domestic demand per capita	1. $El_{G,dom}$ 2. d 3. TD_{dom}	1. $El_{G,dom}$ 2. d 3. TD_{dom}	1. $El_{G,dom}$ 2. $El_{P,dom}$ 3. d	1. $El_{G,dom}$ 2. COV_{wwt} 3. $El_{P,dom}$	1. $El_{G,dom}$ 2. $El_{P,dom}$ 3. cost curve	1. $El_{G,dom}$ 2. $El_{P,dom}$ 3. cost curve
Irrigation demand per hectare	1. $El_{G,irr}$ 2. d 3. TD_{irr}	1. $El_{G,irr}$ 2. d 3. TD_{irr}	1. $El_{G,irr}$ 2. $El_{P,irr}$ 3. d	1. COV_{wwt} 2. $El_{G,irr}$ 3. $El_{P,irr}$	1. $El_{P,irr}$ 2. $El_{G,irr}$ 3. cost curve	1. $El_{P,irr}$ 2. $El_{G,irr}$ 3. cost curve
Industrial demand per dollar	1. $El_{G,ind}$ 2. d 3. TD_{ind}	1. $El_{G,ind}$ 2. d 3. TD_{ind}	1. $El_{G,ind}$ 2. $El_{P,ind}$ 3. iaf	1. COV_{wwt} 2. $El_{G,ind}$ 3. $El_{P,ind}$	1. $El_{G,ind}$ 2. $El_{P,ind}$ 3. cost curve	1. $El_{G,ind}$ 2. $El_{P,ind}$ 3. cost curve
Sea-level rise	1. β Antarctica 2. β Greenland 3. α Antarctica 4. α Greenland	1. β Antarctica 2. β Greenland 3. α Antarctica 4. α Greenland	1. lcf 2. $loss_{evap}$ 3. p 4. φ	1. lcf 2. $loss_{evap}$ 3. p 4. $S_{glac,i}$	1. β Antarctica 2. β Greenland 3. lcf 4. $loss_{evap}$	1. β Antarctica 2. β Greenland 3. lcf 4. $El_{P,ind}$
Groundwater-level decline	1. lcf 2. $loss_{evap}$ 3. p 4. wsf 5. φ 6. v	1. lcf 2. wsf 3. $loss_{evap}$ 4. p 5. $El_{G,ind}$ 6. $S_{fgw,i}$	1. lcf 2. $loss_{evap}$ 3. p 4. φ 5. wsf 6. v	1. lcf 2. $loss_{evap}$ 3. p 4. wsf 5. φ 6. v	1. lcf 2. $loss_{evap}$ 3. p 4. wsf 5. φ 6. $S_{fgw,i}$	1. lcf 2. $El_{P,ind}$ 3. $El_{G,ind}$ 4. wsf 5. p 6. $loss_{evap}$
Potential water supply	1. φ 2. f 3. θ 4. $loss_{evap}$ 5. lcf	1. φ 2. f 3. θ 4. $loss_{evap}$ 5. $AGWR$	1. iaf 2. φ 3. f 4. θ 5. $loss_{evap}$	1. iaf 2. φ 3. f 4. θ 5. $loss_{evap}$	1. $loss_{evap}$ 2. lcf 3. $El_{P,irr}$ 4. wsf 6. $El_{G,ind}$	1. COV_{wwt} 2. $El_{G,ind}$ 3. $El_{P,ind}$ 4. wsf 5. $loss_{evap}$

[1] Declaration of symbols: see the list of symbols in the back of this book.

groundwater parameters appear much more important for future sea-level rise, due to the relatively large contribution of groundwater loss to sea-level rise. A similar statement can be made for potential water supply. Adopting the hierarchist or the egalitarian world-view, one will conclude that the extent of groundwater recharge and the functioning of artificial reservoirs are the most critical factors for the future potential supply, but taking the individualist view, factors which relate to terrestrial evaporation and consumptive water use are most important, due to the positive

Table 6.5. Main list of parameter values in the AQUA World Model.

Parameter	Symbol	Equation	Unit	Hierarchist	Egalitarian	Individual.
Ratio private/public water demand per capita	rpp	3.3	-	0.75	0.75	0.75
Growth elasticity						
- domestic water demand	$El_{G,dom}$	3.7	-	1.2	1.1	1.3
- irrigation water demand	$El_{G,irr}$	3.7	-	0	0	0.1
- industrial water demand	$El_{G,ind}$	3.7	-	0.5	0.4	0.8
Price elasticity						
- domestic water demand	$El_{P,dom}$	3.7	-	0	-0.1	-0.4
- irrigation water demand	$El_{P,irr}$	3.7	-	0	-0.05	-0.2
- industrial water demand	$El_{P,ind}$	3.7	-	0	-0.2	-0.6
Technological diffusion rate	d	3.8	1/yr	0.002	0.001	0
Technological development rate	TD	3.9	1/yr	0.002	0.001	0
Water source fractions	wsf	3.10	-	See text	See text	See text
Fractions consumptive water use	fcwu	3.11	-	See text	See text	See text
Evaporation fraction of irrigation loss	$loss_{evap}$	3.12	-	0.66	0.78	0.54
Hydropower generation capacity per unit of reservoir storage	θ	3.14	MW/kg	0.10×10^{-9}	0.10×10^{-9}	0.10×10^{-9}
Average reservoir depth	d_{res}	3.15	m	14	14	14
Ice sheet sensitivity of Greenland	$\alpha_{Greenland}$	3.17	mm/yr/°C	0.30	0.45	0.20
Ice sheet sensitivity of Antarctica	$\alpha_{Antarctica}$	3.17	mm/yr/°C	-0.20	0.05	-0.35
Initial imbalance of Greenland	$\beta_{Greenland}$	3.17	mm/yr	0	0.2	0
Initial imbalance of Antarctica	$\beta_{Antarctica}$	3.17	mm/yr	0.1	0.4	0.1
Critical temperature increase glaciers						
- Minimum	$T_{glac,min}$	3.18	°C	1.3	1.3	1.5
- Maximum	$T_{glac,max}$	3.18	°C	4.3	3.8	5.0
Response time glaciers	τ	3.18	yr	100	75	125
Sensitivity oceanic evaporation	λ	3.20	1/°C	0.02	0.04	0.01
Sensitivity advective moisture transport	μ	3.21	1/°C	0	0	0
Loss of consumptive water use	ν	3.22	-	0.3	0.5	0.1
Land cover factors	lcf	3.23	-	Table 6.7	Table 6.7	Table 6.7
Soil water-holding capacities	S_{cap}	3.25	mm	Table 6.7	Table 6.7	Table 6.7
Ratio transpiration / total evaporation	-	-	-	Table 6.7	Table 6.7	Table 6.7

moisture recycling effect of increased evaporation on potential water supply.

From the results, one can see that some parameters have 'local' effects, while others influence the whole system. Some model variables thus appear to be sensitive to parameters which, at first sight, do not relate to the variable in question at all. In the case of the egalitarian world-view for example, wastewater treatment coverage appears to be one of the important factors determining water demand. This can be explained by the fact that the water system is regarded as rather vulnerable to pollution

Table 6.5. Continued.

Parameter	Symbol	Equation	Unit	Hierarchist	Egalitarian	Individual.
Ratios direct runoff / net precipitation	φ	3.28	-	Table 6.7	Table 6.7	Table 6.7
Exponent groundwater runoff	p	3.29	-	1.5	1.0	2.0
Ratio delayed runoff / total groundwater runoff	χ	3.31	-	0.83	0.75	0.95
Lag time fresh surface water store	k_{fsw}	3.33	yr	0	0	0
Delay river runoff	T_{fsw}	3.33	yr	0.044	0.044	0.044
Reservoir storage contributing to stable runoff	f	3.35	-	0.65	0.40	0.65
Artificial groundwater recharge	$AGWR$	3.35	kg/yr	0 to 1×10^{15}	0	0 to 1×10^{15}
Water quality parameters	See text	3.36, 3.37	mg/l	Table 6.8	Table 6.8	Table 6.8
Dilution factor	df	4.2	-	10	20	10
Inaccessible fraction of natural stable runoff	iaf	3.38	-	not applicable	0.35	not applicable
Water supply costs	$COST_{ws}$	-	-	Figure 6.4	Figure 6.4	Figure 6.4
Desalination costs	$COST_{desal}$	-	US$/l	See text	See text	See text
Utilization fraction of hydropower generation capacity	uf	3.45	-	0.42	0.42	0.42
Cost elasticity hydropower	Elc	3.46	US$/yr/MW	See text	See text	See text
Potential hydropower gen. capacity	HGC_{pot}	-	MW	2.4×10^6	2.1×10^6	2.7×10^6
Ratios water price / actual cost	rpc	3.50	-	See text	See text	See text
Demanded public water supply coverage	$COV_{pws,dem}$	3.51	-	See text	See text	Figure 6.5
Demanded sanitation coverage	$COV_{san,dem}$	3.52	-	See text	See text	Figure 6.5
Sanitation costs	$COST_{san}$	-	US$/cap	15	15	15
Demanded wastewater treatment coverage	$COV_{wwt,dem}$	3.53	-	See text	See text	Figure 6.5
Domestic wastewater treatment costs	$COST_{wwt,dom}$	-	US$/kg	0.2×10^{-3}	0.2×10^{-3}	0.2×10^{-3}
Industrial wastewater treatment costs	$COST_{wwt,ind}$	-	US$/kg	0.4×10^{-9}	0.4×10^{-9}	0.4×10^{-9}
Constants in sea level - probability relation	γ_1, γ_2	3.47, 3.54	-	See text	See text	See text
Maximum acceptable critical flooding probability	$P_{crit,max}$	3.54	1/yr	See text	See text	See text
Costs of dyke heightening	$COST_{dyke}$	-	US$/m/km	$0.7\text{-}1.4\times10^6$	$0.7\text{-}1.4\times10^6$	$0.7\text{-}1.4\times10^6$

and depletion, resulting in rapidly rising water supply costs if water becomes scarcer and thus in decreased water demand. Another surprise may be that groundwater-level decline and sea-level rise appear to be vulnerable to the land cover factors and evaporation losses from irrigation. This can be understood through the fact that evaporation from land is a critical factor in the water balance of the continents and as a consequence in the storage of water on land. In the individualist world-view groundwater-level decline and sea-level rise are also rather sensitive to growth and price elasticities of water demand, which can be explained by the importance of these parameters for water demand and indirectly for groundwater withdrawals and groundwater depletion.

An overview of parameter values in the calibrated model is presented in Table 6.5. Below I explain how the different values have been obtained, more or less in the order in which they are presented in the table.

Water demand parameters
Water demand is most sensitive to - in sequence of magnitude - growth elasticities (all perspectives), price elasticities (egalitarian and individualist), supply-cost curves (individualist), and technological diffusion and development rates (hierarchist and egalitarian). There is one exception, for irrigation in the individualist world-view, where price elasticity appears to be more important than growth elasticity. The ratio of private to public water demand per capita for domestic water supply is of minor importance within all perspectives. For each perspective, the model has been calibrated separately for domestic, irrigation and industrial water demand. As calibration criteria I have used the estimates of historical sector water supplies as given by Shiklomanov (1997) (Table 6.6). For each perspective, there are too many parameters to find a single best fit. The strategy has been to assume values from the

Table 6.6. Estimates of historical water supply used for calibration.

Water supply (10^{12} kg/yr)	1900	1910	1920	1930	1940	1950	1960	1970	1980	1990
- Domestic[1]	16	n.a.	n.a.	n.a.	37	53	83	130	208	321
- Irrigation[1]	525	n.a.	n.a.	n.a.	891	1124	1541	1850	2191	2412
- Livestock[2]	11	n.a.	n.a.	n.a.	n.a.	18	n.a.	n.a.	60	n.a.
- Industrial[1]	38	n.a.	n.a.	n.a.	127	182	334	548	683	681

[1] Shiklomanov (1997) for 1900 and 1940-1990.
[2] L'vovich and White (1990) for 1900, 1950 and 1980.   n.a. = not available

literature for the parameters for which water demand is least sensitive, and to calibrate the most influential parameter. For all perspectives, growth elasticities were the parameters which were ultimately calibrated; the other values have been estimated on the basis of the literature.

The ratio of private to public water demand per capita for domestic water supply has been assumed at 0.75 for all perspectives, a value suggested by RA (1994). Although this value should be regarded as a very rough estimate, total water demand simulated is fairly insensitive to this parameter, and I therefore did not consider it necessary to expend much effort on getting more accurate estimates.

Price elasticities are only relevant for the egalitarian and individualist (see Section 5.4). Empirical data on price elasticity of domestic water demand per capita show a wide range, with values which generally fall between zero and - 0.7 but in some cases even lower.[1] According to Bower *et al.* (1984), one might expect values between zero and minus one, because water is a commodity which typically cannot be substituted or is difficult to substitute, and is therefore price inelastic. For domestic water demand, the global average price elasticity has been assumed at a value of -0.1 for the egalitarian and -0.4 for the individualist. For industrial water demand per dollar value added, a price elasticity of -0.2 has been taken for the egalitarian and -0.6 for the individualist, which assumes that water use in industry can be reduced more easily than water use for domestic purposes. The price elasticity of irrigation water demand per hectare has been estimated to be small, at - 0.05 for the egalitarian and -0.2 for the individualist, on the assumption that increasing water prices will not only lead to a decreasing demand per hectare but ultimately also to a smaller area of irrigated land.

Non-price-driven technological development and diffusion are only relevant for the hierarchist and the egalitarian. The individualist considers the rate of technological development and the time it takes before people start using new techniques purely a matter of prices. For this reason, efficiency improvements become effective only through price developments in the individualist view. For the domestic and industrial sector, a technological development rate of 0.2 per cent per year has been assumed for the hierarchist (large-scale technology) and 0.1 per cent per year for the egalitarian (small-scale technology). For irrigation, I have assumed

---

[1] Keller and Van Driel (1985): -0.4 in the Netherlands for the period 1953-1981; Kooreman (1993): -0.1 in the Netherlands in 1989; Nieswiadomy (1992): 0.0 to -0.2 in the US in 1984; Renzetti (1992): -0.6 in Canada in 1986; Postel (1992): -0.3 to -0.7 in a number of countries; Groenen *et al.* (1993): -0.1 to -0.2 in the Netherlands in 1989. Gibbons (1986) gives a review of studies in the USA and cites values between zero and -1.6, the lowest value referring to the summer. According to Rogers (1985), the range of typical values for the price elasticity of municipal water demand is from -0.15 to -0.7.

the maximum efficiency improving from 80 per cent in 1900 to 95 per cent in 2000.[1] The diffusion rates for all sectors have been taken at 0.2 and 0.1 per cent per year for the hierarchist and egalitarian respectively. In the domestic sector, these values effectively result in an increase in actual water-use efficiency in the period 1900-2000 of 12 per cent for the hierarchist and 7 per cent for the egalitarian. The actual irrigation efficiency improves by 11 per cent in this period for the hierarchist and 6 per cent for the egalitarian. For the industrial sector, the model simulates an increase in actual water-use efficiency in the period 1900-2000 of 17 per cent for the hierarchist and 10 per cent for the egalitarian. The efficiency improvements in the industrial sector are greater because the initial difference between actual and potential efficiency is greater than in the domestic sector (see initial values Table 6.3).

Using the parameter values discussed above, the model has been calibrated for the three perspectives, resulting in growth elasticities as presented in Table 6.5. The growth elasticities are highest for the individualist and lowest for the egalitarian, which is consistent with the different perceptions of demand in relation to growth described in Chapter 5. The lowest response to economic growth has been found for irrigation water demand per hectare (zero for the hierarchist and egalitarian), which can be explained by the fact that economic growth probably influences the size of the irrigated area rather than water demand per hectare. The latter is instead influenced by natural requirement and technology used. The growth elasticities of domestic water demand established through calibration are rather high if compared to the values found in some national studies.[2] That these studies found smaller responses of domestic water demand per capita to economic growth than my global analysis might be explained by the fact that these studies refer to industrialized countries, where elasticities are probably less than in developing countries. More in general, one might assume that growth elasticities become smaller if gross national product per capita increases. The model takes the calibrated growth elasticity values for a gross world product per capita (GWP_{pc}) in the range from zero to 4000 US$/yr (the 1990 value). For higher values of GWP_{pc}, it is supposed that the growth elasticities decrease linearly to a quarter of their original value if $GWP_{pc} = 10000$ US$/yr, to remain constant again for a $GWP_{pc} > 10000$ US$/yr.

---

[1] It is currently possible to achieve an irrigation efficiency of 95 per cent, using so-called low-energy precision application (LEPA) in combination with water conservation and land preparation methods (Postel, 1992, 1993).

[2] Keller and Van Driel (1985): 0.7 in the Netherlands for the period 1953-1981; Nieswiadomy (1992): -0.2 to 0.4 in the US in 1984; Renzetti (1992): 0.5-0.9 in Canada in 1986; Groenen et al. (1993): 0.1 in the Netherlands in 1989.

Water source fractions

A water source fraction is the ratio of the supply from a particular source to total water supply (Equation 3.10). Water source fractions for the four types of sources considered (fresh surface water, renewable and fossil fresh ground water, and salt water) differ throughout the world. Some countries rely mainly on surface water, while others rely heavily on ground water (e.g. 95 per cent groundwater use in Libya and Saudi Arabia). In a few countries the contribution of desalination has grown to a significant volume (e.g. in Saudi Arabia, Kuwait, the United Arab Emirates, Libya). Water source fractions also differ per sector: according to Shiklomanov (1997), about 50 per cent of global domestic water supply, 20 per cent of agricultural supply and 4 per cent of industrial supply (including mining) is from ground water. Globally, most of the total water use on earth is presently supplied from fresh surface water. The remaining part, except for a small fraction which is taken from salt water, is withdrawn from fresh ground water. According to Kulshreshtha (1993), ground water provided 29 per cent of the global water supply in 1990. Shiklomanov (1997) gives a lower figure of 19 per cent for the same year. The volume of water from desalting plants is very small but growing. In 1960, the worldwide capacity of desalting plants was still insignificant, in 1970 it was about 0.4×10^{12} kg/yr (0.016 per cent of global water supply) and in 1980 about 2.7×10^{12} kg/yr (0.09 per cent of global water supply) (Van der Leeden *et al.*, 1990). The global desalination capacity at the end of 1989 was estimated at 4.85×10^{12} kg/yr (Gleick, 1993a), which corresponds to about 0.14 per cent of global water supply. In the case of groundwater withdrawals, not much is known about the historical development. Certainly absolute withdrawals have been increasing with the increase in total water use, but it is unclear whether the fraction of groundwater use has changed significantly during the 20th century. It is also not clear which part of the groundwater withdrawals is taken from shallow, renewable groundwater resources and which from deep, fossil ground water. Only some case-specific data are available. A well-known example of intensive pumping of deep fossil ground water is the Great Man Made River Project in Libya, where vast quantities of ancient water are withdrawn to irrigate the desert (Pearce, 1992). According to Shiklomanov (1997), non-renewable groundwater resources have been exploited for between a few decades and a century. In the model, it has been assumed that the water source fractions for both renewable and fossil ground water and salt water have increased during the 20th century, at the expense of the water source fraction for fresh surface water. For renewable ground water a linear increase has been assumed from 20 per cent in 1900 to 25 per cent in 2000, for fossil ground water from zero to 1 per cent (a first rough estimate due to lack of data) and for salt water from zero in 1900 to

0.2 per cent in 2000, with the largest increase in the last quarter of the century (this follows the development given by Van der Leeden *et al.*, 1990). For developments in the 21st century, different estimates are used for each of the four perspectives. For the hierarchist and fatalist, it has been assumed that the situation in the year 2000 will remain unchanged throughout the 21st century (see Section 5.4). Because the egalitarian will try to reduce groundwater use, a linear decrease of the groundwater source fraction has been assumed, from 25 per cent in 2000 to 15 per cent in 2100. The individualist will expand the use of fossil ground water and salt water (both to 5 per cent in 2100) and reduce the fraction surface water use (to 65 per cent in 2100).

Consumptive water use fractions
According to L'vovich and White (1990), the global average consumptive fraction for domestic water use decreased from 72 per cent in 1900 to 51 per cent in 1980, due to the increase in pipeline systems of water supply. Korzun *et al.* (1978) suggest a decrease from 25 to 15 per cent in the same period. Shiklomanov (1990, 1993) does not discuss a decrease and uses figures for this period which range between 21 and 27 per cent. For the period up to 2025, Shiklomanov (1997) expects a decrease in the consumptive fraction of domestic water use towards 13 per cent, mentioning the increase in pipeline systems as the main reason. In the model, a linear decrease is assumed from 50 per cent in 1900 to 20 per cent in 2000 and down to 10 per cent in 2100 for the hierarchist. For the egalitarian, the model takes values which are 10 per cent higher (thus from 55 per cent in 1900 to 11 per cent in 2100) and for the individualist values which are 10 per cent lower. For the consumptive fraction of livestock water use, L'vovich and White (1990) assume 100 per cent in 1900 (no return flow at all), decreasing to 83 per cent in 1980. For the hierarchist, a decrease is assumed from 90 per cent in 1900 to 80 per cent in 2000 and 70 per cent in 2100; for the egalitarian and the individualist, the model again takes values which are 10 per cent higher and lower respectively. For industrial water use, L'vovich and White (1990) suggest that the consumptive fraction decreased from 22 per cent in 1900 to 12 per cent in 1980, partly due to the relatively large increase in low-consumptive water use for cooling thermal electricity generating plants. Korzun *et al.* (1978) give a slight decrease, from 7 to 4 per cent. Again, Shiklomanov (1990, 1993) does not mention a decrease and suggests values in the range 7-9 per cent for the period 1900-1980. Shiklomanov (1997) expects that the consumptive fraction of industrial water use will *increase* towards 13 per cent in 2025, due to the growth of recirculation systems which use less water but have, in relation to input, greater evaporation losses. In the model, a decrease is applied for the hierarchist from 20 per cent in 1900 to 10 per cent in 2000, to remain at that level until 2100. Once again,

the model takes values which are 10 per cent higher and lower for the egalitarian and individualist respectively.

For irrigation, the consumptive part of the total water withdrawal is relatively large, but different authors use different estimates. Korzun *et al.* (1978) use a range of 73-79 per cent for the period 1900-1970, Shiklomanov (1990, 1993) applies a range of 76-78 per cent for the period 1900-1980, L'vovich and White (1990) use 86-89 per cent for the period 1900-1985 and Postel *et al.* (1996) assume a value of 65 per cent for the present. In the model, the fraction consumptive water use for irrigation is calculated as that part of the irrigation water which is taken up and transpired by the plant plus the fraction of the remaining part which evaporates during conveyance (see Equation 3.12). The latter fraction has been taken at 66 per cent for the hierarchist, 78 per cent for the egalitarian and 54 per cent for the individualist (which corresponds to consumptive fractions of 78, 86 and 70 per cent in 1900 and slightly higher values in 2000).

Reservoir parameters
Average hydropower generation capacity per unit of reservoir storage has been assumed at a value of 0.10×10^{-9} MW/kg, calculated from a global artificial reservoir storage of 5525×10^{12} kg in 1985 (L'vovich and White, 1990) and a global hydropower generation capacity of 562×10^{3} MW in the same year (UN, 1992b). The average reservoir depth has been estimated as 14 m, assuming an artificial reservoir area in 1985 of 0.40×10^{6} km^2 (L'vovich and White, 1990; Korzun *et al.*, 1978; Shiklomanov, 1993). This is the extent of the reservoirs when full, but excluding backwater areas (surfaces of contributing lake and river waters). Including backwater areas, the total might be 50 per cent more (L'vovich and White, 1990; Korzun *et al.*, 1978). However, when reservoirs are not full the area is smaller, so that it seems appropriate to take, as an average, the area of full reservoirs without the backwaters.[1]

Evaporation parameters
The increase in oceanic evaporation has been modelled as a function of the global mean temperature increase (Equation 3.20). The sensitivity of oceanic evaporation to temperature increase has been assumed at 2 per cent per ^{0}C for the hierarchist, 4 per cent per ^{0}C for the egalitarian and 1 per cent per ^{0}C for the individualist. The uncertainty range of 1-4 per cent has been based on the variation of values found in

---

[1] An improvement in the model could be obtained by accounting for variation in the reservoir area within a year.

the literature. Generally, authors provide estimates of the increase in global evaporation rather than oceanic evaporation, so I have had to rely partly on global estimates. A summary of the results of equilibrium simulations with different General Circulation Models given in Gates *et al.* (1992) suggests that global evaporation will increase by 1.1 to 2.1 per cent per ^{0}C global mean temperature increase. Larger values are found for example by Rind (1988), who gives a global evaporation increase of 2.75 per cent per ^{0}C, and by Wetherald and Manabe (1975) who give a value of 3 per cent per ^{0}C. The empirical relation proposed by Budyko (1982) results in a value of 4 per cent per ^{0}C.

The temperature sensitivity of the advective moisture transport from oceans to land has been taken to be zero for all perspectives, thus ignoring the fact that global warming might result in an increased or decreased net exchange between the oceanic and terrestrial atmosphere. Although there is no reason to exclude this possibility beforehand, I did not feel able to make any rational estimate. General Circulation Models should be able to provide a range of possible values, but these values are generally not reported and therefore unknown to me.

The loss of consumptive water use to the oceans should fall in a range between zero and 100 per cent. As has already been observed in Section 2.5, L'vovich and White (1990) implicitly assume a loss of 100 per cent, but this is very unrealistic. On a global scale, about 65-70 per cent of the natural evaporation from land returns to land as precipitation (estimated on the basis of Korzun *et al.*, 1978), which makes it implausible that none of the anthropogenic evaporation is recirculated on land. According to Budyko (1986), the effect of increased evaporation on precipitation depends on the size of the region within which the evaporation has changed. If this region is small, the effect will be insignificant, but if one considers evaporation increases over entire continents, the effect will be substantial. For the loss of anthropogenic evaporation a value of 30 per cent has been assumed for the hierarchist, a relatively high value of 50 per cent for the egalitarian and a low value of 10 per cent for the individualist.

Land evaporation depends most heavily on the land cover factors, and to a lesser extent on the soil water-holding capacities. Therefore, the latter have been assumed on the basis of different sources (Table 6.7) and the former have been calibrated in such a way that total land evaporation equals 72×10^{15} kg/yr in an equilibrium run (with initial values in 1900 and no changes), the value given by Korzun *et al.* (1978). I realize that soil water-holding capacities in fact depend on soil texture and rooting depth rather than on land cover, but on the supposition that there is some relationship between soil properties and land cover, soil water-holding capacities can be linked to land cover to provide a crude estimate.

Table 6.7. Land cover-specific parameter values.

Land cover type	Land cover factor[1] [-]	Soil water-holding capacity[2] [mm]	Ratio direct runoff / net precipitation[3] [-]	Ratio transpiration to total evaporation[4] [-]
Forest	1.1	200	0.67	0.80
Grassland	0.8	125	0.77	0.65
Desert	0.3	100	0.86	0.50
Wetland	1.1	800	0.67	0.90
Rain-fed cropland	1.0	125	0.77	0.65
Irrigated cropland	1.0	150	0.77	0.65
Urban area	1.0	100	0.91	0.30
Open water	1.0	not applicable	not applicable	not applicable

[1] Relative values based on Van Deursen and Kwadijk (1994). Absolute values resulting from calibration on total land evaporation.
[2] Estimates based on Thornthwaite and Mather (1955), Vörösmarty et al. (1989), Groenendijk (1989), Prentice et al. (1992, 1993), and Van Deursen and Kwadijk (1994).
[3] Relative values based on L'vovich (1979). Absolute values resulting from calibration on total groundwater runoff.
[4] Assumptions on the basis of Rind et al. (1990), Salati and Vose (1984) and Speidel and Agnew (1988).

For the ratio of transpiration to total evaporation on land, Speidel and Agnew (1988) suggest an average value of one third. Rind et al. (1990), however, cite a value of 75 per cent for meadows and 90 per cent for forests and a value for the Amazon rain forest from another study of 85 per cent. Salati and Vose (1984) found a value of 65 per cent for the Amazon rain forest. The model uses different values for each land cover type, as shown in Table 6.7.

Runoff parameters
The division of the net precipitation into direct surface runoff and groundwater runoff depends on factors such as soil, slope and land cover. On sloping soils about 20 per cent of precipitation infiltrates the ground, but on flat, permeable soils the figure can be as high as 80 per cent (Peixoto and Kettani, 1973). For the global average ratio of direct surface runoff to net precipitation, the best reference today is probably L'vovich (1979). He analysed a large number of hydrographs in order to distinguish a groundwater and a direct surface runoff component. For groundwater runoff as a fraction of total runoff, L'vovich found a global average value of 31 per cent, with specific values for the continents ranging between 24 and 36 per cent. L'vovich's estimates are widely used by others, including Korzun et al. (1978) who gives a global groundwater runoff of 13.32×10^{15} kg/yr. It is difficult to find specific

data for different land cover types (which would be preferable), because drainage areas often have more than one land cover type. However, classifying the world into different zones and making some generalizations, L'vovich (1979) finds some values for the ratio of groundwater to total runoff which are useful in this context.[1] The ratios of direct runoff to net precipitation in the model, as presented in Table 6.7, have been obtained by first assuming relative values for the different land cover types on the basis of L'vovich's zonal estimates and then calibrating the absolute values in such a way that the total groundwater runoff in an equilibrium run (starting the simulation in the initial year 1900) is 13.32×10^{15} kg/yr.

Groundwater runoff is determined by two parameters: exponent p and response factor κ_{fgw} (Equation 3.29). If one knows the initial groundwater runoff and assumes certain values for exponent p and the initial groundwater storage, the value of κ_{fgw} can be calculated (if the groundwater store is supposed to be in equilibrium in the initial year). For the initial groundwater runoff, Korzun's value of 13.32×10^{15} kg/yr is used (see above). In an earlier version of AQUA I assumed a linear groundwater store ($p=1$), following Kwadijk (1993), but it was shown that model behaviour is significantly different when another relation is used, for example a quadratic relation ($p=2$) (Hoekstra et al., 1997). The difference is that in the latter case the lag time of the groundwater store is no longer constant, but increases if the store empties. Physically, this can be understood as the disappearance of the upper groundwater layers, which have the fastest response time. A greater value for p means that groundwater withdrawals will affect the groundwater level to a lesser extent, although the response will be faster. Exponent p probably falls somewhere between 1 and 2. Shaw (1994) gives a range of 1.25 to 1.67, but these values are not representative because they refer to a case where Equation 3.29 was applied to the total runoff from a catchment area instead of to the groundwater runoff alone. In the present version of the AQUA World Model, a value of 1.5 is used for the hierarchist, 1.0 for the egalitarian and 2.0 for the individualist. The model outcomes are even more sensitive to the initial groundwater storage than to the value of p. In Hoekstra et al. (1997), I took an initial groundwater storage of 4×10^{18} kg for all perspectives, which is the estimate of L'vovich (1979) for the volume of ground water which is in active exchange with the surface. Although Korzun et al. (1978) give approximately the same figure for the amount of water in active exchange (3.6×10^{18} kg, see Table 2.1), I now consider this a relatively high estimate and use different values for the

---

[1] L'vovich (1979) finds the ratio of groundwater to total runoff to be: 29-50 per cent for different kinds of forests, 20-25 per cent for the temperate steppes, 23-40 per cent for the dry to wet (sub)tropical savannahs, 10 per cent for the desert savannahs in the (sub)tropics and 36 per cent for the sub-arctic tundras.

three perspectives: 2×10^{18} kg for the hierarchist, 4×10^{18} kg for the egalitarian and 1×10^{18} kg for the individualist. The rationale for choosing smaller values is that one can argue for not taking the *whole* zone of active exchange (with an average lag time of $4 \times 10^{18} / 13.32 \times 10^{15} = 300$ years), but only the zone of *most* active exchange (where the bulk of the groundwater flow goes and where lag times are much shorter). Assuming that the groundwater store is initially in equilibrium, κ_{fgw} is found to be 3.6×10^7, 300 and 8.7×10^9 for the hierarchist, egalitarian and individualist respectively. A large p in combination with a large κ_{fgw} (individualist) means that the groundwater store responds relatively fast, but to a lesser extent than in the case of a small p and a small κ_{fgw} (egalitarian). According to Equation 3.30, the initial lag time k_{fgw} of the groundwater store is 150 years for the hierarchist, 300 years for the egalitarian and 75 years for the individualist. In the hierarchist and individualist world, these lag times increase if the groundwater store is being emptied (fastest for the individualist, where the model simulates an increase of 1 per cent in the period 1900-2000).

The ratio of delayed surface runoff to total groundwater runoff has been taken at 0.83 for the hierarchist, calculated on the basis of a total groundwater runoff of 13.32×10^{15} kg/yr and a subsurface runoff of 2.2×10^{15} kg/yr (Korzun et al., 1978). For the latter, L'vovich and White (1990) give a slightly greater value of 2.4×10^{15} kg/yr. However, uncertainties are great, illustrated by the fact that estimates of subsurface runoff range between zero and 12×10^{15} kg/yr (Speidel and Agnew, 1988). For the egalitarian, the model uses a ratio of 0.75, which corresponds to the earlier assumption that ground water is a relatively slow-responding system, where deeper aquifers which connect to remote lowlands and the ocean have a significant role in runoff. The smaller ratio means that less groundwater flow will contribute to river runoff, which is in agreement with the egalitarian perception of water as a precious resource. For the individualist, the model takes a larger ratio of 0.95, a value proposed by for example Nace (1967).

Rivers have an average renewal time of about sixteen days (Korzun et al., 1978). According to Gleick (1987a), for large river basins a lag fraction of half a month is often assumed, that is, half of the water available for runoff in any given month runs off during that month. The remaining fraction becomes available for runoff in the following month. In the model, it is assumed that $k_{fsw}=0$ and $T_{fsw}=0.044$ year (= sixteen days).

The part of the maximum reservoir volume contributing to stable runoff differs per reservoir, making it difficult to obtain a global figure for the role of reservoirs in regulating river runoff. L'vovich and White (1990) estimate that the global average portion of the maximum reservoir volume contributing to stable runoff is about 65 per cent (3500 km^3 of the reservoir capacity of 5500 km^3 in 1985), with continental figures ranging from about 50 to 75 per cent. They consider this a conservative estimate, because the figure is based on the assumption that the useful volume of a reservoir is only filled and emptied once a year, a condition which applies only to rivers with one pronounced wet period. However, the value of 65 per cent is also used by for example Postel *et al.* (1996). Strangely enough, in a different part of their paper L'vovich and White (1990) suggest a value of 40 per cent (2180 km^3 over 5500 km^3, see also note [6] in Table 2.4), a value which is also used by Korzun *et al.* (1978) (2000 km^3 over 5000 km^3). L'vovich (1979) uses a global average value of 45 per cent (1855 km^3 over 4100 km^3). In this study, a value of 65 per cent is assumed for the hierarchist and the individualist (who are optimistic about managing stable runoff) and a value of 40 per cent for the egalitarian (who does not consider an increase in the resource base a real option).

For all perspectives, artificial groundwater recharge in the 20th century has been supposed to be nil on a global scale (although currently there are several projects, see Van der Leeden *et al.*, 1990). For the egalitarian and fatalist, it has been assumed that this will remain so during the 21st century. For the hierarchist and individualist, who will try to expand their resource base, it has been assumed that groundwater recharge will reach a rate of 1.0×10^{15} kg/yr by the year 2100. This is a low estimate compared to the prediction of L'vovich (1979) that the global artificial groundwater recharge rate would have already reached 5.0×10^{15} kg/yr by the year 2000, but as L'vovich's expectation has not been realized and as there are no other global estimates of future groundwater recharge, I use a more conservative estimate.

Sea-level rise parameters
The sea-level rise simulated appears to be sensitive to a whole range of parameters, due to the fact that sea-level rise consists of a number of different components (Figure 2.6). Sea-level rise has been calibrated separately for the three perspectives, using middle response parameter values for the hierarchist, high values for the egalitarian and low values for the individualist. For each of the perspectives the model should simulate a sea-level rise over the 20th century which falls within the observed range of 100-250 mm (Warrick *et al.*, 1996). The hierarchist perspective has been calibrated in such a way that the total sea-level rise over the period 1900-2000 is 150 mm, the egalitarian perspective so that the simulated sea-level rise is about 250 mm, and the individualist perspective so that the simulated sea-level rise is about 100 mm. The three resulting simulations of historical sea-level rise are shown in Figure 6.3.

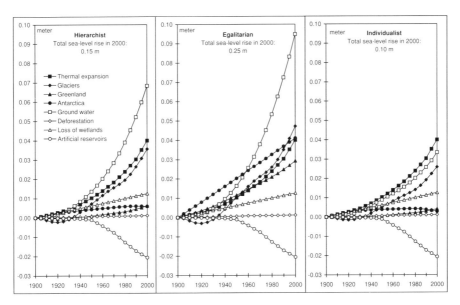

Figure 6.3. Simulated contributions to historical sea-level rise for the hierarchist, egalitarian and individualist perspectives.

For thermal expansion of the ocean, which is an input scenario of the model, the middle estimate of the IPCC, 40 mm, is used for all perspectives (Warrick *et al.*, 1996; Table 2.5). For sea-level rise as a consequence of glacier melting, the glacier parameters have been adjusted so that, for the hierarchist world-view, the model simulates a rise of 35 mm over the period 1900-2000, corresponding to the middle estimate of the IPCC. Because the calibration problem is 'underdetermined', I first assumed a response time of 100 years (the mid-value, given by Wigley and Raper, 1995) and a difference between the minimum and the maximum critical temperature increase of 3.0 °C and then calibrated the middle value for the critical temperature increase. For the egalitarian, I took a response time of 75 years and a range between $T_{glac,\,min}$ and $T_{glac,\,max}$ of 2.5 °C and calibrated the middle value for T_{glac} in such a way that the model simulates a rise of 45 mm over the period 1900-2000. For the individualist, I took a response time of 125 years, a range between $T_{glac,\,min}$ and $T_{glac,\,max}$ of 3.5 °C and calibrated the middle value for T_{glac} using a sea-level rise criterion of 25 mm in the 20th century. The three parameter sets found (Table 6.5) differ slightly from the sets used by Wigley and Raper (1995). One reason for this is that they start their simulation in 1880 and not in 1900 as in this study. In fact, the year for which the glaciers are assumed to be in equilibrium (the initial year of simulation) is another uncertain glacier parameter, which can be assumed as 1880 as well as 1900, because there are no empirical data to indicate which would be more accurate. As

Wigley and Raper (1995) show, the simulated sea-level rise is less sensitive to the choice of initial year if one forces the model with smoothed data (as done in this study, see Section 6.2). A further reason for differences is that Wigley and Raper (1995) take an initial glacier stock of 108×10^{15} kg, while I use a value of 180×10^{15} kg, as given by the IPCC (Warrick et al., 1996).

According to the IPCC, the contributions from the ice sheets of Antarctica and Greenland form the greatest uncertainty in explaining the sea-level rise in the 20th century (Warrick et al., 1996; Table 2.5). The present imbalance of the Greenland ice sheet might be ± 0.4 mm/yr and the imbalance of the Antarctic ice sheet ± 1.4 mm/yr. The sensitivity of Greenland to temperature increase has been estimated at 0.30 ± 0.15 mm/yr/°C and the sensitivity of Antarctica at -0.20 ± 0.25 mm/yr/°C. Although the uncertainties are great, in combination with assumptions on the other contributions to sea-level rise the possibilities are limited, in particular for the initial imbalances, because extreme imbalance values are difficult to correlate with the observed total sea-level rise of 100-250 mm in the past one hundred years. For example, if both ice sheets were growing and caused a sea-level decline of 180 mm, the other contributions would have been extremely large (to compensate for this decline), something which is impossible or at least highly improbable. For the sea-level rise from the Greenland ice sheet, the model takes an initial imbalance of zero and a sensitivity of 0.30 mm/yr/°C for the hierarchist, an imbalance of 0.2 mm/yr and a sensitivity of 0.45 mm/yr/°C for the egalitarian, and an imbalance of zero and a sensitivity of 0.20 mm/yr/°C for the individualist. Regarding the Antarctica ice sheet, the model assumes an initial imbalance of 0.1 mm/yr and a sensitivity of -0.20 mm/yr/°C for the hierarchist, an imbalance of 0.4 mm/yr and a sensitivity of 0.05 mm/yr/°C for the egalitarian, and an imbalance of 0.1 mm/yr and a sensitivity of -0.35 mm/yr/°C for the individualist.

The sea-level rise through groundwater loss on land is a result of the simulated groundwater depletion. The model gives a rise in the period 1900-2000 of 68 mm for the hierarchist, 95 mm for the egalitarian and 35 mm for the individualist. These estimates are considerably higher than the minimum estimate of 8.6 mm given by Sahagian et al. (1994), which is based on the analysis of a number of specific cases of groundwater loss in North America, North Africa, Arabia and the Aral and Caspian basins. However, the estimates are below Korzun's (1978) estimate of 0.8 mm/yr for the period 1900-1964. According to Gornitz et al. (1982), lowering groundwater tables may have caused a sea-level rise of 'a few centimetres' during the period 1880-1980. Warrick et al. (1996) give a range for direct anthropogenic contributions to sea-level rise (including the groundwater contribution) of -50 to 70 mm (Table 2.5).

The simulated sea-level rise in the period 1900-2000 as a result of deforestation is 1 mm for all perspectives. Such a low value corresponds to previous estimates that sea-level rise through deforestation has been nil (Warrick et al., 1996). However, the value is lower than the estimate of Sahagian et al. (1994) who calculate a sea-level rise of 3.4 mm as a result of deforestation in tropical areas in the period 1940-1990. They arrive at such a relatively high value by assuming that 360 mm of water would be released by the destruction of a forest (160 mm of water in living matter plus 200 mm of water in soil and dead organic matter). In this study an average value of 75 mm is used (ignoring the loss of water in living matter and taking a soil moisture loss of 75 mm, which is the difference between the soil moisture content in forests and that in grasslands or croplands). The assumption that the loss of water in living matter can be ignored in terms of sea-level rise has been based on the estimate of Korzun et al. (1978) that the total water content of living matter on earth is 1.12×10^{15} kg which, spread over the ocean surface, is equal to a layer of only 3 mm. It should be noted that this is at the lower end of the uncertainty range, which is $1\text{-}50 \times 10^{15}$ kg according to Speidel and Agnew (1988) and $0.3\text{-}80 \times 10^{15}$ kg according to Korzun et al. (1978).

The simulated sea-level rise through loss of wetlands is 13 mm for all perspectives. Although this is a substantial part of the total sea-level rise observed in the 20th century (100-250 mm), most authors do not regard wetland loss as a significant factor in historical sea-level rise (Warrick et al., 1996). The only calculation known to me is that by Sahagian et al. (1994), who estimate a sea-level rise of 1.3 mm in the period 1900-1990 as a result of wetland loss in the United States alone. If one makes the rough assumption that the wetland area of the United States forms 5 per cent of the global wetland area (Mitsch and Gosselink, 1993) and that the losses elsewhere have been comparable, the estimate by Sahagian et al. (1994) implies a sea-level rise of 26 mm as a result of global wetland loss. This is more than the 13 mm simulated by AQUA, which can be partly explained by the different assumption used for the volume of water lost if wetlands are drained. Sahagian et al. (1994) assume that 1000 mm is lost, while I take an average value of 675 mm (the difference between the soil moisture content in wetlands and the soil moisture content in grasslands or croplands).

The simulated sea-level *decline* as a result of artificial reservoirs is 15 mm for the period 1900-1985, which corresponds to a simulated increase in reservoir volume of 5.5×10^{15} kg (for all perspectives). For the 20th century, the model simulates a sea-level decline due to artificial reservoirs of 21 mm. These estimates roughly correlate with earlier estimates. According to Gornitz et al. (1982) for example, the trapping of water behind dams in the period 1880-1980 may have reduced the sea level by 10 to 20 mm; Gornitz and Lebedeff (1987) give a range of 5 to 10 mm. Newman and Fairbridge (1986) find that reservoirs have caused a sea-level decline of 12.5 mm in the period 1957-1982 (cf. this study: 10 mm). Sahagian *et al.* (1994) give a minimum estimate of sea-level decline due to artificial reservoirs of 5.2 mm in the period 1930-1990, based on a total volume of large dammed lakes of 1900 km^3.

Water quality parameters
The model uses five types of water quality parameters: the average natural concentrations of the substances under consideration, standard deviations, quality standards, the average concentrations in untreated wastewater and transmission coefficients. Global average natural concentrations of nitrate, ammonium, dissolved organic nitrogen and phosphate in rivers have been taken from Meybeck (1982). Standard deviations for the concentrations of these substances have been derived from the statistical distributions of samples as given in Meybeck and Helmer (1989) and Meybeck (1992). They give distributions for both natural and polluted waters (the latter from the GEMS/Water monitoring network). It appears that, except in the case of dissolved organic nitrogen, the standard deviations increase if waters become more polluted (Table 6.8). Each substance-specific standard deviation has therefore been defined as a function of the average concentration, using a linear relationship. The deviation data in Table 6.8 are used for the hierarchist. For the egalitarian, standard deviations are applied which are 25 per cent higher than the values for the hierarchist, and in the case of the individualist standard deviations which are 25 per cent lower.

The average concentrations in untreated wastewater and transmission coefficients have been calibrated in such a way that the model simulates average concentrations at the end of the 20th century which are equal to the average concentrations observed by the GEMS/Water monitoring network. The standards for water quality classes A to C used are presented in Table 6.8. I realize that the choices for specific standards may give rise to critical commentary, because there is little consensus in the discussion on standards. The two main areas of discussion are probably: what are the criteria for establishing certain standards and, given certain criteria, what are the standards. As for the criteria, it is most useful to refer to the intended use or function of the water, for example sustaining aquatic life, drinking or irrigation. However, it is unlikely that one particular function has the highest standard for every substance, and that another function has the second-highest standard for every substance, etc. The real situation is that one function requires the lowest concentration of one substance, while another function has the strictest

Table 6.8. Water quality parameter values.

Substance	Type of water	Statistical distribution of samples[1]			Log-normal distribution[2]		Water quality standards		
		10%	50%	90%	Cavg	Cdev	Cst,A [3]	Cst,B [4]	Cst,C [5]
NO_3 (mg/l)	Natural	0.05	0.10	0.2	0.10	0.058	0.2	10.0	50.0
	Polluted	0.05	0.70	9.0	0.69	5.3			
NH_4^+ (mg/l)	Natural	0.005	0.015	0.04	0.015	0.014	0.04	1.5	5.0
	Polluted	0.009	0.110	1.2	0.108	0.66			
DON (mg/l)	Natural	0.05	0.26	1.0	0.25	0.42	1.0	10.0	50.0
	Polluted	0.25	0.90	1.5	0.95	0.38			
PO_4^{3-} (mg/l)	Natural	0.002	0.010	0.025	0.0095	0.011	0.025	1.0	5.0
	Polluted	0.004	0.025	0.200	0.026	0.078			

[1] Meybeck and Helmer (1989) and Meybeck (1992).
[2] Best fit given the three percentile values.
[3] As a standard for 'natural' water quality, the model uses the concentration which is found in 90 per cent of the samples from natural waters.
[4] Several sources, including WHO (1993) and Van der Leeden et al. (1990).
[5] Own assumptions.

standard for a different substance. Nevertheless, as a rule of thumb, one may assume that natural aquatic ecosystems put the most strict requirements on the quality of a body of water, followed by drinking and finally irrigation and industrial purposes.

For calculating the Water Pollution Index, a dilution factor is needed (Equation 4.2). According to Falkenmark and Lindh (1974), untreated wastewater has to be diluted ten to fifty times before acceptable concentrations are attained. UNEP (1991) appears to use a dilution factor of about 13. Schwarz et al. (1990) and Postel et al. (1996) give a dilution requirement of 8.9×10^5 kg/yr per person which, with a current population of about 6 billion and an untreated wastewater production of about 8×10^{14} kg/yr, means a dilution factor of 7. The model uses a dilution factor of 10 for the hierarchist and individualist and 20 for the egalitarian.

Inaccessible fraction of natural stable runoff
The fraction of the natural stable runoff which is inaccessible for human use is an interesting parameter only in the egalitarian world-view, where it is a determinant of potential water supply. A value of 35 per cent has been assumed, somewhere in between the values given by Ambroggi (1980) and Postel et al. (1996). The former suggests a value of 42 per cent (5.0×10^{15} kg/yr in uninhabited areas over 12.0×10^{15} kg/yr total) while the latter arrive at a value of 19 per cent (2.1×10^{15} kg/yr in

inaccessible areas over 11.1×10^{15} kg/yr total). According to Postel *et al.* (1996), their figure is based on a conservative estimate of inaccessible remote flows, because they only considered the inaccessible remote flows of the Amazon, Zaire-Congo, and northern-tier undeveloped rivers.

Water supply costs

Current water supply costs vary widely from country to country, with values ranging from 0.05 US$/m^3 in the Philippines to 4.65 US$/m^3 in Cape Verde (Gleick, 1993a). RA (1994) estimates the current global average water supply costs for intake water quality A at 0.60 US$/m^3, for quality B at 0.65 US$/m^3, for quality C at 0.73 US$/m^3, and on average at 0.68 US$/m^3. They estimate that about 55 to 60 per cent of this consists of fixed costs. As mentioned in Section 3.7, water supply costs for withdrawals from surface water and renewable ground water supposedly increase as a function of water scarcity. Data on the magnitude of such an increase are difficult to obtain and one has to rely on data in specific cases. Rosegrant (1997) for instance, shows how real capital costs for new irrigation systems in five Asian countries increased in approximately twenty years by a factor of between 2 and 4. A specific example of the increase in water supply costs for some cities in Zimbabwe is given in Chapter 8. In most cases data on increasing costs of supplying water are not linked to scarcity data, not least because 'scarcity' (in the physical sense) is not measured regularly, which in turn is probably due to the lack of an unambiguous definition of water scarcity. As a consequence, the cost curves used in the model are highly uncertain, which has been a reason for assuming different cost curves for the hierarchist, egalitarian and individualist. Figure 6.4 shows the cost curves of public water supply according to three world-views, for water quality class A. The curves for the hierarchist, the egalitarian and the individualist world-view have been calibrated in such a way that current water supply costs are about equal according to the different world-views. The curves for supply from quality classes B to D show a similar cost increase, but lie above the curves shown in the figure, due to additional treatment costs. The cost curves for irrigation, livestock and industrial water supply also show a similar cost increase, but lie somewhat below the curves shown in the figure, due to less strict treatment requirements. Desalination costs depend heavily on the salt content of the intake water and the desalination technique used. Gleick (1993a) gives a range of 0.25 - 1.0 US$/m^3 for brackish water and 1.3 - 8.0 US$/m^3 for sea water. Postel (1996) gives smaller ranges of 0.45-0.70 and 1.0-1.5 US$/m^3 respectively. For the historical development of average desalination costs, data from Van der Leeden *et al.* (1990) have been used. The model assumes that average costs decreased from 5.0 US$/m^3 in 1950 to 1.0 US$/m^3 at present. For the individualist, it

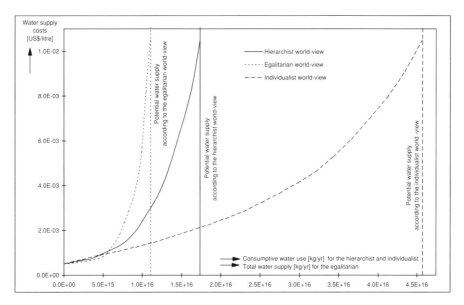

Figure 6.4. Water supply - cost curves per perspective, for public water supply from water quality class A. The values for potential water supply are those for 1990.

has been assumed that costs will decrease slightly during the 21st century as a result of technological improvements. For the hierarchist and egalitarian, desalination costs have been assumed to remain at the same level.

Hydropower parameters

According to Edmonds and Reilly (1985), the global hydroelectricity resource base is 95 EJ/yr, a theoretical maximum representing the amount of energy obtained from continental water runoff per year and calculated from land elevation and runoff data. This theoretical measure does not distinguish between water collected in large rivers which fall rapidly (high-grade resources) and dispersed, intermittent runoff flows, nor does it account for efficiency of capturing and converting the energy. About 80 per cent of the theoretical maximum would be available for development, i.e. 76 EJ/yr, which corresponds to a hydropower generation capacity of 2.4×10^6 MW. L'vovich and White (1990) cite a study from the 1970s giving comparable figures. In the model, it has been assumed for the hierarchist, egalitarian and individualist that they can use 80, 70 and 90 per cent of the theoretical resource base respectively. Data from the UN (1992b) on global hydroelectric generation capacity and generation for the period 1970-1990 show that average capacity utilization varied between 0.38 and 0.46, with an average of 0.42. The model takes this average value as a time-constant for all perspectives. Total average costs of hydropower generation have been assumed to be 50000 US$/yr/MW at present, a value based on data from Edmonds and Reilly (1985). These costs include both capital costs and operational and maintenance costs. The cost elasticity of hydropower has been assumed to decrease from 100 if none of the potential hydropower generation capacity is utilized to 0.1 if the capacity is fully utilized (intermediate values assumed according to the relation $E_C = 0.1 \times 2 \wedge (10 - 10 \times HGC / HGC_{pot})$).

Water price
Empirical data on water prices and actual costs are generally case-specific and it is difficult to obtain a global picture, let alone a picture of the historical development during the 20th century. Nevertheless, there is clear evidence that, in most parts of the world and in all sectors, the price of water - the tariff charged to the consumer - is lower than the actual supply costs (Gleick, 1993a; Serageldin, 1995). Historically, water has been a free resource (Bower *et al.*, 1984). According to the World Bank (1993), water prices charged in municipal water supply projects which were financed by the World Bank in recent years cover only 35 per cent of the actual supply costs, and charges in many irrigation systems are much lower. FAO (1995a) adds that these World Bank figures are probably better-than-average. As an example, FAO (1995a) cites a study from the mid-1980s showing that average subsidies to irrigation in six Asian countries covered 90 per cent of the total operational and maintenance costs. On average, Serageldin (1995) estimates that the cost recovery in the water sector is about 25 per cent at present, which is much lower than in other infrastructure sectors. In the model, it has been assumed that in the period 1900-1950 water prices were 15 per cent of actual costs. For the period from 1950 until the present, the model assumes that water prices increased to 25 per cent of actual costs (with the largest increase during the last two decades). For the 21st century, a policy of limited water price increase has been assumed for the hierarchist (water prices increase towards 75 per cent of actual costs by 2025), a water-taxing policy for the egalitarian (prices growing to 110 per cent by 2025) and a market-pricing policy for the individualist (prices going up to 100 per cent by 2025). In the case of the fatalist, prices remain at 25 per cent during the whole of the 21st century.

Public water supply and sanitation
As discussed in Chapter 5, the hierarchist and egalitarian actively manage public water supply and sanitation. Coverage demand is regarded as an input of the model. The values presented in Table 6.9 have been taken as historical input data for the hierarchist and egalitarian. For the hierarchist, the model assumes an increasing coverage demand for both public water supply and sanitation towards 100 per cent

The AQUA World Model   181

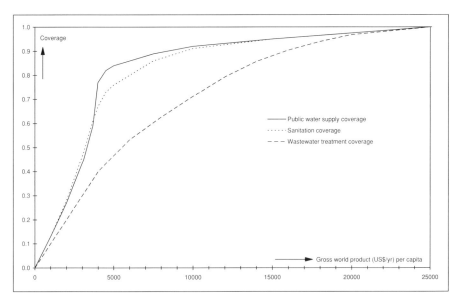

Figure 6.5. Public water supply, sanitation and wastewater treatment as a function of gross world product per capita. These curves are only used in the individualist world-view.

by the year 2050. For the egalitarian, 2025 is taken as the full coverage target year. In the individualist world-view, public water supply and sanitation coverage improve as a function of economic growth. The model uses the curves as presented in Figure 6.5, which have been correlated with the public water supply and sanitation data given in Table 6.9. It has been assumed that improvement of public water supply and sanitation conditions takes place particularly in a specific phase of economic growth (up to 4000 US$/cap). After this phase, the majority of people will have access to safe water and proper sanitation facilities, but further improvements towards full coverage will be slow. For the fatalist, it has been assumed that the number of people with public water supply and sanitation will remain constant from 2000, which implies that coverage will decline if population grows. The assumptions for public water supply costs have been discussed above. Annual sanitation costs have been assumed to remain constant at an average level of 15 US$ per capita, including both investment and operational and maintenance costs. According to Christmas and De Rooy (1991), capital investment costs for sanitation vary from 20 US$ per capita for low-cost technology in rural areas to 350 US$ per capita for high-cost technology in urban areas. The annual cost estimate has been obtained by assuming a depreciation time of twenty years and annual operational and maintenance costs which equal the annual investment costs.

Wastewater treatment

For the hierarchist and egalitarian, domestic and industrial wastewater treatment are regarded as policy variables. For the period 1900-1990 the model uses the values presented in Table 6.9. The historical development assumed is merely a rough estimate, because data are scarce, especially for the past and for developing countries. Through lack of data, the development of industrial wastewater treatment has been supposed to be similar to the development of domestic wastewater treatment. According to Postel *et al.* (1996), current domestic wastewater treatment in the countries of the Organisation for Economic Co-operation and Development is estimated to cover about 60 per cent of the population. A similar figure is given by the European Environment Agency, which estimates that in 1990 about 60 per cent of the European population benefited from sewage treatment (Stanners and

Table 6.9. Public water supply, sanitation and wastewater treatment estimates used for calibration.

	1900	1910	1920	1930	1940	1950	1960	1970	1980	1990
Public water supply coverage[1]	0.10	0.13	0.16	0.19	0.22	0.25	0.32	0.45	0.59	0.77
Sanitation coverage[1]	0.10	0.13	0.16	0.19	0.22	0.25	0.32	0.48	0.61	0.67
Wastewater treatment coverage[2]	0.05	0.07	0.10	0.13	0.16	0.20	0.24	0.28	0.32	0.36

[1] Data for 1980 and 1990 based on coverage data for developing countries from Christmas and De Rooy (1991) and the assumption of 100 per cent coverage in developed countries. Data for 1970 estimated on the basis of WHO (1984) and Gleick (1993a). Data before 1970: own estimates.
[2] Estimates on the basis of UNEP (1991), Stanners and Bourdeau (1995), Eurostat (1995) and Postel *et al.* (1996).

Bourdeau, 1995; Eurostat, 1995). During the last few decades, wastewater treatment facilities in industrialized countries have improved significantly. According to UNEP (1991), the treatment coverage in Belgium for instance increased from 5 per cent in 1970 to 33 per cent in 1980, in the Netherlands from 46 per cent in 1975 to 64 per cent in 1983, and in Sweden from 78 per cent in 1970 to 99 per cent in 1983. Information on wastewater treatment in developing countries is sparse, but treatment coverage is certainly far more limited, often estimated at about 5 per cent. This is also the value which has been taken for global treatment coverage at the beginning of the 20th century. In the individualist world-view, wastewater treatment is a function of economic growth. For both domestic and industrial wastewater treatment the model uses the curve as presented in Figure 6.5, which has been correlated with the data in Table 6.9. As in the case of public water supply and sanitation, it has been assumed that improvement of wastewater treatment takes place particularly in a specific phase of economic growth, but wastewater treatment is supposed to follow public water supply and sanitation after a certain delay. For the fatalist, it has been assumed that the amount of wastewater treatment will remain constant after the year 2000, which implies that coverage will decline if total wastewater production increases. According to Gleick (1993a), wastewater treatment costs range from 0.07 to 1.8 US$/m^3, depending on the type of process. Costs of secondary treatment, which may be taken as indicative of average costs, range from 0.2 to 0.4 US$/m^3. These costs include both capital costs and operational and maintenance costs. In the model, average domestic wastewater treatment costs have been assumed at 0.2 US$/m^3 and average industrial wastewater treatment costs at 0.4 US$/m^3.

Coastal defence parameters

The constants in the relationship between sea level and flooding probability (Equations 3.47 and 3.54) are different for each cluster and taken from RA (1994). For the maximum acceptable critical flooding probabilities, which also differ per coastal cluster, I have used as a reference the critical flooding probabilities in 1990, provided by RA (1994). The coastal defence strategy of hierarchists is to actively reduce flooding probabilities. This strategy has been implemented by assuming that hierarchists aim at critical flooding probabilities by the end of the 21st century which are one tenth of the probabilities in 1990. For individualists, a reduction in maximum acceptable critical flooding probabilities depends on economic growth (see Section 5.4). It has been tentatively assumed that the maximum acceptable levels reduce by one tenth if gross world product per capita increases by a factor 10. For egalitarians, the equality principle has been applied: they will aim at reducing the critical flooding probabilities in the areas in the world most at risk. In these areas, maximum acceptable critical flooding probabilities will be reduced to a level of once per ten years by the end of the 21st century. For fatalists, it has been assumed that no coastal improvements will be made. In areas where the critical flooding probability was already less than once every ten years in 1990, maximum acceptable probabilities are kept at 1990 levels. The costs of raising coastal protection by one metre over a length of one kilometre have been assumed at 1.4×10^6 US$/m/km for stone-protected sea dykes, 0.7×10^6 US$/m/km for clay-covered sea dykes and 1.0×10^6 US$/m/km for sand dunes (RA, 1994). Data on population and capital densities in the coastal clusters for the year 1990 have been taken from RA (1994). For other years, population densities are calculated according to the ratio of *POP(t)* to *POP(1990)* and capital densities according to the ratio of $GWP_{pc}(t)$ to $GWP_{pc}(1990)$.

6.4 Discussion of results

The AQUA World Model simulates an increase in total expenditure on public water supply and sanitation in the world (including both investment costs and operational and maintenance costs and expressed as a fraction of gross world product) from 0.2-0.3 per cent in 1900 to 0.5-0.6 per cent in 1960, 0.8-0.9 per cent in 1980 and 1.1-1.2 per cent today. By comparison, the World Bank has estimated that the proportion of gross domestic product invested in public water supply and sanitation in developing countries rose from about 0.25 per cent in the 1960s to about 0.45 per cent in the 1980s (Serageldin, 1994). Assuming that annual investment costs are in the same order of magnitude as annual operational and maintenance costs, the simulation results and the World Bank estimates are relatively close to each other. It should be noted, however, that the World Bank estimates refer to developing countries, and my estimates to the world as a whole.

The model produces an interesting result with respect to sea-level rise through water loss on land. In the individualist explanation of 20th century sea-level rise, water loss on land contributes 28 mm, in the hierarchist explanation 61 mm and in the egalitarian explanation 88 mm. This range of 28-88 mm is significantly above the range of -50 mm to 70 mm given by the Intergovernmental Panel on Climate Change (IPCC). According to the middle estimate of the IPCC, 75 mm of the observed 180 mm sea-level rise during the period 1890-1990 is explained by climate change, only 5 mm by water loss on land and the remaining 100 mm is left unexplained (Warrick *et al.*, 1996; see also Table 2.5). The results reported in this chapter suggest that a significant fraction of this unexplained sea-level rise can be ascribed to water loss on land. However, uncertainties on all separate contributions remain great, including the contribution from water loss on land.

7 A global water assessment

Can the world population be provided with an adequate and sustainable supply of clean water? How will the global hydrological cycle be changed in the future? There are too many uncertainties to give simple answers to these questions. However, one can do some interesting exploratory work if certain assumptions are made. In this chapter, I present such an exploration, using the AQUA World Model as a basis. I report on experiments with this model and explain the role of population and economic growth, agriculture, technology, water pricing and other types of water policy. Different perspective-based scenarios are presented for future global water supply and sea-level rise, distinguishing between utopias and dystopias. I will review low, medium and high risk developments and consider the implications of various water policy strategies. The changing interaction between water and development is explained in terms of a three-phased 'water transition'. The chapter concludes with an analysis of strategies to safeguard a sustainable supply of clean water.

7.1 Introduction

During the 20th century, human pressures on the water system have grown to such an extent that the hydrological cycle has been changed in many regions of the world. In several places water scarcity has become a critical issue. Given the expected growth of the world's population and the ongoing economic growth, it is likely that hydrological processes will continue to change and that water will become scarcer in the future. The aim of this global water assessment is to obtain a picture of the changes which can be expected, to get a feeling for the relative importance of the different mechanisms of change and to assess possibilities of anticipating the changes through public policy. I try to go further than just a presentation of possible water futures and will put emphasis on the explanation of *how* different futures can come into being. Through an understanding of the mechanisms behind certain developments, it is possible to get a clear idea of where people could interfere in the process of change, in order to redirect the future.

To develop scenarios of possible water futures, experiments have been carried out with the AQUA World Model, following the perspective-based scenario approach discussed in Section 5.1. This approach means that several possible water futures will be presented, each of which is characterized by:

- a specific socio-economic and environmental context,
- a specific world-view on water, and
- a specific style of water management.

The 'context' refers to exogenous developments which are not part of the study and for which assumptions have to be made. The study distinguishes between a hierarchist, an egalitarian and an individualist context, which will be described in the next section. The 'world-view' denotes the way in which various water issues are perceived (e.g. issues such as water demand, availability, scarcity, and robustness of the water system). The 'management style' refers to a particular perception of how water should be managed. As discussed in Chapters 5 and 6, three world-views (hierarchist, egalitarian and individualist) and four management styles (hierarchist, egalitarian, individualist and fatalist) are distinguished. Considering all possible combinations of context, world-view and management style, one arrives at 3×3×4 = 36 possible water futures. To present all of these would not only be laborious, but would also result in a dry and opaque review. In addition many of these futures are quite similar, so that I have decided to discuss only a few specific futures in detail. I start with the three so-called utopian futures, in which context, world-view and management style all correspond to the same perspective. These utopian futures represent the 'ideal world' according to the hierarchist, the egalitarian and the individualist.[1] After the three utopias I present three dystopias, which only differ from the utopias in respect of the management style applied. By analysing the effects of a fatalist management style, where public policy intervention is non-existent or ineffective, and contrasting the dystopian with the utopian worlds, it is possible to form an impression of the true importance of public policy in each of the utopias. After this detailed discussion of the three utopias and three dystopias, a further thirty dystopias remain. I will examine the full range of these possible futures by focusing on a limited number of variables and showing how different combinations of assumptions will result in different forecasts for the year 2100.

After the presentation of the various possible futures, an important policy question remains unanswered: what action taken would be best? What are the risks of any specific management style, given the fact that it might fit within one particular utopian world, but will be inappropriate if the world does not appear to behave like this utopia? Which chances will be lost if people spend money in the egalitarian way and the world turns out to be as the individualist had told us? Or, what will happen to the environment if mankind follows the individualist approach and the world behaves as the egalitarian had warned us it would? A risk assessment will be carried out by considering the effect of different management styles under

---

[1] The term 'ideal world' has been put in quotation marks, because an ideal world is here understood to be a world which evolves *if people consistently follow a certain perspective,* and not some imagined perfect world.

various conditions. The chapter concludes with an analysis of how high-risk futures may be avoided, in order to safeguard a sustainable supply of clean water.

7.2 The context

In this study three types of context are considered: hierarchist, egalitarian and individualist. Each type of context refers to a particular development path in respect of population, economics, land use, energy use, and global warming. The three types of context have been defined in such a way that they roughly correspond to the integrated hierarchist, egalitarian and individualist utopian futures described in Rotmans and De Vries (1997). In this section I limit myself to a description of the quantitative assumptions for each context, to give the reader an insight into the figures on which the water scenarios which will be presented in the next sections are based, and to enable the reader to make comparisons with scenarios developed by other authors.

The hierarchist context is strongly based on 'middle estimates' in the literature. The world's population will continue to grow, but the growth rate will decline, resulting in 11.3×10^9 people in the year 2100 (Figure 7.1), in correspondence with the medium scenarios of the UN (1990) and World Bank (1991). There will be continued economic growth, resulting in a gross world product of about 250×10^{12} 1990-US$/yr by 2100, comparable to the IS92a-scenario of the IPCC (Leggett *et al.*, 1992). This scenario means that economic production will grow by a factor of about 10 during the 21st century (cf.: in the 20th century, production has grown by a factor of about 20). Livestock growth rates have been assumed as equal to population growth rates, which results in livestock numbers of 3.7×10^9 ox equivalents in 2100, about double the current numbers. Deforestation will continue in the tropics, but there will be some reforestation in the temperate zone. In 2100 the global forest area will be 30×10^6 km^2, i.e. about three quarters of the area today (Figure 7.2). The area of rain-fed cropland will increase in the developing world but decrease in the developed world, so that the extent of rain-fed cropland in the world as a whole will change only slightly. In both the developing and the developed world, there will be continued but decreasing growth in irrigated cropland, approaching a global area of about 3.5×10^6 km^2 in 2100. The global urban area will grow more rapidly than the population, because the average global level of urbanization will increase from 45 per cent today towards 70 per cent in 2100. The global urban area will thus reach 1.6×10^6 km^2 by 2100, which is about three times the present urban area. The global expanse of wetlands will decrease further as a result of land reclamation, but the loss rates will be less than in the 20th century. About one third of the current global

188 · Perspectives on water

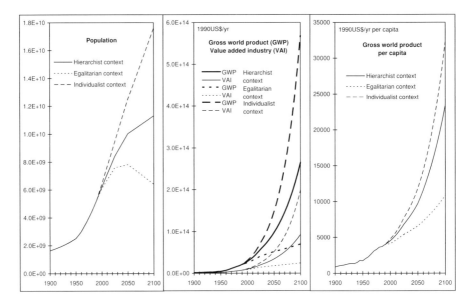

Figure 7.1. Population, gross world product and value added in the industrial sector.

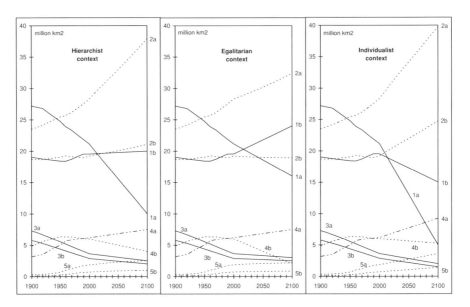

Figure 7.2. Land cover changes: 1. forests, 2. grasslands, 3. wetlands, 4. rain-fed croplands and 5. irrigated croplands. A distinction is made between a. developing world and b. developed world. Not shown in the figure are: deserts, urban areas and fresh surface water areas.

wetland area will be lost during the 21st century, resulting in an area of 4.5×10^6 km^2 by 2100. Hydropower generation capacity will be extended to 1.0×10^6 MW by 2100, which is about 40 per cent more than the capacity today. By the end of the 21st century the global mean surface air temperature will be 2.5 °C above the value in 1900 (2.0 °C above the value in 1990), which corresponds to the middle estimate of the IPCC for the IS92a-scenario (Kattenberg et al., 1996).

The egalitarian context is characterized by modest growth rates. The world's population will reach a peak of 7.8×10^9 around the year 2050 and then decrease in the second half of the 21st century, a development path comparable to the medium-low scenario of the UN (1990). Economic growth is much smaller than in the hierarchist world, both in an absolute sense and on a per capita basis (Figure 7.1). By the end of the 21st century gross world product will amount to 70×10^{12} 1990-US$/yr, which is about three times the present level of economic production. Deforestation in the developing world will be neutralized by reforestation in the developed world, so that the total global forest area will remain at approximately 40×10^6 km^2 (Figure 7.2). Expansion of rain-fed cropland in the developing world is equal to that in the hierarchist world (although the type of production is extensive rather than intensive). The area of rain-fed cropland in the developed world will be reduced, as in the hierarchist world, but more sharply (fewer people and a relatively low-calorie and more vegetarian diet). The area of irrigated cropland will increase in both the developing and developed world, but by less than in the hierarchist world. By the end of the 21st century, the global irrigated cropland area will have reached 3.0×10^6 km^2, which is about 20 per cent more than today (cf.: in the 20th century the irrigated cropland area increased by a factor of more than 5). The level of urbanization will increase less than in the hierarchist world, towards 60 per cent in 2100. The global urban area will grow to 0.8×10^6 km^2, which is about 40 per cent more than at present. The loss of wetlands will be smaller than in the hierarchist world, resulting in a global area of 5.5×10^6 km^2 by the year 2100. Hydropower generation capacity will be extended towards 0.9×10^6 MW in 2100. This increase will be largely obtained from small-scale hydropower plants, not from building large new dams. Although counter-intuitive, the increase in global mean temperature in the egalitarian world will be greater than in the hierarchist world (see also Rotmans and De Vries, 1997). This is the paradox of the egalitarian: a relatively low pressure on the environment will have a relatively great impact, due to the relatively high vulnerability of the earth system assumed. It has been assumed that in the egalitarian world global mean temperature in the year 2100 will be 3.0 °C above the value in 1900, which is 2.5 °C above that in 1990.

In contrast to the egalitarian context, the individualist context is characterized by high growth rates. By the year 2100 the earth will be populated by 17.6×10^6 people, which corresponds to the medium-high population scenario of the UN (1990). This scenario implies an increase in population during the 21st century by a factor of about 2.8, which is much more than in the hierarchist or the egalitarian world, but not extreme if compared to the growth factor of nearly 4 during the 20th century. Gross world product will increase to about 550×10^{12} 1990-US\$/yr in 2100, corresponding to the IS92e-scenario of the IPCC (Leggett et al., 1992). The production per capita will increase during the 21st century by a factor of nearly 7 (the growth factor during the 20th century was about 5). In the individualist world, half of the current global forest area will be lost during the 21st century. The area of rain-fed cropland in the developed world will decrease, but to a lesser extent than in the hierarchist or egalitarian worlds. In the developing world there will be a strong increase in the area of rain-fed cropland, because the largest percentage of population growth will take place in this part of the world. By 2100, the global rain-fed cropland area will be 14.5×10^6 km^2, which is 20 per cent more than at present. The global irrigated area will increase to 5.0×10^6 km^2, which is about twice the present area. The total cropland area will grow by a factor of about 1.3 only. The world population can be fed due to a large increase in food production efficiency. More wetlands will be lost than in the hierarchist or egalitarian worlds: by 2100 a global wetland area of 3.5×10^6 km^2 will remain, which means a loss of nearly half of the current area. The global urban area in the individualist world will grow much faster than in the hierarchist or egalitarian worlds, because both population and global average urbanization level grow faster. By the end of the 21st century, the urbanization level in the world as a whole will be 75 per cent, about the level presently found in the developed world. The total urban area in the world will then be 2.6×10^6 km^2, which is approximately five times the present urban area. Hydropower generation capacity will be extended to 1.2×10^6 MW in 2100. Global warming will be less than in the hierarchist or egalitarian world, because of the robust climate system assumed. Global mean temperature in the year 2100 will be 1.8 °C above the value in 1900 (1.3 °C above that in 1990).

Although I have put the emphasis on figures in the above presentation of the three contexts, they differ more fundamentally than merely in the fact that one context has larger or smaller growth rates than the others. Each context is characterized by a particular life-style. For a more elaborate discussion of the characteristics of the hierarchist, egalitarian and individualist, the reader is referred to Chapter 5.

7.3 Three water utopias

In this section three so-called 'water utopias' are discussed: a hierarchist, an egalitarian and an individualist utopia. Each water utopia is based on one particular, coherent perspective on how the world may evolve in general (context), on how water and human beings interact (world-view on water), and on how to manage water resources (management style).

A hierarchist water utopia
In the hierarchist water utopia, a strong growth in water supply infrastructure during the 21st century is made possible by constructing new artificial surface reservoirs and by implementing artificial groundwater recharge on a large scale. In the second half of the century, water quality will improve because of intensive investment programmes to increase the coverage of wastewater treatment. There is a certain amount of water conservation, but not as much as some people (egalitarians and individualists) thought would be necessary. Domestic water demand per capita will grow by a factor of about 2 in the 21st century (Figure 7.3). In developed countries,

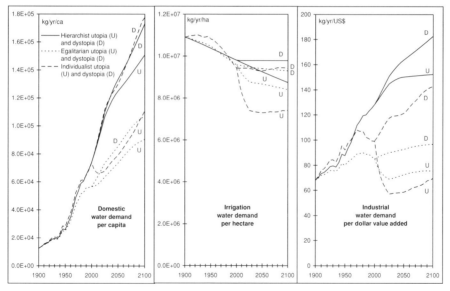

Figure 7.3. Specific water demand per sector. In the hierarchist utopia, water conservation is less than in the egalitarian or the individualist utopia, which corresponds to the hierarchist idea that water scarcity is a problem of supply rather than of demand. In the three dystopias, active measures to improve water-use efficiency are lacking, so water-use intensities reduce only as a consequence of the 'autonomous' increase in costs (which is a result of increased scarcity). The growth of water-use intensities in the domestic and industrial sectors is a result of economic development.

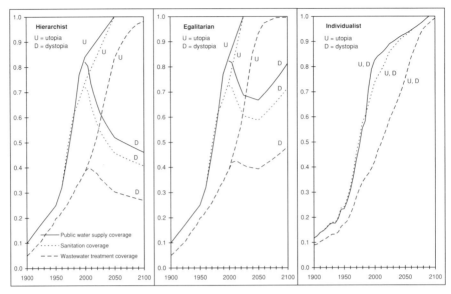

Figure 7.4. Coverage of public water supply, sanitation and wastewater treatment. In the hierarchist and egalitarian utopias, coverage will increase during the 21st century as a result of intensive development programmes. In the hierarchist and egalitarian dystopias, the lack of adequate policies will impede the expansion of infrastructure; as a result of population growth, coverage will decline. In the individualist world-view, improvements in public water supply, sanitation and wastewater treatment depend on economic growth rather than public policy, so - in this respect - the individualist dystopia does not differ from the individualist utopia.

water use per capita will often remain at the same level or even become less, but water-use intensities in developing countries will increase strongly, due to improved economic conditions and living conditions. In irrigation, the average water application factor will decrease steadily, through the use of more efficient techniques. Water-use intensities in industries will grow in developing, but decrease in developed countries. The entire global population will have public water supply and proper sanitation facilities by the year 2050 (Figure 7.4). Full coverage of wastewater treatment will follow after some decades. As shown in Figure 7.5, global water supply will continue to grow during the 21st century, with industry becoming the largest water-use sector instead of agriculture. Domestic and agricultural water use continue to increase, but the growth rates will level off. Industrial water use will grow exponentially as a result of the increasing growth of industrial production. The growth of global consumptive water use[1] will be considerably lower than the growth

---

[1] Consumptive water use is that part of a freshwater withdrawal which is lost through evaporation. The volume of consumptive water use often gives a better indication of the possible effect of water use on the terrestrial hydrological cycle than the volume of total water use (see Chapter 2).

A global water assessment 193

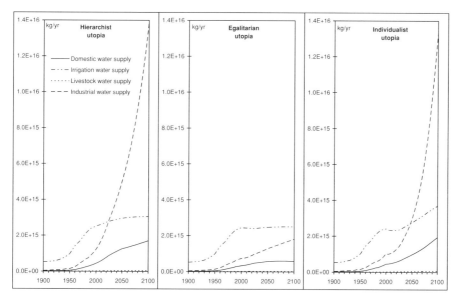

Figure 7.5. Water supply per sector in each of the three utopias.

of total water supply and will level off well below potential water supply (Figure 7.6). This is caused by the fact that the sector with most consumptive water use, the agricultural sector, shows relatively little growth. The exponential growth of industrial water use can be realized relatively easily, because most of the water returns directly to streams or other bodies of fresh water, so making it available for re-use.

Water scarcity, expressed as the percentage of potential water supply taken by consumptive water use, will grow from 12 per cent in 2000 to 20 per cent by 2100. This, together with a worsening water quality situation, will result in a significant increase in global average water supply costs during the first half of the 21st century (Figure 7.7). However, in the second half of the century average water supply costs will stabilize, as a result of improved water quality. In the period 2000-2025 water prices will show a relatively rapid increase (but less so than in the other two utopias, as will be explained below), due to the new pricing structure introduced. Water users will have to pay total operational and maintenance costs and about half of investment costs, which together is generally more than users pay today. In the 21st century, total expenditure in the water sector will continue to grow, but expressed as a fraction of gross world product, it will decrease steadily (Figure 7.8). However, future expenditure differs per item: relative expenditure in irrigation will fall sharply, expenditure in public water supply, sanitation and domestic wastewater treatment will show a moderate decline, relative expenditure in industrial water

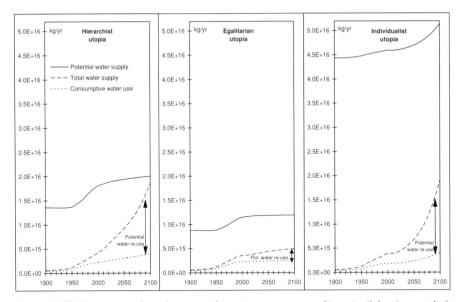

Figure 7.6. Total water supply and consumptive water use compared to potential water supply, in each of the three utopias. In the hierarchist utopia, potential water supply increases as a result of new artificial surface reservoirs and artificial groundwater recharge. Potential water supply in the egalitarian utopia will increase only a few per cent during the 21st century. The increase in potential water supply in the individualist utopia is largely due to the beneficial effect of consumptive water use on water recycling on land.

supply will stabilize, but relative expenditure in industrial wastewater treatment will greatly increase.

Until 2025 the water quality situation in the world will worsen, but thereafter quality starts to improve again, due to increased coverage of wastewater treatment (Figure 7.9). The water balance on land will change as a result of different processes. In the hierarchist utopia, the total evaporation from land is estimated to increase by 12.6 per cent in the period 1900-2100, the largest part of this in the 21st century (Figure 7.10). This increase is made up as follows: 15 per cent as a result of global warming, 5 per cent from consumptive water use, 0.4 per cent from artificial reservoirs and -7.8 per cent through land use changes (deforestation being the most important factor). As a result of the increased evaporation precipitation will increase as well, but to a lesser extent, resulting in a decrease in net precipitation. An area of great uncertainty is the net advective moisture transport from oceans to land, for which some specific assumptions have been made (see Section 6.3). Although net precipitation will decrease (about -3 per cent in the period 1900-2100), total runoff will *increase* (about +3 per cent). The difference between total runoff and net precipitation represents a loss of water on land, caused by groundwater abstractions.

A global water assessment

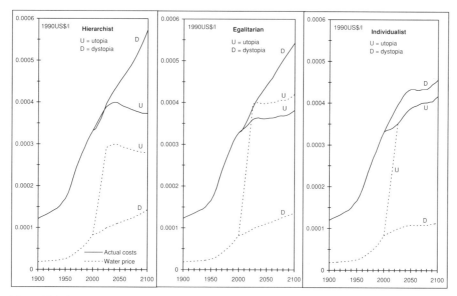

Figure 7.7. Average water costs and prices. In the 20th century, water costs and prices have increased as a result of growing scarcity and pollution. The water-pricing policies in each of the utopias will result in a large price increase in the first quarter of the 21st century, most pronounced in the egalitarian utopia because of the introduction of water taxes. The increasing prices will slow down the growth in water demand and thus reduce the increase in water scarcity and costs. Improved water quality in the second half of the 21st century will also have a beneficial effect on cost development. In the three dystopias, water-pricing policies are lacking, which will keep prices relatively low. However, as a result water will be used inefficiently, water scarcity will increase, and actual water costs will rise more than in the utopias. A further driving force behind the cost increase in the dystopias (except the individualist dystopia), is the worsening water quality situation.

As is be discussed below, this will result in declining groundwater levels, most noticeably in densely populated areas.

Stable runoff has changed and will continue to change due to a number of factors. However, a crucial issue in the 21st century will be the loss of stable runoff due to groundwater withdrawals (Figure 7.11). The consequences of land cover changes and climate change on stable runoff will be noticeable in certain regions of the world, but are less omnipresent than the effects of groundwater exploitation. The reduction in stable runoff as a result of groundwater withdrawals will be partly compensated for by building new large dams, to create additional freshwater reservoirs, but the number of new reservoir projects will be limited. During the 20th century artificial reservoirs could effectively increase stable runoff, but in the 21st century their effect will not be sufficient even to compensate for the loss of stable

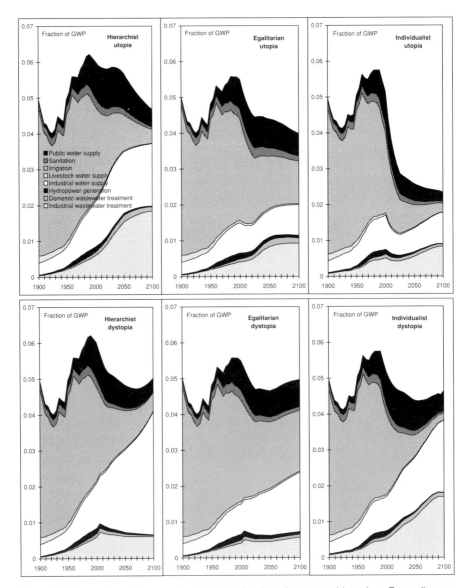

Figure 7.8. Expenditure in the water sector as a fraction of gross world product. Expenditure on coastal defence is not shown, because this is relatively small.

runoff from intensive groundwater exploitation. For this reason, people will start to implement artificial groundwater recharge on a large scale. In the period 2020-2030, global stable runoff will reach its maximum, after which it will steadily decrease. This decrease can only be reversed if the number of artificial recharge projects is increased, or if groundwater exploitation itself is reduced. Intensive groundwater use in the hierarchist utopia will not only reduce stable runoff, but also lower groundwater tables (Figure 7.12). Worldwide, groundwater tables will on average drop by a water column of one metre during 2000-2100, which is about three metres in real terms if one assumes an average porosity of one third.

By the end of the 21st century, total sea-level rise in the hierarchist utopia will reach about 1 metre (measured from the reference year 1900, see Figure 7.12). Considering the rise in sea level in the period 1900-2100 as a whole, about 31 per cent is due to thermal expansion of the oceans, 22 per cent to melting of glaciers and 5 per cent to melting and calving of the Greenland ice sheet (see Table 7.1). Antarctica has a small negative effect on the sea level (-1.3 per cent), as has the construction of new reservoirs (-2.6 per cent). The contributions of deforestation and loss of wetlands are 0.3 and 1.6 per cent respectively. A large contribution of about 44 per cent is from loss of ground water, largely due to groundwater withdrawals but also to decreased percolation as a result of land use changes. As for the climate-related components of sea-level rise, the hierarchist projection roughly corresponds to the middle values given by the IPCC (Warrick *et al.*, 1996). On the assumption that the net direct anthropogenic contribution to sea-level rise (i.e. the net contribution from ground water, land use changes and reservoirs) is zero, the IPCC provides a medium estimate for sea-level rise of 490 mm in the period 1990-2100. Without direct anthropogenic contributions, sea-level rise in the hierarchist utopia will be 525 mm over the same period (505 mm for 2000-2100). Despite the rising sea level, it will be possible to effectively reduce the risk to population and capital (Figure 7.13). Raising and reinforcing coastal defences will require annual investment growing to about $4-5\times10^9$ US$/yr (total investment during the 21st century being about 0.4×10^{12} US$). These costs include the expenditure required for a worldwide reduction in critical flooding probabilities to a level of one tenth that of 1990. Excluding this increase in safety levels, total investment during the 21st century would be about 0.24×10^{12} US$.

An egalitarian water utopia
The egalitarian water utopia is characterized by water conservation and a stabilization of water use. Additionally, the world population attains full coverage of public water supply and sanitation by the year 2025, which is a quarter of a century earlier than in the hierarchist utopia (but more easily reached due to lower population growth). Even water-poor countries maintain a high level of water self-sufficiency, including not importing water-intensive products. Due to successful management of demand, supply side measures (such as building large dams) need not be implemented. Water quality will improve greatly and stable runoff can be

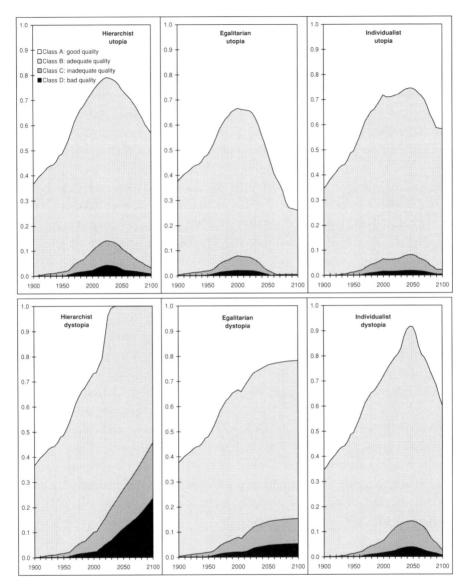

Figure 7.9. Surface water quality. In the hierarchist and egalitarian dystopias water quality improvements will fail to materialize, due to the increasing volume of untreated wastewater disposals (wastewater production increases while treatment coverage declines). In the individualist dystopia wastewater production increases, but increasing treatment coverage results in water quality improvements in the second half of the 21st century.

sustained at today's level. It is not feasible that groundwater-level decline and sea-level rise will be halted in the 21st century, due to the long response times involved, but everything possible will have been done to prevent a worse situation occurring (e.g. through a shift toward using less ground water and more surface water).

By the end of the 21st century, domestic water demand per capita will be 40 per cent lower than in the hierarchist utopia, thanks to extensive conservation programmes (Figure 7.3). In industry water-use intensities will decrease significantly, not increase as in the hierarchist utopia. The volume of industrial water use will remain below the quantity used for irrigation (Figure 7.5). During the 21st century, total water supply in the world will increase by a factor of about 1.4 (in the hierarchist utopia: 4.4), and consumptive water use by a factor of about 1.07 (in the hierarchist utopia: 1.8). Water scarcity, expressed as the ratio of actual to potential supply, will grow from 31 per cent in the year 2000 to 41 per cent in 2100. However, by the end of the 21st century there is very little growth in water supply and water scarcity, and a stable situation will have been reached (Figure 7.6). In the 22nd century water scarcity might even diminish again, due to a decrease in the world's population. The conservation of water during the 21st century will be partly the result of a restructuring of the water-charging system during the first quarter of the century. In the new charging system, which will be fully operational worldwide from the year 2025, full water costs plus an additional water tax of 10 per cent of the actual costs will be charged to the consumer (Figure 7.7). Water is only subsidized in cases where people would not otherwise have access to clean water. However, the total volume of subsidized water is relatively small. Actual water costs will stabilize at about the same level as in the hierarchist utopia. The annual expenditure in the water sector will be significantly less than in the hierarchist utopia, but expressed as a fraction of gross world product, it will only be slightly less (Figure 7.8).

Water quality will improve more than in the hierarchist utopia, because total production of wastewater is less and full coverage of wastewater treatment is attained earlier (Figure 7.9). Terrestrial evaporation will increase by about 17 per cent in the period 1900-2100, the largest increase as a result of global warming during the 21st century (Figure 7.10). The increase in terrestrial precipitation will be slightly less than that in evaporation, which implies that net precipitation will decrease slightly. Total runoff will become less accordingly. However, there will be a small difference between total runoff and net precipitation, resulting in a continued loss of water on land (this difference cannot be seen in Figure 7.10, because it is small compared to the total quantities of net precipitation and runoff). A reduction in groundwater use cannot prevent a continuing decline in groundwater tables during the 21st century, due to the high vulnerability and slow response of the system.

Table 7.1. Estimates of historical and future sea-level rise according to three utopias.

Sea-level rise (in mm)	Past (1900-2000)			Future (2000-2100)		
	Hierarchist	Egalitarian	Individualist	Hierarchist	Egalitarian	Individualist
Component contributions						
Thermal expansion	40	40	40	280	340	220
Glaciers	35	45	25	200	270	128
Greenland ice sheet	6	29	4	44	92	26
Antarctica ice sheet	6	41	3	-19	48	-35
Groundwater loss	68	95	35	393	243	290
Deforestation	1	1	1	2	0	5
Drainage of wetlands	13	13	13	3	1	5
Artificial surface reservoirs	-21	-21	-21	-7	-4	-12
Total sea-level rise	148	243	100	896	990	627
Excl. the last four components[1]	87	155	72	505	750	339

[1] For a comparison with the data from the Intergovernmental Panel on Climate Change: see Table 2.5.

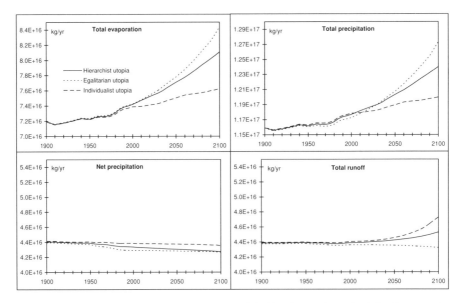

Figure 7.10. Change in the water balance of the global land area, in each of the three utopias. The changes in the three dystopias are rather similar (the hierarchist dystopia corresponds to the hierarchist utopia, etc.).

However, the rate of decline will lessen (in contrast to the hierarchist and individualist utopias, see Figure 7.12). The contribution of ground water to sea-level rise in the 21st century is less than in the other utopias, but total sea-level rise will be greater, due to a bigger contribution from the climate-related components (Table 7.1). The contributions from thermal expansion, glaciers and ice sheets will be larger because the global mean temperature will increase more and glaciers and ice sheets will be more sensitive. As shown in Figure 7.13, the number of people at risk will decrease and the amount of capital at risk will stabilize as a result of worldwide improvements to coastal defences. Including expenditure to reduce critical flooding probabilities in the most vulnerable areas in the world to a maximum of once every ten years, annual global investment costs for coastal defence vary between 2×10^9 and 5×10^9 US$/yr (total investment during the 21st century being about 0.32×10^{12} US$). Excluding the increase in safety levels total investment costs during the 21st century would be about 0.27×10^{12} US$.

An individualist water utopia

The individualist water utopia is characterized by a combination of efficiency improvements and growth. A strategy of market-pricing of water will lower water-use intensities in all sectors. Total demand and supply will increase strongly in all sectors, due to improved economic and living conditions, especially in the developing world. Parallel to economic growth, coverage of public water supply and sanitation will improve. The worldwide introduction of the 'polluters pay' principle will make wastewater treatment and water re-use cost-effective measures. Water use will be globally distributed in an efficient way, which means that water will be used most intensively in places where it is most abundant and cheapest. Water-poor countries will import water-intensive products if possible.

By the end of the 21st century, total water supply and consumptive water use will be as large as in the hierarchist utopia. In the individualist utopia, too, industrial water supply will exceed agricultural water supply. However, there is an important difference between the individualist and hierarchist utopias. In the hierarchist utopia total water supply serves a medium-sized population with medium-range economic development, while the same amount of water is used in the individualist utopia to sustain a larger population and a higher level of economic development, realized through highly efficient types of water use. As can be seen from Figure 7.3, specific water demands in the individualist utopia are much lower than in the hierarchist utopia. Specific water demand in the industrial sector for instance is 50 per cent less than in the hierarchist utopia. Despite the great increase in water demand, water scarcity will remain low, due to the high assessment of potential water supply

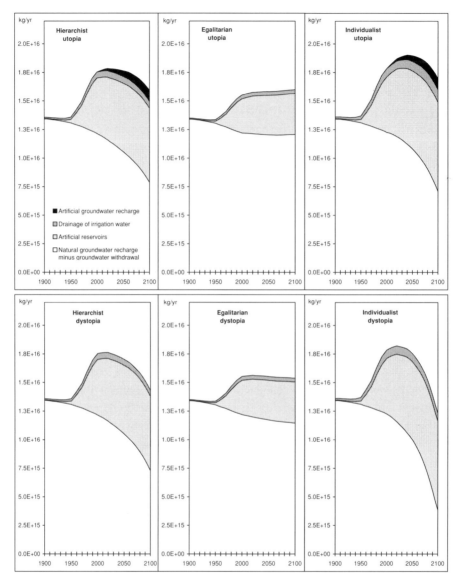

Figure 7.11. Stable runoff. In the hierarchist and individualist utopias, stable runoff is reduced by intensive groundwater use. During the next few decades, this effect is neutralized by measures to increase stable runoff, such as artificial groundwater recharge and the construction of artificial surface reservoirs. In the long term these measures cannot prevent a decrease in stable runoff. In the hierarchist and individualist dystopias, where measures are lacking, stable runoff will decrease much more sharply than in the corresponding utopias. In the egalitarian utopia, reduced groundwater use means that measures such as artificial groundwater recharge and building new dams are not necessary. The egalitarian dystopia is worse than the utopia, but much better than the other two dystopias.

(Figure 7.6). A further reason is that potential water supply will increase as a result of greater water recycling on land (reappearance of anthropogenically evaporated water as precipitation). Water scarcity, expressed as the ratio of consumptive water use to potential water supply, will grow from 4.1 per cent in 2000 to 7.8 per cent in 2100. Water supply costs increases will be of the same order of magnitude as in the hierarchist and egalitarian utopias (Figure 7.7). In absolute terms total expenditure in the water sector in the individualist utopia is larger than in the other two utopias, but the fraction of gross world product spent in the water sector continuously decreases and becomes much smaller than in the other utopias (Figure 7.8). As a result of an increase in wastewater treatment, water quality will start to improve in the second half of the 21st century (Figure 7.9). Changes in the terrestrial water balance will be less than in the other two utopias, mainly because of a smaller temperature increase during the 21st century. However, total runoff will increase by about 7 per cent, due to a negative balance of the groundwater store. This negative balance, which means an annual loss of ground water, is mainly caused by intensive exploitation of the groundwater stock. The volume of artificial groundwater recharge assumed in the individualist utopia will not be sufficient to neutralize the loss of stable runoff as a result of groundwater withdrawals (Figure 7.11). Despite the relatively heavy pressure on the groundwater system, groundwater tables will decline less than in the other two utopias, due to the assumption of a relatively robust system (responding reasonably fast, but to a lesser extent than in the other two utopias). However, by the end of the 21st century the rate of decline will become as high as in the hierarchist utopia (much higher than in the egalitarian utopia; see Figure 7.12). Sea-level rise in the individualist utopia will be lower than in the other two utopias for two reasons. The contribution of ground water will be relatively small for the reason mentioned, as will the climate-related contributions, due to a slight increase in global mean temperature only and comparatively insensitive glaciers and ice sheets. Despite the rising sea level and the growth in coastal population, the number of inhabitants at risk will decrease during the 21st century by about 40 per cent. Capital at risk will increase by a factor of nearly 1.5, which is, however, not much if compared to the growth in total coastal capital (with a growth factor of 6.8 in the period 2000-2100). In terms of capital loss this means, by the year 2100, an annual loss which amounts to 0.03 per cent of gross world product. Investment in coastal defence will amount to a total value of 0.28×10^{12} US$ during the 21st century. By the year 2100 annual investment will be about 0.01 per cent of gross world product. Higher investment could reduce the annual loss, but this investment would have to be made in less developed areas, where investment capacity is relatively low.

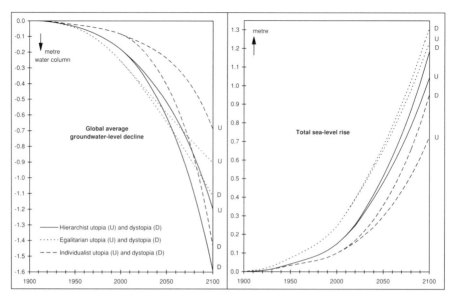

Figure 7.12. Groundwater-level decline and sea-level rise. Of the three utopias, the hierarchist shows the greatest groundwater-level decline by the end of the 21st century, due to the combination of medium sensitivity of the groundwater system and intensive exploitation. In the individualist utopia, the groundwater system has been assumed to be robust, but there is also intensive exploitation of the system. The egalitarian utopia is a future with reduced exploitation, but a comparatively vulnerable system. In the dystopias the pressure on the system becomes greater in all cases. Of the three utopias the egalitarian shows the highest sea-level rise, for two reasons: the world will become warmer than in the other two utopias and the climate sensitivity of glaciers and ice sheets has been assumed to be greater.

7.4 Three water dystopias

What will happen to each of the three utopias if the fatalist management style is applied instead of the utopian management style? Fatalist management style essentially means that no new water policy measures are implemented and current practice remains more or less unchanged (see Chapters 5 and 6). Today's water-pricing structures will be taken into the 21st century, so that water will still be subsidized; there will be no water conservation programmes; no extensive investment programmes to expand public water supply and sanitation infrastructure; no expansion of the irrigated cropland area; no wastewater treatment policy; no artificial groundwater recharge to alleviate the effect of intensive exploitation of the groundwater store; and no coastal defence strategy. In each of the dystopias, the fatalist management style has been assumed to become operational in the year 2000.

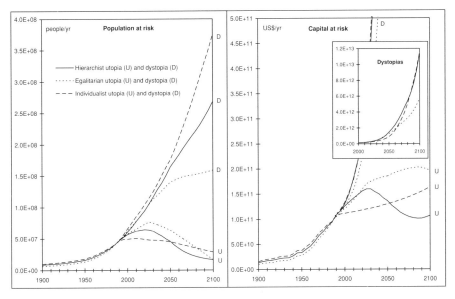

Figure 7.13. Coastal population and capital at risk. Although coastal defence strategies in the three utopias differ fundamentally, they all succeed in stabilizing or reducing the total of population and capital at risk. In the dystopias, where coastal defence strategies are lacking, risks will increase exponentially, most severely in the case of capital at risk (see inset). This increase is due to the rise in sea level, as well as population growth and an increase in economic capital in coastal areas.

A hierarchist water dystopia

The hierarchist water dystopia differs from the hierarchist utopia in many respects. Due to a lack of adequate policies, coverage of public water supply and sanitation will decline, there will be less water conservation, total water use will be greater, water costs will become higher, total wastewater production will increase, treatment coverage will diminish, and water quality will continue to get worse. Global average public water demand per capita will increase towards 475 litres per person per day by 2100, about 15 per cent higher than in the hierarchist utopia. Water demand per hectare irrigated land will remain at the current level and not continue to decrease. Industrial water demand per dollar value added will keep on increasing instead of stabilizing. As a consequence, total water use will become 12 per cent more than in the hierarchist utopia. The percentage of the world population with public water supply and sanitation will drop dramatically, returning to 1970 levels by the year 2100. The same applies to global coverage of wastewater treatment (Figure 7.4). Average water supply costs will become about 50 per cent higher than in the hierarchist utopia, but a smaller fraction will be charged to the consumer (Figure

7.7). Total expenditure in the water sector will be of the same order of magnitude as in the hierarchist utopia, but the allocation will be different: more expenditure on water supply (particularly industrial water supply), and less on wastewater treatment (Figure 7.8). Water quality will deteriorate severely: by the year 2030 there will be no natural, pristine lakes or rivers left anywhere in the world. Nearly 50 per cent of the fresh surface waters will be of inadequate or bad quality (Figure 7.9). The loss of stable runoff due to groundwater exploitation will be greater than in the hierarchist utopia. This, together with the absence of artificial groundwater recharge, will result in a rapid decline in stable runoff, and by 2100 stable runoff will be 10 per cent lower than in the hierarchist utopia (Figure 7.11). The worldwide decline in groundwater levels will be much more severe, with a global average groundwater-level decline of -1.6 metre in the period 1900-2100, which is 33 per cent more than in the utopian world (Figure 7.12). Total sea-level rise in this period will be 1.19 metre, 14 per cent higher than in the utopia. Due to a lack of adequate coastal defence measures, population and capital at risk will increase greatly (Figure 7.13). By the end of the 21st century capital at risk will reach a level of about 11.4×10^{12} US$/yr, which means an annual capital loss amounting to 4.3 per cent of gross world product.

An egalitarian water dystopia
The egalitarian dystopia resembles the hierarchist dystopia, but is in some respects less extreme. An important difference is that public water supply and sanitation coverage decline in the first half of the 21st century, as in the hierarchist dystopia, but start to increase again in the second half. The egalitarian will point out the advantage of low growth: 'mismanagement' will be less harmful in a world where mankind puts relatively little pressure on the environment. Water quality will become worse in the egalitarian dystopia as well, but not by as much as in the hierarchist dystopia. Stable runoff will decrease only slightly, and not exponentially as in the hierarchist dystopia.

Due to the lack of water conservation programmes and the absence of a water-taxing policy, water-use intensities will be higher than in the egalitarian utopia (Figure 7.3), and total water use will increase by 10 per cent. Water costs will rise as a result of increasing water scarcity and pollution (Figure 7.7). By the end of the 21st century total expenditure in the water sector will amount to 5 per cent of gross world product, compared to 4 per cent in the egalitarian utopia (Figure 7.8). Global average groundwater-level decline in the period 1900-2100 will be 1.1 metre (cf. the egalitarian utopia: 0.9 metre). Total sea-level rise in the same period will be 1.3 metre (cf. the egalitarian utopia: 1.2 metre). Population at risk will continue to increase during the first half of the 21st century, but begin to stabilize in the second half, due to a decline in the population living in coastal areas. Regarding population at risk the situation will be much better than in the hierarchist dystopia (Figure 7.13). However, capital at risk will grow towards 5.7×10^{12} US$/yr, which implies an annual loss of capital of about 8.2 per cent of gross world product, nearly twice as much as in the hierarchist dystopia.

An individualist water dystopia
The differences between the individualist dystopia and utopia are much smaller than the differences between the hierarchist or egalitarian dystopia and utopia. One reason for this is that in the individualist world-view many processes are 'autonomous', regulated by the economic mechanisms of demand and supply. As a result, the type of water policy (in this case: either individualist or fatalist management style) influences the individualist world less than the hierarchist or egalitarian world. The individualist world is also comparatively robust, so 'mismanagement' is less disastrous than in the more vulnerable worlds of the hierarchist and egalitarian.[1] Probably the most essential difference between the individualist dystopia and utopia is the absence of a market-pricing policy in the dystopia. This means that water prices will remain at about current levels during the whole of the 21st century (Figure 7.7). As a result, people will not be forced to conserve water in some way, so that water-use intensities will be much greater than in the utopian case (Figure 7.3). By the end of the 21st century, total water use will be about 75 per cent more than in the individualist utopia, but consumptive water use will be only 12 per cent more. Water scarcity in the individualist dystopia, expressed as the ratio of consumptive use to potential water supply, will grow from 4.1 per cent in 2000 to 8.4 per cent in 2100 (cf. 7.8 per cent in the individualist utopia). The change in the global water balance will not be very different from what has been presented for the individualist utopia (Figure 7.10), apart from the fact that stable runoff will be reduced much more than in the utopia, due to a more intensive exploitation of the groundwater store and the absence of artificial groundwater recharge (Figure 7.11). Because public water supply, sanitation and wastewater treatment coverage are supposed to develop as a function of economic growth, the individualist dystopia does not differ from the utopia in this respect (Figure 7.4).

---

[1] Note the fundamental difference between the individualist and egalitarian rationale. The egalitarian advocates low growth in a vulnerable world (because mismanagement will be less harmful in a low-growth than in a high-growth world), while the individualist argues for high growth in a robust world (because the effects of mismanagement can be better overcome in a high-growth than in a low-growth world).

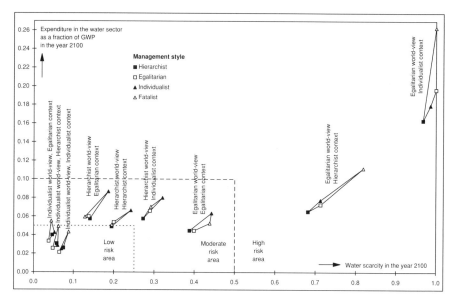

Figure 7.14. Possible water futures with respect to water scarcity and water costs. Each of the thirty-six possible futures is characterized by a specific context, world-view and management style. The nine clusters of futures show the effects of alternative management styles given a particular combination of context and world-view. Within the scope of possible futures, areas of low, moderate and high risk have been distinguished.

Total expenditure in the water sector, however, will become twice as high as in the utopia, as a result of a larger water supply, higher water supply costs, and a greater flow of wastewater to be treated (Figure 7.8). Global average groundwater-level decline during the 21st century will be more than twice as much as in the utopia; total sea-level rise will be 1.4 times as much (Figure 7.12). The absence of adequate coastal defence measures will result in an exponential growth of population and capital at risk (Figure 7.13). In terms of capital loss, this means that by 2100 about 2.0 per cent of gross world product will be lost annually, due to flooding.

7.5 Possible water futures: the full range

In the previous sections, a limited selection of possible water futures has been discussed, namely three utopias and three dystopias. In this section a larger set of possible futures will be presented, by considering all possible combinations between context, world-view and management style. For ease of survey, I have chosen to show four indicators for just one point in the future, the year 2100. Water scarcity and water expenditure in 2100 have been plotted in Figure 7.14 and groundwater-

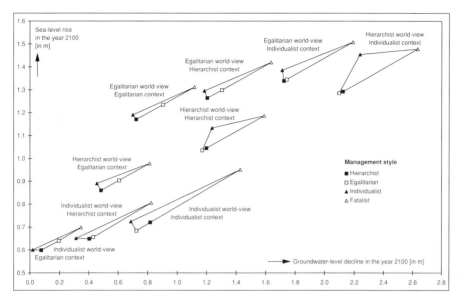

Figure 7.15. Possible water futures with respect to groundwater-level decline and sea-level rise. The thirty-six futures shown correspond to the futures presented in Figure 7.14. The nine clusters of futures show the effects of alternative management styles, given a particular combination of context and world-view.

level decline and sea-level rise in Figure 7.15. From the results, one can see that it is not superfluous to look at 'the other' possible water futures in addition to the three utopias and three dystopias already discussed.

Two dystopian combinations of context and world-view in particular appear to result in extraordinarily high water scarcity and expenditure by the end of the 21st century. This is the case if socio-economic growth is relatively high (the individualist or hierarchist context) while water demand and potential supply are governed by the egalitarian world-view (Figure 7.14). In the most extreme case, water scarcity grows by up to 100 per cent worldwide, while expenditure in the water sector amounts to 25 per cent of gross world product. This is the egalitarian nightmare of rapid growth in a vulnerable environment. Assuming high instead of low growth in the egalitarian utopia (i.e. shifting from the egalitarian to the individualist context), water demand will become about 2.5 times greater and average water supply costs per litre will increase by a factor of about 20. The increase in demand will occur despite lower water-use intensities in all sectors, a side effect of the greater economic growth and increasing water prices. Water quality will still improve during the course of the 21st century as a result of the increasing wastewater treatment coverage, but to a lesser extent, due to the larger volume of total wastewater production. In an extreme (but possible) scenario such as this, the style of water management can substantially worsen or relieve the crisis, but the influence of water policy is far from enough to really solve the crisis or prevent catastrophes such as food shortages, rivers and lakes drying up and groundwater depletion. Series of relative dry years will have disastrous consequences for regional economies as a whole. In fact, the only real solution if the water system is as vulnerable as the egalitarian believes is to prevent a high growth rate.

Groundwater level and sea level are equally very vulnerable to high economic growth (Figure 7.15). In the worst case global average groundwater-level decline will reach 2.6 metre water column by 2100, while sea-level rise will amount to 1.5 metre. Assuming an average soil porosity of one third, actual groundwater levels would on average be 8.7 metres below the levels of 1900 (or 8.2 metres below today's levels). This will result in subsidence of soils and reduced natural stable runoff. Examples of such a scenario can already be observed in many regions of the world, especially in large cities. The decline of water tables by tens of metres, sometimes more than 100 metres, has been observed locally in a number of European countries, China, the USA and several Arab countries (Shiklomanov, 1997). Subsidence, as a result of falling groundwater tables, also has already been observed in many places. In Mexico City for example, subsidence of 10 metres has been reported, and according to the World Bank (1993) over-pumping of ground water causes a subsidence rate of 0.14 m/yr in some areas of Bangkok.

7.6 Risk assessment

What are the circumstances which will create a global water crisis? In other words: under which circumstances will water put a significant constraint on human health and development on a global scale? And what are the necessary conditions to prevent such a crisis? Which developments can be characterized as risky? Because hierarchists, egalitarians and individualists hold fundamentally different views, I will start to consider these questions from each perspective.

In the hierarchist world-view, water scarcity is defined as the ratio of consumptive water use to stable runoff. I tentatively assume that a water scarcity of 0-25 per cent implies a low risk, 25-50 per cent a moderate risk and 50-100 per cent a high risk, an approach based on some earlier classifications of vulnerability to water scarcity (Kulshreshtha, 1993; Gleick, 1993b; Raskin *et al.*, 1995, 1996, 1997; Falkenmark and Lindh, 1976). High risk means that a major impact of water scarcity on population and economic development will probably occur. Moderate risk means that this impact will occur, but to a lesser extent and not worldwide. Low risk means that the future will not bring water scarcity problems which significantly exceed the problems currently encountered. As shown in Figure 7.14, similar risk areas have been assumed with respect to expenditure in the water sector. If this expenditure exceeds 5 per cent of gross world product, mankind enters the moderate risk area, and if it exceeds 10 per cent of GWP one enters the high risk area, where people experience insurmountable obstacles to providing their water needs. The twelve experiments with the hierarchist world-view show that water scarcity at the end of the 21st century will be in the range of 13 to 32 per cent while expenditure will be between 5 and 9 per cent of GWP (Figure 7.14). According to these results, hierarchists typically move within the moderate risk area, which corresponds to their world-view (see Chapter 5). From these experiments, it can be concluded that the main risks are either that growth is larger than foreseen (the individualist context), or that the individualist management style is being applied. However, the risks will probably stay within limits and can be managed reasonably well. From the results, there appears no reason to advocate such an explicit water-pricing policy as that of the egalitarians and individualists. With respect to groundwater-level decline and sea-level rise the hierarchist should also be attentive to the consequence of high growth (see Figure 7.15). If economic growth appears to be larger than expected (the individualist context), one will have to consider a reduction in groundwater use as advocated by the egalitarian or an increase in artificial groundwater recharge, the individualist solution.

For the egalitarian I have defined a low, moderate and high risk water scarcity area by using the same 25 and 50 per cent criteria as in the hierarchist world-view. However, these risk lines do not have the same physical meaning, due to the fact that egalitarians use a different definition of water scarcity. In the egalitarian view one enters the high risk area if *total water supply* becomes more than half of *stable runoff in inhabited areas*, whereas according to the hierarchist view this only happens if *consumptive water use* exceeds half of *total stable runoff*. As a result, it is not surprising that the egalitarian world appears much more vulnerable than the hierarchist world, with a realistic chance of entering the high risk area (see Figure 7.14). Low growth appears to be essential for the egalitarian, much more urgently so than for the hierarchist. In addition particularly the fatalist style of water management should be avoided at all costs, because this could worsen the situation significantly, as the fatalist is the only one who would leave water underpriced to the extent it is now. The scarcer water becomes, the more disastrous such an approach would be, because there will continue to be no strong stimulus to conserve water. For this reason the egalitarian even advocates the introduction of water taxes. According to the egalitarian, intensive efforts are needed to remain within the low risk area, to avert unbridled growth and its inevitable consequences, including an unacceptable decline in groundwater levels (Figure 7.15).

For the individualist one could apply the same 25 and 50 per cent risk lines for water scarcity as for the hierarchist and egalitarian. Within the individualist perception, however, this means that one enters the high risk area only if *consumptive water use* becomes more than half of *total runoff*. The simulation results show that, within his own perspective, the individualist typically moves within the low risk area. The most perilous development would be if fatalists prevented the establishment of a market-oriented policy. This would impede efficient water use and result in unnecessary large expenditure in the water sector (Figure 7.14). Furthermore, the lack of any effective response to groundwater depletion (e.g. artificial recharge) would lead to an unnecessarily large decline in groundwater level and rise in sea level (Figure 7.15). Despite these possible disturbances, the individualist world is so robust that it is not very difficult to stay within the low risk area. However, this does not mean that the individualist world is a low risk world from every point of view. As discussed above, from the egalitarian and hierarchist perspectives, individualist influences are equal to high risk, which is the paradox of the individualist. This paradox can be explained by introducing two levels of risk, the first level referring to the risks as perceived *within* a certain world-view and the second to the risks which emerge if one adopts alternative world-views. In this terminology, the individualist utopia is a low-risk world on the first risk level, but a high-risk world on the second level. In the concluding section of this chapter I will analyse second-level risks and consider how these could be minimized.

7.7 Water and development: the water transition

The conclusion that a rapid growth in water demand carries a high risk for both environment and development, corresponds to the recent findings of other authors (see for instance Young *et al.*, 1994). For this reason, a number of recent official documents have proposed a global shift from a supply-oriented towards a demand-oriented policy (UN, 1997a, 1997b). From the model experiments reported in this chapter we can learn that the call for such a shift is not just a result of the recent awareness of the environment, but can also be understood if we consider the changing interaction between water and development. Interpretation of the model experiments shows that this relationship can be characterized according to three phases. In the first phase the originally more or less equilibrium state of the water system is disturbed by increasing water withdrawals and rates of pollution. A characteristic of this disturbance is that it is self-reinforcing (there is positive feedback): the development of water resources supports socio-economic development and this in turn drives a further exploitation of and pressure on the water system. This is the phase of exponential curves. In the second phase, increasing water scarcity and water supply costs induce competition between water users. Water supplies may become less reliable, temporarily falling short, yields in irrigated agriculture may be affected, pollution starts to kill aquatic life and fishing industries begin to be threatened. Growth rates will be moderated. In the third phase water scarcity has reached a level where people are forced to increase water-use efficiency and reduce pollution. Under favourable socio-economic conditions (a stabilizing population, reasonable living standards), the third phase can be concluded with a stabilization of water demand and a recovery of the quality of the natural water system.

These three phases of the changing interaction between water and development will hereafter be referred to as the water transition (see also Rotmans and De Vries, 1997; Hoekstra, 1995). The basic idea behind this concept is that increasing water demand will result in increasing water scarcity, which in turn will force the stabilization of water demand. However, a water transition is not necessarily as smooth as described above. The second phase can be accompanied by serious problems if socio-economic conditions are unfavourable (e.g. a combination of high population growth rates and low economic growth). In such a case, people would be unable to mitigate negative impacts. As a result the third phase will not happen, water demand will not stabilize but continue to grow, water supply will approach its limits and a water crisis becomes inevitable. An unexpected succession of dry years will be disastrous, and serious repercussions for public health and the economy of a country as a whole are inevitable. The worsening conditions might act as a trap: water scarcity and water supply costs increase (because water demand grows and available clean water diminishes) in a situation where better water supply is an essential precondition for socio-economic improvement. Only favourable demographic, economic or technological developments, or outside help, can free people from such a trap.

A smooth water transition requires suitable water policies in each phase. Generally, the type of water policy which accompanies the first phase of rapid expansion will be inappropriate in the second phase of increasing scarcity and might even be harmful in the third phase of stabilization. In the first phase, a well-established water policy is often lacking. There is no great need for this either, because a self-regulating mechanism is at work: water demand and supply increase as a result of socio-economic development, which in turn is supported by the increased exploitation of water. This process is facilitated by continued technological development. During the first phase, governments generally become increasingly involved, particularly in infrastructure projects and technological innovation. In the second phase, the exploitation of the water system has become quite intensive and water users begin to compete, prompting governments to formulate a more balanced policy. In order to supply the needs of all, governments initiate ambitious water resources development projects, and they begin to promote wastewater treatment. This phase often includes different forms of water subsidization. In the third phase, water scarcity reaches a level at which exploitation limits become evident, requiring a transformation from supply-oriented towards demand-oriented policy.

The water transition concept is intended to be a general framework for thought, useful for positioning different perspectives in relation to each other. Each of the three phases of the water transition is likely to be dominated by one particular perspective. The individualist perspective is likely to be dominant in the first phase of exponential growth. The hierarchist perspective will probably prevail in the second phase of finding a balance between the desirable and the possible. Finally, in the third phase of stabilization, the egalitarian perspective is likely to become the ruling force. One can also put it the opposite way: if the individualist point of view dominates, the way society is perceived accords with the first phase of the transition. From this point of view, it is even possible to deny the likelihood of a second or third phase. When the hierarchist point of view dominates, people apparently recognize their world in the description of the second phase of the transition. In this perception it is accepted that the first phase of unbridled growth is over, but the need to enter a third phase of stabilization is not recognized. Once the egalitarian point of view dominates, people appear to feel the need to move from the second to the third phase. The egalitarian perspective is not necessarily the end of the line, which will be reached sooner or later. If, at any time in the stabilization phase, it becomes clear that more water is available than was thought (e.g. through new desalination techniques) or if a way is found to use water much more efficiently (e.g. through new techniques of water re-use), a shift will probably occur from the egalitarian to the individualist perspective. In this way, the third phase of a transition can lead to the first phase of the next transition, which will then take place under the new conditions.

The transition concept can be used not only to put the different perspectives into a broader context, as shown above, but also to position different regions in the world. One could say for instance that many developing countries are still in the first phase of water transition. In these regions, where waterborne diseases often account for 80 per cent of all illness (Postel, 1992), the wish to develop water resources dominates all other interests, so that forces for water conservation are few. One could argue that a few Western countries have reached the third phase of transition. In the Rhine basin for instance, the rate of growth in water demand is now only small and river water quality has started to improve. However, it is too simplistic to position world regions on the water-transition scale merely on the basis of their development status. In fact, the individualist perspective, which typically correlates with the first phase of water transition, can be found not only in developing countries, but in industrialized countries as well (see for instance Anderson, 1995). Those who adhere to the individualist view in industrialized countries suppose that continued growth is possible through a continuing increase in water-use efficiency (e.g. through desalination of sea water, stabilization of natural runoff by means of dams and artificial groundwater recharge, and concentration of water-intensive production processes in water-rich areas). Similarly, the egalitarian perspective is not just represented in the industrialized word, but can also be found in developing countries.

That there is room for divergent points of view within particular regions is due to the great uncertainties regarding upper limits to water supply and the abilities of mankind to develop water-saving technology. According to the individualist perspective, continued growth can take place because there are still many possibilities of more efficient water use and a large volume of water remains still unexploited. The most important condition for further growth is a proper functioning of the market mechanism: if markets do not function well, incentives to develop and refine new technologies such as desalination will not exist and efficiency improvements will fail to materialize. The hierarchist clearly perceives limitations and is more prudent in assessing future possibilities. Hierarchists try to extend the second phase of the transition, by increasing supply possibilities and attempting to minimize negative impacts. Their strategy is not to reduce water demand, but to increase supply potential at the same rate as demand expands. Typical elements of such a strategy are for instance the construction of large surface reservoirs, artificial groundwater recharge, wastewater treatment, and land and soil management. According to the egalitarian perspective reducing water demand has grown into a sheer necessity, to prevent such future catastrophes as droughts and absolute water scarcity. Egalitarians are inclined to emphasize that the second phase cannot continue. Efforts to increase supply possibilities are regarded as a temporary postponement of the real measures which have to be taken: conservation of water, reduction of wastewater production, introduction of water taxes and education of people to increase 'water literacy'.

The central idea in the transition hypothesis is that the interaction between water and development can be schematized into three phases: increasing exploitation, balancing of the desirable and the possible, and finally stabilization. In the third phase conditions may change, so initiating the first phase of a new transition cycle. Each phase in the transition requires its own type of water policy, which implies that there is not one ideal type of water policy. To move from one type of policy to another is generally difficult, but a probably even greater problem in formulating appropriate water policy is uncertainty about the current phase, and thus about the *necessity* for change. In the concluding section of this chapter I will use the concept of risks to analyse how such uncertainty can be handled.

7.8 Conclusion

Let us return to the main question in this chapter: how can the world population be provided with sufficient clean fresh water, in a sustainable way? The experiments show that there is not one single answer, but it has become clear that under some conditions risks are smaller than under others. In Section 7.6, it has been argued that risks can be perceived on two different levels: the level of a single perspective and a level which exceeds the perspectives. With regard to risks on the first level, it has become clear that each perspective has a distinctive perception of risks, which means that preferred policy strategies differ per perspective. Risks on the second level are not apparent within one particular world-view, but emerge if one adopts alternative world-views. In this section I will analyse how people might avoid high risks on the second level.

The set of possible futures presented in this chapter shows that the type of water policy is only one factor determining future water scarcity and changes in the water system. Another, more influential, factor is the type of future socio-economic development (the 'context'). This has been illustrated in Figures 7.14 and 7.15, where one can see that, given a particular combination of context and world-view, the style of water management has a limited influence on the type of development. This means that 'water problems' can only partially be solved through 'water policy'. Socio-economic policies which influence population and (the type of) economic growth can be much more effective. Futures with high population growth and high economic growth are typically high-risk futures. If one values risk avoidance, this requires efforts to extend birth control, particularly in regions with a high population growth rate, for example through stimulating socio-economic development of the poor. In regions with strong economic growth, it requires efforts to discourage activities or production processes which put great pressure on the environment, and to stimulate activities which are environmentally friendly such as many activities in the service sector. It is beyond the scope of this study to go into further details on population and economic policy, but I feel it is important to make this point: if there is no shift in the future towards lower growth, the possibility of water policy effectively managing risks of water scarcity, depletion of aquifers, high water costs, etc. is rather limited.

Given certain contextual conditions, the possibilities of managing future water scarcity through water policy are restricted, but the situation will become worse if these possibilities are not used. Of all the management styles, the fatalist one carries the highest risks. Fatalists themselves will probably consider this an ironic twist of fate. However, as fatalists in any case regard the future as a lottery, they will have ambivalent feelings towards the more active style of management proposed by the hierarchist, the egalitarian or the individualist. Hierarchists, egalitarians and individualists should avoid - while trying to attain their own ideal worlds - obstructing each other to such an extent that none of their strategies in fact becomes really effective. It would make sense to search for approaches which fit into all perspectives (although possibly for different reasons). Some form of improved water pricing is for instance an essential element in any desirable future. If water subsidies were not removed at all, as is the case with the fatalist management style, this would result in 10 per cent more water use globally by the year 2100 in the egalitarian utopia and 12 per cent in the hierarchist utopia. In the individualist utopia water use would even increase by 75 per cent.

If one wished to aim at minimizing risks, I would recommend a change in policy priorities as presented in Table 7.2. A major shift should take place from water supply towards water demand policy. In developing countries, where the greatest extension of water supply infrastructure is likely to occur, any supply policy should at least be accompanied by a demand policy. An important policy ingredient should be to drastically reduce water subsidies in all water-use sectors. Although for different reasons, this agrees with the preferences of hierarchists, egalitarians and individualists. An active water supply policy contrary to the market mechanism could be recommended only in some cases of drinking-water supply. Subsidies for public water supply can be defended only if they benefit people who would otherwise not be able to afford safe water. The extent to which subsidies should be removed in all other cases needs to be debated: partial removal (hierarchists), complete removal (individualists and egalitarians) or even adding a tax (egalitarians).

In the past, more effort has probably been put into improving water supply technology than into educating people to use water more efficiently. In the future these priorities should change, because there is a long interval between technological possibilities and the actual use of these possibilities. This is supported by the

Table 7.2. A change in water policy priorities needed to safeguard a sustainable supply of clean water.

Policy	Past	Future
General		
- water demand policy	+	+++
- water supply policy	+++	+
Water demand policy		
- water pricing (removing subsidies)	+	+++
- water supply technology	+++	+
- water education	+	++
Water supply policy		
- supply infrastructure policy	+++	+
- water quality policy	+	+++
- land and soil policy	+	++
- climate policy	~ 0	+

Note: The priorities are intended to be indicative and merely have a comparative value: from relatively low priority (+) to relatively high priority (+++).

sensitivity analysis in the previous chapter, which showed that future water use depends more strongly on the technological *diffusion rate* (which relates to 'learning') than on the technological *development rate*. The diffusion of efficient techniques of water use into households could be accelerated by improving 'water literacy'. Leaving efficiency improvements to the functioning of the price mechanism, as advocated by individualists, carries the risk of not being effective, as illustrated in the hierarchist worlds (see Figure 7.14). One of the possibilities for increasing water-use efficiency is to stimulate water re-use, especially in industries. Industrial water use has had least attention from policy makers to date. In many of the possible water futures, however, the industrial sector will become the largest water user. Regarding irrigation, the application of more efficient techniques can reduce consumptive water use to some extent, but it is more important to critically evaluate plans for new irrigation schemes. A central theme should be the weighing of irrigation against other ways of enlarging food production.

Because domestic water use is small compared to other types of water use, the aim of increasing public water supply coverage in developing countries causes little conflict with the aim of saving water. The target of 'drinking water for all' justifies high investment in public water supply. The individualist strategy of improving public water supply and sanitation through economic growth carries a high risk, because the situation will become worse if economic growth appears to benefit the rich and not the poor or if it stagnates. If the aim is to minimize second-level risks, the egalitarian and hierarchist strategies are to be preferred to the individualist strategy. The individualist management strategy equally carries a high risk regarding improvements in water quality. The individualist strategy serves well within the individualist utopia, but is much less appropriate in the hierarchist or egalitarian world. The opposite, applying the hierarchist or egalitarian management style in the individualist world, does not worsen the situation, but can at best be regarded as 'overdone' or an inefficient spending of money.

In the future more attention should be paid to the effects of intensive land use and erosion on water availability. As has already been explained, deforestation and the drainage of wetlands reduce water recycling on land. Land clearing, intensive land use and erosion increase flooding and reduce stable runoff. The issue of climate change also deserves more attention in the future, particularly the question of how greenhouse gas emissions can be reduced, but this comes under energy rather than water policy.

A final observation is that - to whatever utopia one aspires - risks could be minimized by an adequate response to early signs of dystopian trends and by flexibility in adapting world-views and management styles to new circumstances. The three utopias have been formulated in such a way that they are plausible, but they have to be readjusted continuously in the light of new insights. For instance, according to current scientific knowledge, the groundwater system may be either robust and fast-responding (as in the individualist world-view) or vulnerable and slow-responding (the egalitarian view). If a choice is made to intensify the exploitation of groundwater resources to accommodate rapid socio-economic growth, thus taking the risk of the 'egalitarian nightmare', nothing would undermine this growth more than the neglect of possible signs that the groundwater system behaves according to the egalitarian and not the individualist world-view. In other words, it is essential that people are aware of the second risk level, the risk that one's world-view, including its perception of risks, is a fallacy. This is what sociologist Ulrich Beck (1997) means when he says that the greatest danger in current societies is that people regulate for small-scale risks (first-level risks) but neglect the large-scale risks (second-level risks).

8 The AQUA Zambezi Model

The AQUA Zambezi Model has been developed to support an integrated assessment of freshwater use in the Zambezi basin. This chapter discusses how the Zambezi basin has been schematized and which input data have been used. Processes which relate to human pressures on the basin, socio-economic impacts of water scarcity and societal responses are simulated separately for each of the eight riparian countries. Hydrological processes and water quality are simulated separately for eight sub-basins. Runoff per sub-basin has been calibrated in four consecutive steps, to get a feeling for the importance of each of the various hydrological parameters.

8.1 Introduction

The Zambezi basin in Southern Africa is one of the great international river basins in the world. Eight nations have part of their territory in the basin: Zambia, Angola, Zimbabwe, Mozambique, Malawi, Tanzania, Namibia and Botswana. In the past, water resources development in the basin has been dominated by national single-purpose projects. These projects have rarely taken into consideration the interests of other users, other countries or the consequential environmental impacts. Despite this lack of comprehensive planning, there have been no major conflicts in the utilization of the Zambezi river system, probably due to the fact that many parts of the basin still offer sufficient scope for further development. However, if all current plans for such development are realized, different interests will inevitably begin to clash. At present, the installed hydropower generation capacity in the main stream of the Zambezi amounts to nearly 3500 MW, but there are plans in more or less advanced stages of development for at least another 6000 MW.[1] At the same time there are ambitious plans throughout the basin to extend irrigation. In the Caprivi Strip, the Namibian territory in the upstream part of the Zambezi basin, for instance, water demand for irrigation has been estimated as likely to grow by a factor of about 200 in the period 1990-2020, extending the present irrigated area of 1.6 km^2 to about 300 km^2 (Heyns, 1995). Downstream, in Mozambique, plans have been made to extend the irrigated area within the Zambezi basin from the present 100 km^2 to 2400 km^2 (World Bank and UNDP, 1990h). In Zambia, Zimbabwe and Malawi too, irrigation development has high priority (World Bank and UNDP, 1990d,e,g). Furthermore,

---

[1] Currently installed capacity: Victoria Falls 108 MW, Kariba 1266 MW, Cahora Bassa 2075 MW. Possible future installations: Batoka Gorge 1600 MW, Devil's Gorge 1240 MW, Mupata Gorge 1000 MW, extension Cahora Bassa 550 MW, Mapanda Uncua 1600 MW (Dale, 1995).

given continuing population growth, an increasing level of urbanization and an expected rise in average living standards, domestic water demand will increase significantly throughout the basin. In order to avoid possible future conflicts, which are likely to arise if water use in the basin continues to intensify, and to stimulate environmentally sound management of water, the Zambezi countries initiated the Zambezi Action Plan (ZACPLAN), assisted by the United Nations Environment Programme (Dávid et al., 1988). In 1987, the plan was implemented as a programme of the Southern African Development Co-ordination Conference (SADCC), later the Southern African Development Community (SADC). Despite all efforts, progress of the Zambezi Action Plan is behind original expectations, partly because of a lack of funding, but also due to failures within the funded projects (SADC-ELMS, 1994). In 1993, UNEP proposed the carrying out of a rapid integrated water assessment for the Zambezi basin, in addition to the ongoing and long-term ZACPLAN activities. Such a rapid assessment could provide a focus for the long-term ZACPLAN activities, by signalling the most urgent future water supply issues within the basin. The rapid assessment project was established in 1994 as a joint initiative of SADC and UNEP, supported by RIVM as a collaborating centre of UNEP. I was asked to contribute to the assessment by expanding the generic AQUA model for the Zambezi basin and applying it to support the assessment. During the project, two workshops with regional experts have been organized, one in January 1995 in Lilongwe, Malawi, and one in November 1996 in Harare, Zimbabwe. The aim of the first workshop was to identify and analyse the most important water policy issues in the Zambezi basin and to gather basic information on these issues (UNEP, 1995b). At this workshop a preliminary version of the AQUA Zambezi Model was presented and discussed. The outcome of the first workshop gave direction to subsequent adaptation and improvement of the model. The aim of the second workshop was to explore possible water futures and to discuss and select the most promising water policy strategies (UNEP, 1996). During this second workshop the AQUA Zambezi Model was used interactively, to support the development of scenarios of future water supply and to analyse the consequences of various water policy strategies. In this chapter I describe the final version of the model, including some improvements made after the workshop in Harare.

The purpose of the AQUA Zambezi Model follows from the above: the model should support a policy analyst in exploring possible future developments with respect to water use in the Zambezi basin and in examining the effects of various water policy strategies. In the assessment project of SADC/UNEP, the year 2025 was chosen as the time horizon. However, I implemented the model in such a way that explorations can readily be made up to 2050, the time horizon used in the analyses

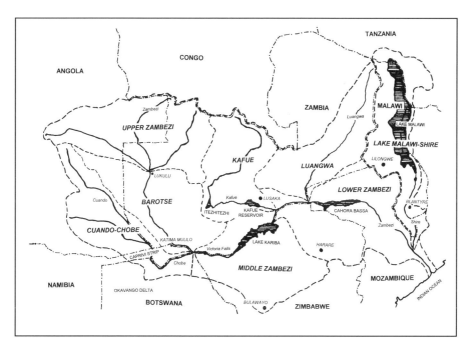

Figure 8.1. Map of the Zambezi basin.

in the next chapter.[1] The year 1990 has been chosen as the initial year of simulation. Theoretically it could have been useful to start the simulation further in the past, to test the performance of the model by simulating historical developments, but in practice this appeared to be pointless, as historical data are too scarce to serve such a purpose.

For the description of hydrological processes, the Zambezi basin has been schematized into eight sub-basins: the Upper Zambezi, Barotse, Cuando-Chobe, Middle Zambezi, Kafue, Luangwa, Lake Malawi - Shire, and Lower Zambezi (Figure 8.1). For the description of water demand and supply, the basin has been schematized into eight socio-economic regions, corresponding to the national territories which constitute the basin. As described in Chapter 3, the AQUA model consists of four sub-models: the pressure, state, impact and response sub-models. The pressures on the Zambezi water system per country are translated into pressures per sub-basin in order to calculate changes in the state of the water system for each sub-basin (Figure 8.2). The changes per sub-basin are then translated back into changes per country, so that impacts and societal response can be calculated for each

---

[1] In fact, the model can be run beyond 2050 as well, but it has been tested only for the period 1990-2050.

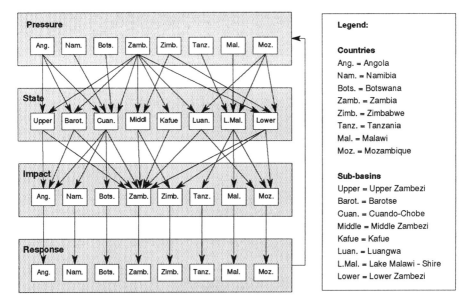

Figure 8.2. Model structure of the AQUA Zambezi Model.

country.[1] River flow in the basin has been schematized as shown in Figure 8.3. The Zambezi river rises in the Upper Zambezi basin and flows via the Barotse basin, Middle Zambezi basin and Lower Zambezi basin towards the Indian Ocean. The Cuando-Chobe basin connects to the Middle Zambezi basin at the Chobe confluence, just upstream of the Victoria Falls. The Kafue, Luangwa and Lake Malawi - Shire basins drain into the Lower Zambezi basin.

The basic assumptions within the model can be varied according to three 'world-views' - hierarchist, egalitarian and individualist - as described in Chapter 5. Each world-view consists of a specific set of equations and initial and parameter values, representing a coherent perception of how the world works. In addition four management styles have been pre-defined - hierarchist, egalitarian, individualist and fatalist. Each management style consists of a particular set of parameter values representing a certain policy strategy. Once a pre-defined management style has been chosen, the user of the model can adjust particular elements if preferred.

The AQUA model has already been described in general terms in Chapters 3 to 5. In this chapter, I will discuss which data have been used to make the generic AQUA model operational for the Zambezi basin. A separate section describes how

---

[1] Note that calculations are not made for the countries as a whole, but only for those parts of national territories which lie within the Zambezi basin.

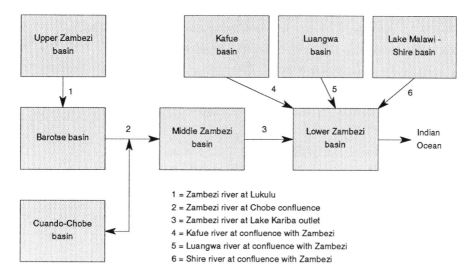

Figure 8.3. River flow schematization in the AQUA Zambezi Model.

the hydrological parameters in the model have been calibrated. The chapter concludes with a discussion of some model results regarding the present situation in the Zambezi basin. Results with respect to possible futures will be reported in the next chapter.

8.2 Input data

Although it might seem easy to find accurate data on the size of sub-basins, this is not in fact the case. Even data on the size of the Zambezi basin as a whole vary widely from source to source, as illustrated in Table 8.1. Most estimates of the size of the Zambezi basin fall within a range of about 1200-1400×10^3 km^2. This variation may be partly due to the absence of accurate data on the borders of the basin. I can find no explanation for the relatively low estimate of 540×10^3 km^2 by Probst and Tardy (1987). The relatively high estimate of 1870×10^3 km^2 by SADC-ELMS and DANIDA (1994) is caused by the fact that they consider the Okavango Delta part of the Zambezi basin. Most studies, including this one, exclude the Okavango Delta, because it can largely be regarded as an internal drainage area, where all incoming water collects and evaporates again without net outflow. However, depending on the climatic conditions in a particular year, there can be some minor interaction between the Okavango Delta and the Zambezi basin through the Selinda Spillway, which for some researchers is a reason to include the Okavango Delta in their definition of the Zambezi basin. Unfortunately, reliable data on the exchange through the Selinda

Table 8.1. Some basic data for the Zambezi basin from a number of sources.

Author(s)	Total area (10^3 km^2)	Annual runoff (10^{12} kg/yr)	Total population (10^6)
Balon and Coche (1974)	1190	-	-
Meybeck (1976, 1988)	1340	224	-
Korzun et al. (1977, 1978)	1330	106	-
UN (1978)	1420	-	-
Czaya (1981)	1330	79	-
Szestay (1982)	1295	221	5.6 (in 1970), 12 (in 2000)
Bolton (1983)	1300	-	-
Milliman and Meade (1983)	1200	223	-
Probst and Tardy (1987)	540	77	-
Pinay (1988)	1400	-	-
Speidel and Agnew (1988)	1330	269	-
Gandolfi and Salewicz (1991)	1300	-	20 (in the early 1980s)[1]
Vörösmarty and Moore (1991)	1220	98	20 (in the early 1980s)[1]
Howard (1994)	1420	110	22 (in 1983)
SADC-ELMS and DANIDA (1994)	1870	103	-
Masundire and Matiza (1995)	1400	106	-
UNEP (1995a)	1250	103	-
This study[2]	1360	110	25.5 (in 1994)

[1] Original source: UNEP (1987).
[2] Area based on UNEP (1984); runoff based on simulation results AQUA Zambezi Model; population based on Deichman (1994).

Table 8.2. Spatial schematization of the Zambezi basin (areas in 10^3 km^2).

Sub-basin	Ang.	Nam.	Bots.	Zamb.	Zimb.	Tanz.	Mal.	Moz.	Total
Upper Zambezi	109	-	-	96	-	-	-	-	205
Barotse	37	-	-	115	-	-	-	-	152
Cuando-Chobe	92	16	12	15	-	-	-	-	135
Middle Zambezi	-	-	-	31	135	-	-	-	166
Kafue	-	-	-	158	-	-	-	-	158
Luangwa	-	-	-	147	-	-	-	4	151
Lake Malawi - Shire	-	-	-	-	-	27	107	21	155
Lower Zambezi	-	-	-	20	80	-	-	138	238
Total	238	16	12	582	215	27	107	163	1360

Source: sub-basin boundaries have been taken from UNEP (1984) and country boundaries from ESRI (1992).

Spillway are lacking. An explanation for some of the lower estimates of the Zambezi basin area might be that a number of researchers, such as Vörösmarty and Moore (1991), exclude not only the Okavango Delta but also the Cuando-Chobe basin. In this study the Cuando-Chobe basin is included, because under some circumstances interaction between the Chobe river and the Zambezi river can be significant, particularly in relatively dry or relatively wet years (see also Section 8.3). Here the Zambezi basin is estimated to cover 1360×10^3 km^2, on the basis of data from UNEP (1984). The areas of the sub-basins are shown in Table 8.2.

Calculations of the average annual runoff from the Zambezi basin range between 77×10^{12} and 269×10^{12} kg/yr (Table 8.1). One reason for the large variation in estimates might be the imprecise definition of the Zambezi basin mentioned above. However, a more important reason is probably that measuring total runoff from the Zambezi basin into the Indian Ocean is difficult due to the large flood plain in the delta of the river, so that estimates depend either on the interpretation of different discharge measurements or on indirect calculations on the basis of precipitation measurements and estimated evaporation. A third reason is that the different estimates probably represent averages for different periods. Unfortunately, most authors do not mention the period to which their data refer. Average annual runoff data may vary considerably if one takes different periods, due to the large climatic fluctuations in the region. As shown in Figure 8.4, the average Zambezi river flow at the Victoria Falls for the period 1911-1990 was for instance about 1060 m^3/s, but for the period 1911-1920 an average of 710 m^3/s was recorded and for the period 1961-1970 an average of 1520 m^3/s (Mukosa *et al.*, 1995). In the AQUA Zambezi Model runoff is simulated per sub-basin on a monthly basis. The hydrological parameters in the model have been calibrated as discussed in Section 8.3.

The potential water supply of a country[1] is divided into two components: potential supply from internal and from external sources. The first refers to the available amount of water due to precipitation within the country. The second consists of the water flow entering the country from upstream. Because runoff is calculated per sub-basin (in the state sub-model) and not per country (as needed for the calculation of potential water supply in the impact sub-model), a translation is required from catchment-specific information to country-specific information (see also the text box in Section 3.4). To translate internal sources per sub-basin to internal sources per country, the model uses an area distribution matrix, based on the data presented in Table 8.2. It is slightly more difficult to assess the external water sources per country. One difficulty is how to account for a 'diffuse' contribution of

---

[1] 'Country' refers here to 'that part of a country which lies within the Zambezi basin'.

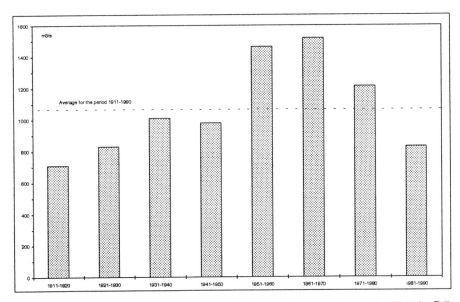

Figure 8.4. Ten-year averages of annual runoff for the Zambezi river at the Victoria Falls (Mukosa et al., 1995).

Table 8.3. Land cover in the Zambezi basin (areas in 10^3 km^2).

Sub-basin	Forest	Grassland	Rain-fed cropland	Irrigated cropland	Surface water	Total
Upper Zambezi	0	198	7	0.06	0	205
Barotse	1	142	9	0.06	0	152
Cuando-Chobe	42	91	2	0.03	0	135
Middle Zambezi	84	70	6	0.16	5	166
Kafue	5	139	13	0.08	1	158
Luangwa	21	119	11	0.07	0	151
Lake Malawi - Shire	10	92	22	0.22	31	155
Lower Zambezi	56	169	10	0.31	3	238
Total	219	1020	80	1	40	1360

Sources: Olson et al. (1985) and FAO (1996). The digital land cover map from Olson has been used to obtain data for three categories: forest, grassland and agricultural land. The latter category has been split into cropland and permanent pasture on the basis of the national ratios of cropland areas to areas of permanent pasture (national ratios taken from FAO). The permanent pastures have been added to the grasslands. The cropland areas have been subdivided into rain-fed and irrigated areas on the basis of the national ratios of rain-fed to irrigated areas (national ratios taken from FAO). In the Tanzanian part of the Zambezi basin, the irrigated cropland area has been assumed to be nil, in correspondence with data from Kivugo and Nnunduma (1994) and the World Bank and UNDP (1990f). Surface water areas: see Table 8.4.

water from upstream, as for instance in the case of Malawi, where a diffuse flow of water enters from Tanzania (see Figure 8.1). In the state sub-model, the Tanzanian part of the Zambezi basin is not a separate entity, but is part of the Lake Malawi - Shire basin, so there is no basis in the model on which to estimate the external water sources of Malawi. Although this is primarily a technical problem, due to the schematization chosen, something similar exists in reality: there are no direct measurements available to assess the diffuse flow from Tanzania into Lake Malawi. In this case, the external flow from Tanzania is estimated to be small and has been neglected. The same assumption has been made with respect to the Angolan water contributions to Zambia, Namibia and Botswana. Other difficulties arise where the Zambezi river forms the border between two countries: between Zambia and Namibia and, further downstream, between Zambia and Zimbabwe. In these cases, it has been assumed that the riparian countries can both draw on the entire river flow.[1] An exception has been made for Botswana, which abuts on the Zambezi river at just one point. In this case, the Zambezi river flow at this point is not considered a possible external source for Botswana.

Let me summarize now, per country, how the external sources have been defined. No external sources have been distinguished for the Angolan, Botswanese, Zambian, Tanzanian and Malawian territories within the Zambezi basin. The external water source of the Namibian territory has been defined as the runoff from the Barotse basin (which is generated in Angola and Zambia). The external water sources of Zimbabwe are formed by the runoff from the Barotse and Kafue basins. The latter fully originates in Zambia, but is available to Zimbabwe after the Kafue has joined the Zambezi. Theoretically, Zimbabwe can also draw on the runoff from the Cuando-Chobe basin, but this flow is generally negligible. The external sources of Mozambique consist of the runoff from the Middle Zambezi basin (which is generated mainly in Angola, Zambia and Zimbabwe), the runoff from the Kafue and Luangwa basins (generated in Zambia), and the runoff from the Lake Malawi - Shire basin (descending mainly from Malawi and Tanzania).[2]

Land cover
Accurate land cover maps of the Zambezi basin are rare and the classifications of land cover types on these maps do not generally correspond to the classification

---

[1] The fact that two riparian states both have access to the entire flow of the river does not mean that they have the *right* to use it all. In the simulation model, the concept of potential water supply has been defined in the physical and the economic sense, not in the political sense.

[2] 'Runoff' refers in this paragraph either to stable or to total runoff, depending on the world-view chosen (see Chapter 5).

Table 8.4. Major lakes, reservoirs and waterfalls in the Zambezi basin.

Lake, reservoir or waterfall	Sub-basin	Country	Surface area[1] (10^3 km^2)	Storage[2] (10^{12} kg)	Present hydropower generation capacity[3] (MW)
Victoria Falls	Middle Zambezi	Zambia	-	-	108
Lake Kariba	Middle Zambezi	Zambia / Zimbabwe	5.36	116-185	1266
Itezhitezhi reservoir	Kafue	Zambia	0.37	0.7-5.7	-
Kafue reservoir	Kafue	Zambia	0.81	0.14-0.84	900
Lake Malawi	Lake Malawi - Shire	Malawi / Mozambique	30.9	7720	-
Nkula Falls	Lake Malawi - Shire	Malawi	-	-	104
Tedzani Falls	Lake Malawi - Shire	Malawi	-	-	40
Lake Cahora Bassa	Lower Zambezi	Mozambique	2.74	12.5-72.5	2075

[1] Surface areas from SADCC and UNEP (1991), Lake Malawi surface from Korzun et al. (1978).
[2] Dead storage and maximum capacity for Kariba, Itezhitezhi, Kafue and Cahora Bassa from Vörösmarty and Moore (1991). Lake Malawi storage from Korzun et al. (1978).
[3] SADCC (1990).

Table 8.5. Major wetlands in the Zambezi basin.

Wetland	Sub-basin	Surface area[1] (10^3 km^2)	Water storage[1] (10^{12} kg)	Water depth[2] (m)
Barotse Flood Plain	Barotse	7.5 - 7.7	8 - 9 during normal flood	1.0 - 1.2
Linyanti Swamp	Cuando-Chobe	n.a.	n.a.	n.a.
Chobe Swamp	Cuando-Chobe	2.5	n.a.	n.a.
Busanga Swamp	Kafue	1.0	n.a.	n.a.
Lukanga Swamp	Kafue	2.1 - 2.6	7.4 - 8.4 storage capacity	2.8 - 4.0
Kafue Flats	Kafue	5.4 - 7.0	10.5 during average peak	1.5 - 1.9
Delta	Lower Zambezi	n.a.	n.a.	n.a.

[1] From Vörösmarty and Moore (1991).
[2] Calculated by dividing storage by surface area.

n.a. = not available

preferred in a water assessment. In particular 'irrigated cropland' and 'wetland' are missing on most maps. In this study, the digital land cover map from Olson *et al.* (1985) has been used as primary data source, supplemented by national data from FAO (1996). Five land cover types are distinguished: forest, grassland, rain-fed cropland, irrigated cropland and open water (Table 8.3). Due to lack of data in the available databases, it has not been possible to distinguish wetlands as a separate land cover type.[1] By using additional, independent data on wetlands (e.g. from Vörösmarty and Moore, 1991, see Table 8.5), it would have been possible to include wetlands as a separate land cover type and thus improve the data set used, but to do this was beyond the scope of this project. The figures presented in Table 8.3 are used in the model as initial land cover data. The extension of cropland areas at the expense of forests and grasslands is considered an exogenous scenario, to be chosen by the user of the model according to the type of future to be explored (see next chapter).

Initial water stocks
Obtaining reliable estimates of water stocks in the Zambezi basin is relatively easy for surface waters, but more difficult for ground water. The greater part of surface waters is formed by the lakes and man-made reservoirs shown in Table 8.4 and the wetlands shown in Table 8.5. Water stocks in rivers have been estimated per sub-basin on the basis of river runoff, stream velocity and river length, using data from Kaltenbrunner (1992), but these are small compared to the water stocks in lakes and wetlands. In the Kafue and Luangwa basins, a groundwater storage of 1.6 m water column has been reported (World Bank and UNDP, 1990d). The remaining part of Zambia, in as far as it is situated in the Zambezi basin, has an average groundwater storage of 3.4 m water column. Provisionally, it has been assumed that the groundwater layer actively contributing to groundwater runoff is 2.5 m water column in all sub-basins.

Population and domestic water demand
On the basis of a population density map of Africa (Deichman, 1994), the population in the Zambezi basin has been estimated at about 25.5 million people in 1994. The distribution of these people over the basin is shown in Table 8.6. The

---

[1] From an hydrological point of view, this means that an element of error is introduced in the simulation of evaporation, particularly in the Barotse Flood Plain and the Cuando-Chobe and Kafue basins, where most wetlands are situated. The fact that wetlands are not distinguished as a separate land cover type does *not* mean that the stabilizing effect of wetlands on river runoff is disregarded. This effect is accounted for in the form of the lag time parameter k_{fsw} and the delay parameter T_{fsw} (see Section 8.3).

Table 8.6. Population in the Zambezi basin in 1994 (in millions).

Sub-basin	Ang.	Nam.	Bots.	Zamb.	Zimb.	Tanz.	Mal.	Moz.	Total
Upper Zambezi	0.19	-	-	0.36	-	-	-	-	0.55
Barotse	0.07	-	-	0.64	-	-	-	-	0.71
Cuando-Chobe	0.09	0.09	0.01	0.07	-	-	-	-	0.26
Middle Zambezi	-	-	-	0.38	3.10	-	-	-	3.49
Kafue	-	-	-	3.15	-	-	-	-	3.15
Luangwa	-	-	-	1.98	-	-	-	0.04	2.02
Lake Malawi - Shire	-	-	-	-	-	0.60	8.40	0.28	9.28
Lower Zambezi	-	-	-	0.76	3.40	-	-	1.86	6.02
Total	0.35	0.09	0.01	7.34	6.51	0.60	8.40	2.18	25.5

Source: based on Deichman (1994).

highest population density is found in Malawi, with an average of about 90 people per km^2 of land, which is nearly five times the average in the Zambezi basin and roughly four times the average in Africa. After Malawi, the Zimbabwean and Tanzanian parts of the Zambezi basin are most densely populated and then the areas of Mozambique and Zambia. Sparsely populated areas can be found in the upstream part of the basin, in Angola, Namibia and Botswana. In the Namibian Caprivi Strip for example, a density of about 5 people per km^2 is found (see also Heyns, 1995). The populations of the basin states have been assumed to grow according to either the low, medium or high scenarios of the United Nations (UN, 1993). Depending on the type of future one would like to explore, one of these UN scenarios can be chosen (see next chapter). For the current distribution of the Zambezi basin population over rural and urban areas, national data have been used as shown in Table 8.7, derived from the FAO (1995b). It has been assumed that the urbanization level over the whole basin will increase over the 21st century, following the same trend as has been noticed in developed countries (see Berry, 1990). At present, the urbanization level in the basin is about 30 per cent, which has been assumed to grow towards 50 per cent by the year 2050. To initialize the pressure sub-model, data are needed on specific domestic water demand in rural and urban areas, in both sectors distinguishing between public and private water supply. Direct estimates of specific demand on the basis of an 'expert judgement' are probably more reliable than indirect calculations on the basis of data on total demand and population size (see text box). After considering a number of direct estimates for different countries in the Zambezi basin, it has been decided to use the following values: 150 kg/day per

How much water do people use for domestic purposes?

It seems to be more difficult to answer such a simple question than one would expect. Some sources give direct estimates of domestic water demand per capita, often distinguishing between urban and rural areas, but most sources primarily give data on total domestic water demand. In that case, demand per capita can be estimated indirectly if population data are also given. Direct estimates, which are generally no more than an 'expert judgement', appear to be relatively rare, but they are probably more useful than the indirect calculations, as will be shown below.

World Bank and UNDP (1990g,h,d) give some direct estimates for Malawi, Mozambique and Zambia. For Malawi, they report a public water demand of 15-40 kg/day per capita in rural areas, 120 kg/day in small urban areas and 200 kg/day in the cities of Lilongwe and Blantyre. For Mozambique, a public water demand is reported of 26 kg/day per capita in rural areas and 100 kg/day in urban areas. For Zambia they give a public water demand of 15 kg/day per capita in rural areas, 120 kg/day in small urban townships and 200 kg/day in large urban areas. JICA (1995) assumes a specific demand in Zambia of 35 kg/day per capita in rural areas, 150 kg/day in small urban areas and 180 kg/day in large urban areas. According to JICA, the present demand in Lusaka amounts to 130-150 kg/day. Whereas these direct estimates give a relatively consistent picture, a more diffuse picture appears if indirect calculations are made on the basis of data on total domestic water demand and population size.

WRI (1994) gives an average specific domestic water demand of 19 kg/day per capita for Malawi and 36 kg/day for Mozambique. Data from FAO (1995b) give an average demand of 24 kg/day per capita for Malawi and 9 kg/day for Mozambique. These averages seem rather low, which might be due to the fact that the figures for 'total' domestic water demand used by WRI and FAO refer to the demand from *public* supply systems only, and exclude *private* supply facilities. On this basis domestic demand per capita will be underestimated if it is calculated by dividing 'total' domestic water demand in a year by the population in that year, particularly if water supply coverage is relatively limited. In fact, one should divide the figure for 'total' domestic water demand by the part of the population with public water supply. Following this approach, the data from WRI (1994) give an average demand of 36 kg/day per capita in Malawi (assuming a public water supply coverage of 52 per cent, see Table 8.10) and 150 kg/day in Mozambique (assuming a coverage of 24 per cent); the data from FAO (1995b) give an average demand of 46 kg/day per capita for Malawi and 39 kg/day for Mozambique. These calculations appear to yield more plausible estimates, except for the questionably high WRI figure for Mozambique. Indirect calculations for the other Zambezi countries on the basis of data from WRI and FAO also yield various unexpected and questionable values for specific water demand. According to WRI data for instance, the specific domestic demand in Zambia would be 148 or 265 kg/day per capita on average (according to the first and second calculation method respectively), very high values if one takes into account that about 60 per cent of the Zambian population lives in rural areas. There may be different reasons for errors, the most important ones being inconsistencies between population and total demand data and misinterpretations of the term 'total demand' (does it include private demand, rural demand, etc.). The indirect calculation method is probably so vulnerable to errors that it is better to make a direct rough estimate of specific domestic water demand on the basis of an 'expert judgement', than to rely on total demand and population data. In this study, an initial specific water demand has been assumed of 25 kg/day per capita for public water supply in rural areas and 150 kg/day for public water supply in urban areas. For private water supply, an initial specific demand of 25 kg/day per capita has been used for both rural and urban areas.

Table 8.7. Population, livestock, irrigation, economic production and hydropower in the Zambezi basin countries.

	Ang.	Nam.	Bots.	Zamb.	Zimb.	Tanz.	Mal.	Moz.	Total
Population[1] (10^6)									
In country	10.0	1.78	1.30	8.45	9.71	27.3	8.75	15.7	83.0
In Zambezi basin	0.35	0.09	0.01	7.34	6.51	0.60	8.40	2.18	25.5
Rural fraction of population[2] (%)									
In country	67	73	79	62	70	78	80	73	71
In Zambezi basin	67	73	79	62	70	78	80	73	71
Cattle[3] (10^6)									
In country	3.10	2.09	2.70	2.88	6.22	13.1	0.84	1.38	32.3
In Zambezi basin	0.11	0.11	0.02	2.50	4.17	0.29	0.81	0.19	8.20
Sheep, goats and pigs[3] (10^6)									
In country	2.54	5.21	2.43	0.89	3.47	12.4	1.22	0.68	28.8
In Zambezi basin	0.09	0.27	0.02	0.77	2.32	0.27	1.17	0.09	5.00
Irrigated cropland[4] (km^2)									
In country	750	40	20	300	1000	1440	200	1050	4800
In Zambezi basin	60	1.6	0.7	248	251	0	200	233	994
Gross national product[5] (10^6 US\$/yr)									
In country	6000	1500	1100	3100	6100	3100	1500	1200	23600
In Zambezi basin	210	76	9	2700	4100	68	1400	170	8700
Value added industry[6] (% of GNP)									
In country	44	35	57	55	40	12	20	15	
In Zambezi basin	44	35	57	55	40	12	20	15	
Hydropower generation capacity[7] (MW)									
In country	400	n.a.	n.a.	1963	666	333	146	2360	5868
In Zambezi basin	0	0	0	1608	666	0	144	2075	4493

[1] National population figures in 1990 from WRI (1992); population figures for the Zambezi basin in 1994 based on Deichman (1994).

[2] Rural population fraction per country from FAO (1995b). Provisionally, the national fractions have been assumed for the Zambezi basin as well. The FAO data per country differ from the data given by WRI (1994), but the WRI average for the basin as a whole comes to 71 per cent as well.

[3] National livestock figures in 1990 from FAO (1996); fraction in Zambezi basin calculated on the basis of the population distribution.

[4] National irrigated areas in 1990 from FAO (1996). The irrigated area of Namibia in the Zambezi basin from Heyns (1995). The irrigated area of Tanzania in the Zambezi basin assumed to be zero on the basis of Kivugo and Nnunduma (1994) and World Bank and UNDP (1990f). Other areas in Zambezi basin calculated on the basis of the total cropland areas in the basin and the national ratios of rain-fed to irrigated areas (see footnote in Table 8.3).

[5] National GNP data in 1989 from WRI (1992); fraction of GNP in Zambezi basin calculated on the basis of the population distribution.

[6] National data for 1990 from UNEP (1993). For Namibia, a regional average of 35 per cent has been assumed. Provisionally, the national ratios have been assumed to be applicable for the Zambezi basin as well. Although UNEP data refer to fractions of gross *domestic* product (GDP), they are applied here to gross *national* product (GNP), an inconsistency in the model due to the fact that no separate input data are used for GNP and GDP.

[7] National capacities in 1989 from IWPDC (1991), see also Gleick (1993a); capacity in Zambezi basin from SADCC (1990).

capita for public water supply in urban areas, and 25 kg/day per capita for public water supply in rural areas and for private water supply in both rural and urban areas. These figures have been adopted for all countries.

Livestock

The initial livestock data used are shown in Table 8.7. From these data it appears that Namibia and Botswana have most livestock in relation to size of population. Within the Zambezi basin, the highest density of livestock in terms of head per km^2 can be found in Zimbabwe. Growth rates for livestock have been assumed to be equal to the population growth rates. Specific water demand for cattle has been assumed to be constant at 12,000 kg/yr/head, which is equal to 33 kg/day/head. This estimate has been made on the basis of several sources and corresponds to the value used in the AQUA World Model (see Chapter 6). The specific water demand for sheep, goats and pigs has been assumed to be eight times less per head.

Irrigation

The total irrigated area in the Zambezi basin in 1990 is estimated at about 1000 km^2, equal to 1.2 per cent of the total cropland area in the basin, which is not much if compared to the percentage of total cropland area which is irrigated in Africa as a whole (6.0 per cent) or in the world (16.5 per cent). It is generally noted that there is a large irrigation potential in the basin still undeveloped (SADCC, 1992). It is unclear which parts of the irrigated areas in Angola and Mozambique are in fact still irrigated at present, because many schemes have been disrupted during the long civil wars in these countries (SADCC, 1992). Initial irrigation water demand has been assumed at 10×10^6 kg/yr/ha for all countries, an average value which is often used (see also Chapter 6).

Economics and industrial water demand

The economic data presented in Table 8.7 show that, in the initial year of simulation, Namibia and Botswana have the highest gross national product per capita of all basin states, about 850 US$/yr. Zimbabwe and Angola follow with about 600 US$/yr per capita. Mozambique has the lowest economic output, of about 75 US$/yr per capita. In the calculation of gross basin product (GBP) in the Zambezi basin, it has been assumed that the gross national products of the basin states contribute to GBP in the same proportion as the national populations contribute to the basin population. This is a very rough estimate and it is questionable, but there were no data to make a more intelligent estimate. GBP calculated in this way amounts to 340 US$/yr per capita. The growth rates of the gross national products of the basin states are regarded as external scenarios, to be chosen by the user of the model according to the type of future to be explored (see next chapter). Data on the value added in the industrial sector show that Botswana and Zambia are the most industrialized countries of all basin states, followed by Angola and Zimbabwe. The national ratios of value added in the industrial sector to GNP have been taken as constant over the whole simulation period, assuming no structural changes in the national economies. For all countries, specific industrial water demand has been assumed at 70 kg/yr per US$ value added in the industrial sector, a value corresponding to the global average at the beginning of the 20th century (see Chapter 6). As an alternative, specific industrial water demand could have been calculated separately for each country, as a quotient of the total industrial water supply reported and the value added in the industrial sector. However, one then encounters the same sort of data inconsistencies as with domestic water demand (see discussion above). By applying the indirect calculation method and using total industrial demand data from FAO (1995b), values for specific industrial water demand are found to vary between 15 and 100 kg/yr per US$ value added, but I can find no well-argued explanation for these variations. Why, for instance, would Malawi use six to seven times more water per value added than Angola? One can question the hypothesis that there is a relationship between value added in the industrial sector and industrial water demand, but as an alternative hypothesis based on empirical data is not at hand, I will stick to this simple approach (see Chapter 3, Equation 3.6).

Hydropower generation
The present hydropower generation capacity in each of the Zambezi states is shown in Table 8.7. Hydropower plants in the Zambezi basin are major contributors to the national capacities. The largest hydropower plants are situated in the main Zambezi river: the Cahora Bassa plant in Mozambique and the Kariba plants on the North and South Banks, in Zambia and Zimbabwe respectively. A smaller plant further upstream in the Zambezi river is situated at the Victoria Falls. In the Kafue basin in Zambia, there is a large plant at Kafue Gorge and a smaller one at Itezhitezhi. In the Shire river, the outlet of Lake Malawi, two small plants can be found, at the Nkula and Tedzani Falls (see also Table 8.4).

Water demand parameters
Important parameters in the model which could not be calibrated due to a lack of historical data, are the growth and price elasticities of domestic, irrigation and industrial water demand, and the technological development and diffusion rates. For this reason, it has been chosen to use the same values as in the AQUA World Model,

taking different values for the hierarchist, egalitarian and individualist (see Section 6.3). Per country, water source fractions have been used as shown in Table 8.8. In general surface water makes a much larger contribution to total water supply than ground water. In most Zambezi states, ground water is used mainly for rural domestic water supply. In Angola groundwater use for irrigation has been estimated to be zero (FAO, 1995b). The development of groundwater resources in Angola is primarily related to drinking water supplies (World Bank and UNDP, 1990b). In Namibia, it has been estimated that about 14 per cent of the irrigated area is supplied from groundwater resources (FAO, 1995b), a value which has been taken as a measure for total water supply as well. In Botswana ground water has been and will continue to be a major source of water, supplying some 80 per cent of the population and most livestock (World Bank and UNDP, 1990c). About 44 per cent of the irrigated area is supplied with ground water (FAO, 1995b). For Zambia JICA (1995) finds a groundwater fraction of 6 per cent for agricultural water supply. FAO (1995b) finds an almost similar value of 5 per cent. In the Copperbelt area in Zambia large groundwater abstractions occur to drain the mines. JICA (1995) reports a groundwater fraction of 17 per cent for domestic and industrial water supply, but this value is clearly too low, because from other data presented by JICA one can conclude that at least 58 per cent of industrial water use consists of groundwater abstractions (due to the mining sector). Furthermore, according to the World Bank and UNDP (1990d), about 50 per cent of current domestic water use in urban areas in Zambia is supplied from groundwater resources, something which has been as high as 100 per cent in the 1960s. In this study, it has been assumed that irrigation depends on ground water for 6 per cent, domestic water supply for 50 per cent and industrial water supply for 60 per cent, resulting in an average value of 17 per cent. In Zimbabwe, total groundwater use might be estimated at about 6 per cent, assuming 25,000 boreholes with a total supply of 70×10^9 kg/yr (World Bank and

Table 8.8. Water source fractions per country.

	Ang.	Nam.	Bots.	Zamb.	Zimb.	Tanz.	Mal.	Moz.	Zambezi basin
Surface water (%)	98	86	56	83	94	75	98	98	92
Ground water (%)	2	14	44	17	6	25	2	2	8

Sources: country data based on World Bank and UNDP (1990b-h), JICA (1995) and FAO (1995b). Data for the Zambezi basin have been calculated on the basis of the national data, weighted according to the relative contribution of a nation to the total water use in the Zambezi basin.

UNDP, 1990e) and a total water supply of 1220×10^9 kg/yr (FAO, 1995b). In Tanzania, ground water from boreholes, shallow wells and springs is estimated to contribute more than 25 per cent to domestic water consumption (World Bank and UNDP, 1990f). Irrigation in Tanzania is mostly based on surface water resources, but irrigation is of minor importance or even absent in the part of Tanzania which lies within the Zambezi basin. The fraction of land irrigated from groundwater sources in Malawi amounts to only 0.05 per cent (FAO, 1995b). According to World Bank and UNDP (1990g), the development of groundwater resources in Malawi has been primarily for rural domestic water supplies, which form about 2 per cent of the total water supply. In Mozambique, surface water is relatively abundant and the potential use of ground water is limited, due to large areas with rocky soils which are not very permeable and to high salinity in many of the available aquifers (World Bank and UNDP, 1990h). It has been estimated that irrigation currently depends entirely on surface waters. Only rural domestic water supply and a minor part of urban domestic water supply and industrial water supply depend on ground water, which means that about 2 per cent of the total water supply in Mozambique comes from ground water. The water source fractions as shown in Table 8.8 have been assumed to be constant over time. The model uses the same 'fractions consumptive water use' as in the AQUA World Model. For the hierarchist world-view these are: from 20 per cent in 1990 to 15 per cent in 2050 for domestic water use, from 80 per cent in 1990 to 75 per cent in 2050 for livestock water use, constant at 10 per cent for industrial water use and dependent on water-use efficiency improvements in the case of irrigation. For the egalitarian world-view 10 per cent higher values have been assumed, and for the individualist world-view 10 per cent lower values.

Water supply costs and prices
As described in Chapter 3, the costs of water supply are assumed to increase if water becomes scarcer. This is a result of the fact that people always look for the cheapest possibilities for water supply. When water demand outgrows its supply capability, the next cheapest source will be chosen. Durham (1995) describes this mechanism in relation to urban water supply in Zimbabwe's part of the Zambezi basin. As shown in Figure 8.5, the average urban water supply costs will increase steadily during the next few years, as a result of increasingly expensive extensions to the urban water supply systems. Each new extension of the urban water supply system will require higher investment than the previous one, due to the increasing distance between the new source and the urban centre. As with Durham (1995), most studies look only a few years ahead, so it is difficult to form an impression of the increase in costs in the long term. As a consequence, we have to rely on very rough estimates. Different

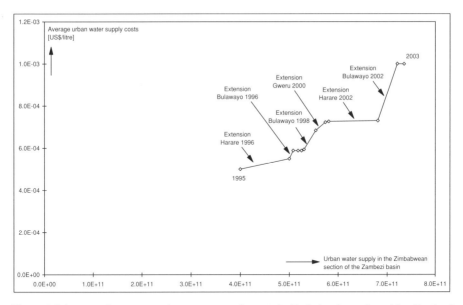

Figure 8.5. Increase in average urban water supply costs in Zimbabwe's section of the Zambezi basin. Data on sizes and investment costs of planned extension projects from Durham (1995). Assumptions: depreciation time of thirty years, interest rate of 10 per cent per year, annual operational and maintenance costs equal to annual fixed costs, and average urban water supply costs in 1995 of 0.5×10^3 US$/litre.

curves have been assumed for the hierarchist, egalitarian and individualist worldviews, using the same relationships between costs increase and water scarcity as in the AQUA World Model. Figure 8.6 provides an example of the public water supply-cost curves for Zimbabwe's section of the Zambezi basin.

As in most regions in the world, water prices in the Zambezi basin are generally lower than the actual costs. However, it is difficult to obtain an accurate picture of the true subsidies, due to a lack of data. As current situation, a charge of 25 per cent of actual costs has been assumed, a rough estimate taken from Serageldin (1995). For the 21st century, a policy of limited water price increase has been assumed for the hierarchist (the percentage of actual costs charged increases towards 75 per cent by 2025), a water-taxing policy for the egalitarian (the percentage growing to 110 per cent in 2025) and a market-pricing policy for the individualist (the price going to 100 per cent of costs by 2025). If the fatalist management style is used, the percentage remains at 25 per cent during the whole of the 21st century.

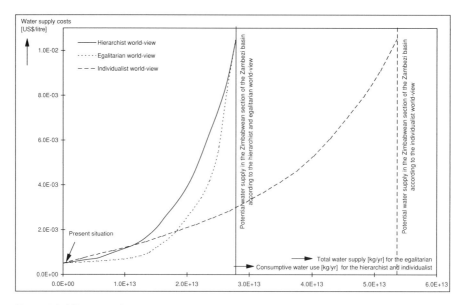

Figure 8.6. Water supply-cost curves per perspective for public water supply and water quality class A in Zimbabwe's section of the Zambezi basin. The values for potential water supply taken are those for the year 1990. The curves are similar for other water-demanding sectors and other water quality classes. The only difference lies in the baseline costs (at zero scarcity), which are lower for irrigation and industrial water supply and higher for water quality classes B to D.

Public water supply, sanitation and wastewater treatment
Recent figures for public water supply and sanitation coverage are shown in Table 8.9. In the hierarchist and egalitarian world-views, public water supply and sanitation coverage improve as a function of active government policy (see Chapter 5). In conformity with the assumptions in the AQUA World Model, it has been assumed that coverage demand for both public water supply and sanitation increases towards 100 per cent by the year 2050 in the case of the hierarchist management style. With the egalitarian management style, 2025 has been taken as the full coverage target year. In the individualist world-view public water supply and sanitation coverage improve as a function of economic growth. For the fatalist it has been assumed that the number of people with public water supply and sanitation remains constant from 1990, which implies that coverage will decline if the population grows. Public water supply costs will change as discussed above. Annual sanitation costs have been assumed to remain constant at an average level of 15 US$ per capita, including both investment and operational and maintenance costs. Current wastewater treatment coverage has been assumed to be 5 per cent throughout the river basin, for both domestic and industrial wastewater. For both the

Table 8.9. Public water supply and sanitation coverage in the Zambezi basin.

	Ang.	Nam.	Bots.	Zamb.	Zimb.	Tanz.	Mal.	Moz.	Zambezi basin[3]
Public water supply coverage[1]									
- Rural areas	0.20	0.37	0.88	0.43	0.80	0.46	0.49	0.17	0.52
- Urban areas	0.73	0.90	1.00	0.76	0.95	0.75	0.66	0.44	0.76
- Total	0.37	0.51	0.91	0.56	0.85	0.52	0.52	0.24	0.59
Sanitation coverage[2]									
- Rural areas	0.20	0.11	0.85	0.34	0.22	0.77	0.81	0.11	0.47
- Urban areas	0.25	0.24	1.00	0.77	0.95	0.76	1.00	0.61	0.85
- Total	0.22	0.15	0.88	0.50	0.44	0.77	0.83	0.25	0.57

[1] Data for 1990 for Angola, Namibia and Botswana, and for 1988 for Zambia, Zimbabwe, Tanzania, Malawi and Mozambique from WRI (1994). The total values have been calculated on the basis of FAO data on rural and urban population (see Table 8.7).

[2] Data for 1990 for Angola, Namibia and Botswana, and for 1988 for Zambia, Zimbabwe, Tanzania and Mozambique from WRI (1994). Data for 1980 for Malawi from WRI (1992).

[3] Data for the Zambezi basin have been calculated on the basis of the national data, weighted according to the relative contribution of a nation to the total population in the Zambezi basin.

hierarchist and the egalitarian world-view, the year 2050 has been chosen as the target year for 100 per cent coverage. According to the individualist world-view wastewater treatment coverage is a function of economic growth. For the fatalist, it has been assumed that the coverage will remain at 5 per cent. As in the AQUA World Model, average domestic wastewater treatment costs have been assumed at 0.2 US$/m^3 and average industrial wastewater treatment costs at 0.4 US$/m^3. These costs include both capital costs and operational and maintenance costs.

Water quality parameters

For the water quality parameters in the model, the same values have been taken as in the AQUA World Model (see Section 6.3), due to a lack of specific data for the Zambezi basin. Although most countries have some formal, nationwide water quality monitoring network, actual records of water quality are very rare, due to poor facilities, unskilled staff and a lack of funding (World Bank and UNDP, 1990). As in the World Model, four substances are considered: nitrate, ammonium, dissolved organic nitrogen and phosphate. Transmission coefficients of 0.9 have been assumed for all sub-basins. This means that the Barotse basin receives 90 per cent of the wastes emitted in the Upper Zambezi basin, the Middle Zambezi basin receives 90 per cent of the wastes emitted in the Barotse basin and 81 per cent of the wastes emitted in the Upper Zambezi basin, etc.

8.3 Calibration of hydrological parameters[1]

River runoff has been calibrated separately for each of the eight sub-basins distinguished. River runoff from the upstream sub-basins was calibrated first, followed by river runoff from the more downstream sub-basins. The calibration was carried out for an 'average year' with respect to climatic conditions and under water-use conditions in the year 1990. Monthly precipitation and temperature data per sub-basin and land cover type have been derived from IIASA's climate database (Leemans and Cramer, 1991). These data represent averages for the period 1931-1960. The calibration has been based on long-term averages of observed monthly river runoff data for the sub-basins distinguished, derived from SADC-ELMS and DANIDA (1994) and referring to different periods (see Table 8.10). The average hydrograph[2] at Lukulu, Zambia, for the period 1968-1991, has been used for the calibration of river runoff from the Upper Zambezi basin. The average hydrograph at the Victoria Falls, for the period 1907-1994, has been used as a measure of the river runoff from the Barotse basin, on the assumption that the major contribution to river runoff at the Victoria Falls comes from this basin (and indirectly from the Upper Zambezi basin) and not from the Cuando-Chobe basin. For the Cuando-Chobe basin no time series of runoff could be obtained, but it has been estimated that the Cuando-Chobe basin does not make a major contribution to the Zambezi river. In some years the 'contribution' could even be negative, due to a reverse flow from the Zambezi into the Chobe river. SADC-ELMS and DANIDA (1994) estimate that in the period from 1959/60 to 1990/91, there was an average net runoff of 0.63×10^{12} kg/yr from the Zambezi into the Chobe, which means a reduction in the annual Zambezi river flow at the Victoria Falls of about 1.5 per cent. It has been estimated that in a wet year, such as 1977/78, the relative volume of reduction can increase to 4 per cent, because the higher water level in the Zambezi means that more water enters the Chobe. In a dry year however, such as 1972/73, the Chobe feeds the Zambezi, and its relative contribution to the total flow at the Victoria Falls can be as much as 9 per cent.

Monthly measurements at the outflow from Lake Kariba in the period 1962-1988 have been used to calibrate river runoff from the Middle Zambezi basin. The Kariba

---

[1] This section is partly based on Vis (1996).

[2] A hydrograph or runoff curve shows the river discharge of a particular river as a function of time. Here, hydrographs are not drawn for one specific year, but for an 'average' year in a certain period.

Table 8.10. River runoff data sets used for calibration.

Sub-basin	Hydrometric station	Period
Upper Zambezi	Lukulu, Zambia	1968-1991
Barotse	Victoria Falls, Zambia / Zimbabwe	1907-1994
Cuando-Chobe	-	-
Middle Zambezi	Lake Kariba Outflow, Zambia / Zimbabwe	1962-1988
Kafue	Kafue Gorge, Zambia	1971-1988
Luangwa	Luangwa Bridge, Zambia	1955-1988
Lake Malawi - Shire	Chiromo, Malawi	1953-1981
Lower Zambezi	-	-

dam was completed in 1960, and therefore only flow data after this date have been considered. For the calibration of river runoff from the Kafue basin, monthly river runoff data at Kafue Gorge for 1971-1988 have been employed (data set after completion of the Kafue reservoir in 1970). The average hydrograph for the period 1955-1988 for Luangwa Bridge has been used to calibrate river runoff from the Luangwa basin. For calibration of river runoff from the Lake Malawi - Shire basin, the average hydrograph for 1953-1981 at Chiromo, Malawi, was chosen (see also UNESCO, 1995). There are no adequate river runoff data for the Lower Zambezi basin, because there are no hydrological stations which measure the full flow in the delta. Records of the Lower Zambezi flow are available for Matundo-Cais, near Tete in Mozambique, and for Lupata, about 80 km downstream of Tete, but both locations are upstream of the Shire confluence, so these data cannot be used as a measure of the total outflow from the Lower Zambezi basin into the Indian Ocean.

It should be noted that the period covered by the climatic data does not correspond to the periods covered by the river runoff data. More recent climatic data than for 1931-1960 are available, but only for specific meteorological stations, not translated into spatial patterns. Composing a new database of climatic data which would correlate more closely with the available runoff data would require considerable effort and goes beyond the scope of this study. The large climatic variations in the Zambezi basin, illustrated by the ten-year averages for the annual runoff of the Zambezi at the Victoria Falls in Figure 8.4, indicate that this mismatch of periods covered can have significant effects on the results. The largest inconsistency might be expected for basins where the runoff records cover relatively short periods, such as the Upper Zambezi. This could lead to some model inaccuracy with regard to the simulation of runoff from these basins, as will be discussed below.

Per sub-basin, the model has been calibrated by minimizing the difference between simulated and observed monthly river runoff values. As a measure of 'agreement', it has been chosen to use the so-called Willmott index *WI* (Willmott et al., 1985), which has also been used for this purpose by Vörösmarty and Moore (1991):

$$WI = 1 - \frac{\sum_{i=1}^{12} |R_{riv,sim}[i] - R_{riv,obs}[i]|}{\sum_{i=1}^{12} |R_{riv,sim}[i] - R_{riv,obs,avg}| + |R_{riv,obs}[i] - R_{riv,obs,avg}|} \quad (8.1a)$$

in which $R_{riv,sim}[i]$ represents the simulated river runoff for month i, $R_{riv,obs}[i]$ the observed river runoff in this month and $R_{riv,obs,avg}$ the average of observed monthly runoff values in a year, defined as:

$$R_{riv,obs,avg} = \sum_{i=1}^{12} R_{riv,obs}[i] / 12 \qquad [kg/month] \quad (8.1b)$$

The larger the Willmott index, the better the simulated hydrograph corresponds to the observed one. The index, which has a range from zero to one, attains one if the simulated hydrograph exactly equals the observed one. As optimization algorithm, the Powell method has been used, a so-called 'direction set method' in multidimensional optimization (Press et al., 1987). The set of parameters which has been varied in order to find optimal solutions has not been the same for all sub-basins, due to the different characteristics and the different availability of data for each of the basins. The basins of the Upper Zambezi, Barotse, Kafue and Luangwa have been calibrated first, and the basins of the Cuando-Chobe, Middle Zambezi, Lake Malawi - Shire and Lower Zambezi subsequently.

The Upper Zambezi, Barotse, Kafue and Luangwa basins
For the Upper Zambezi, Barotse, Kafue and Luangwa basins, the Willmott index has been maximized by calibrating four calibration parameters: *lcf* (the land cover factors in the Thornthwaite equation for potential evaporation), φ (the fraction of net precipitation which forms direct runoff), T_{fsw} (the delay in river runoff) and k_{fsw} (the lag time in the surface water store).[1] Some other parameters have been assumed at

---

[1] For an explanation of the parameters: see Chapter 3, Equations 3.23, 3.28 and 3.33. See also the list of symbols at the end of this book.

fixed values.[1] To get some feeling for the importance of the different calibration parameters, four calibration steps have been distinguished, with each step achieving better simulation results. In the first step, river flow itself has been supposed to be instantaneous, without any delay or lag time (which means that T_{fsw} and k_{fsw} are assumed to be zero). In this step, the land cover factors and ratios of direct runoff to net precipitation have been calibrated simultaneously. Changing the land cover factors means changing the area under the simulated hydrograph. From a physical point of view, land cover factors are empirical constants in the relationship between temperature and potential evaporation, but from a mathematical point of view they can be regarded as the 'handle' to correlate simulated and observed *annual* runoff.[2] The ratios of direct runoff to net precipitation do not influence annual runoff, but influence the pattern throughout the year. If these ratios were one, the rainfall pattern would be translated straightforwardly into a similar river runoff pattern. Smaller ratios result in increasing groundwater recharge and, as a consequence, stabilization of river runoff. The ratios of direct runoff to net precipitation can thus be considered the 'handle' to correlate the magnitude of stable runoff. The simulation results after the first calibration step were clearly inadequate, with correlation values in the range 0.3-0.7. This first step has been reported here to illustrate that a water balance approach is not sufficient to explain river runoff, because river flow dynamics are ignored.[3] In the second calibration step, a delay in river runoff T_{fsw} has been introduced, to account for the time it takes for water which enters the river to leave the basin. From a mathematical point of view, the delay parameter can be considered a 'handle' to synchronize simulated and observed peak runoff. The other parameters were held constant at the values obtained in the first step. Although the results improved, simulated peak discharges were still too large and intensive, just as after the first step, showing that a simple river flow delay is also not enough to explain observed river runoff. Apparently, river flows are not only delayed, but also stabilized. In the third calibration step, the effect has been studied

---

[1] For soil water-holding capacities, the same values have been assumed as in the AQUA World Model (see Table 6.7). Exponent p in the groundwater runoff equation has been assumed at one. The groundwater layer actively contributing to stable runoff has been assumed at 2.5 metres. Subsurface runoff from the basins has been ignored.

[2] The relative magnitudes of the land cover factors for the different land cover types have been kept constant, so that the whole set of land cover factors formed just one handle, in order to avoid an 'underdetermined' optimization function (a function with too many parameters to find one best analytical solution).

[3] See also for example Gleick (1987a), who introduces a river flow delay in addition to a water balance model of the Sacramento basin, or Vörösmarty and Moore (1991), who use a more advanced water transport model in addition to their water balance model of the Zambezi basin.

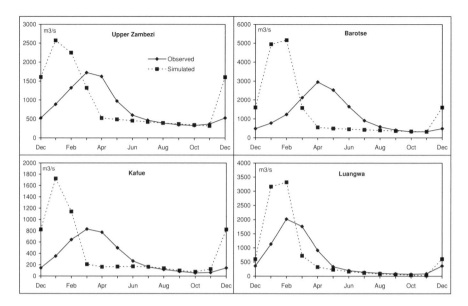

Figure 8.7. Simulated and observed hydrographs for an average year (simulation result after calibration step I).

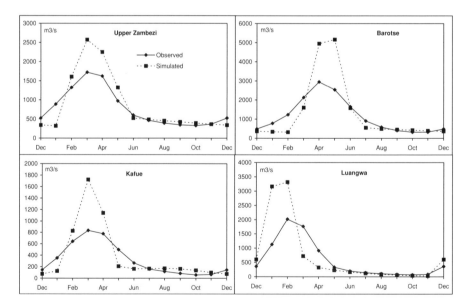

Figure 8.8. Simulated and observed hydrographs for an average year (simulation result after calibration step II).

of applying a residence time k_{fsw} instead of a delay T_{fsw}. In this step, the parameters lct, φ and k_{fsw} have been calibrated simultaneously, because one might expect that particularly the values of φ found in the first calibration step will change (river flow is now a result of stabilization in both the groundwater and the river system and not only in the groundwater system). The results after the third calibration step were much better than after the second step in respect of the form of the hydrographs, but not in the synchronization of simulated and observed runoff peaks. Therefore, a fourth calibration step was carried out, in which a river flow delay T_{fsw} was introduced, to finally achieve synchronization of runoff peaks. The parameter values found in the third step were kept constant. The simulated hydrographs after calibration steps I to IV are shown in Figures 8.7 to 8.10 respectively. One can easily see that the differences between simulated and observed hydrographs become smaller after each step. This can also be seen from the improvement in the Willmott index values, presented in Figure 8.11. The parameter values obtained in each calibration step are shown in Table 8.11, while the results of each step are discussed below.

The hydrographs resulting from the first calibration step are not very satisfactory, except for the fact that the areas under the simulated and observed hydrographs are about equal (Figure 8.7). For all sub-basins, relatively high land cover factors were found (if compared to values found in the literature, see e.g. Chapter 6, Table 6.7), particularly for the Upper Zambezi and Kafue basins. An explanation for the high values for the Kafue basin might be that the large wetland areas in the basin have not been taken into account, which implies an underestimation of evaporation, to be compensated for by evaporation from the other land cover types. An explanation for the high values for the Upper Zambezi basin might be the inconsistency between the climate and runoff data used for calibration. As mentioned above, the period covered by the climatic data does not correspond to the periods covered by the river runoff data. Taking annual runoff data at the Victoria Falls as a measure, 1931-1960 (the period for the climate data) was on average about 10 to 15 per cent wetter than 1968-1991 (approximately the period for the runoff data). Calibration of the model on the basis of relatively high precipitation data for the period 1931-1960 and relatively low runoff data for the period 1968-1991 logically leads to overestimates of the land cover factors which determine evaporation. As a result, the calibrated model will systematically overestimate evaporation from the Upper Zambezi basin and underestimate runoff. Nevertheless, the parameter values obtained by calibration are used for further calculations, because the uncertainties in the data sets used are already too great to justify minor corrections. For the ratios of direct runoff to net precipitation, relatively low values

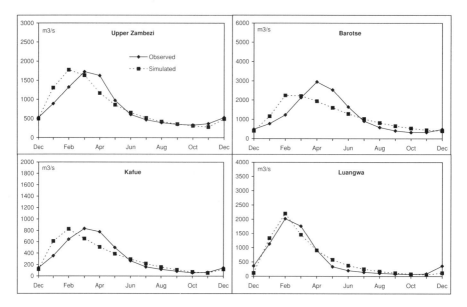

Figure 8.9. Simulated and observed hydrographs for an average year (simulation result after calibration step III).

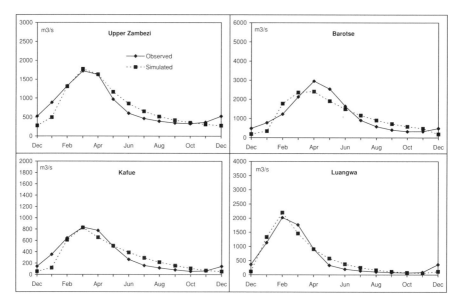

Figure 8.10. Simulated and observed hydrographs for an average year (simulation result after calibration step IV).

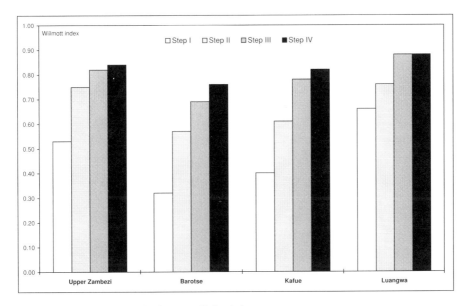

Figure 8.11. Willmott index for river runoff simulations.

were found for the Upper Zambezi, Barotse and Kafue compared to general estimates of such ratios in the literature (see for instance Chapter 6, Table 6.7). This would imply that these basins have relatively extensive groundwater recharge, which seems unrealistic. A more realistic suggestion might be that groundwater recharge is in fact less, and that part of the stabilization of runoff is due to lags in the river system (this will be shown in the third calibration step).

In the second calibration step, a river flow delay was introduced. Only river flow delays of an integer number of months were considered. The hydrographs resulting from this calibration differ from the previous hydrographs only in the fact that the simulated runoff peaks now coincide with the observed runoff peaks (Figure 8.8). Nevertheless, the correspondence of the simulated with the observed runoff curve, measured by the Willmott index, improves significantly with the introduction of the river flow delay. However, the delay of three months found for the Barotse basin is significantly longer than the flood-wave translation period of three to six weeks suggested by other authors (cited in Vörösmarty and Moore, 1991). Estimates in the literature of the flood-wave translation period in the Kafue Flats range from two to three months (again cited in Vörösmarty and Moore, 1991), which corresponds to the delay of two months in the Kafue basin found in this calibration step.

The results after the third calibration step (Figure 8.9) show that significant improvements can be obtained in the simulation of peak runoff if the river system is imagined as a linear store with a certain lag time. Compared to the previous calibration

Table 8.11. Hydrological parameter values.

Sub-basin	Land cover factor[1] [-]	Ratio of direct runoff to net precipitation[2] [-]	Delay river runoff [yr]	Lag time fresh surface water store [yr]	Willmott index [-]
Step I (without stabilization or delay in surface water store)					
Upper Zambezi	2.34, 1.70, 2.13	0.43, 0.53	0	0	0.53
Barotse	1.55, 1.13, 1.41	0.40, 0.50	0	0	0.32
Kafue	2.49, 1.81, 2.26	0.35, 0.45	0	0	0.40
Luangwa	1.38, 1.00, 1.25	0.62, 0.68	0	0	0.66
Step II (without stabilization in surface water store, but with delay)					
Upper Zambezi	2.34, 1.70, 2.13	0.43, 0.53	2/12	0	0.75
Barotse	1.55, 1.13, 1.41	0.40, 0.50	3/12	0	0.57
Kafue	2.49, 1.81, 2.26	0.35, 0.45	2/12	0	0.61
Luangwa	1.38, 1.00, 1.25	0.62, 0.68	1/12	0	0.76
Step III (without delay in surface water store, but with stabilization)					
Upper Zambezi	2.34, 1.70, 2.13	0.64, 0.70	0	0.19	0.82
Barotse	1.55, 1.13, 1.41	0.61, 0.67	0	0.18	0.69
Kafue	2.49, 1.81, 2.26	0.70, 0.75	0	0.28	0.78
Luangwa	1.39, 1.01, 1.26	0.60, 0.66	0	0.11	0.88
Step IV (with stabilization and delay in surface water store)					
Upper Zambezi	2.34, 1.70, 2.13	0.64, 0.70	1/12	0.19	0.84
Barotse	1.55, 1.13, 1.41	0.61, 0.67	1/12	0.18	0.76
Kafue	2.49, 1.81, 2.26	0.70, 0.75	1/12	0.28	0.82
Luangwa	1.39, 1.01, 1.26	0.60, 0.66	0	0.11	0.88

[1] The first value refers to forests, the second to grassland and the third to (rain-fed or irrigated) cropland.
[2] The first value refers to forests, the second value to grassland and (rain-fed or irrigated) cropland.

step, the correlation values have improved. The lag times in the fresh surface water system vary from 1.3 to 3.4 months, with the lowest value for the Luangwa basin and the highest for the Kafue basin. This is consistent with the Kafue basin having most wetlands and the Luangwa basin least. The ratios of direct runoff to net precipitation are much higher than in the two previous calibration steps, which can be explained by the fact that stabilization of runoff in those two steps could only happen through ground water, while surface water now also acts as a buffer. The simulated maximum for surface water storage in the Barotse basin is 13×10^{12} kg, which is of the same order of magnitude as some values cited in Vörösmarty and

Moore (1991). According to their sources, the Barotse Flood Plain would store 8-9×10^{12} kg during a normal flood, but 17×10^{12} kg during a high flood.

As a last calibration step, a delay in river flow was introduced, to simulate peak flow delays. As in the second step, only delays of an integer number of months were considered. The other parameter values were kept at the values which resulted from the third step. After this fourth calibration step, simulated and observed hydrographs correspond reasonably well (Figure 8.10), the Willmott index values varying between 0.76 and 0.88. Calibrating all parameters simultaneously would probably improve the correlation slightly more.

The Cuando-Chobe basin
Runoff from the Cuando-Chobe basin could not be calibrated, due to the absence of a reference hydrograph. Only estimates of annual runoff were available, as discussed above. It may be assumed that, in an average year, the Cuando-Chobe basin does not generate runoff (SADC-ELMS and DANIDA, 1994). The land cover factors have been calibrated in such a way that, in an average year, annual evaporation equals annual precipitation. For the other parameters, averages have been assumed of the values found for the Upper Zambezi, Barotse, Kafue and Luangwa basins.

The Middle Zambezi basin
River runoff from the Middle Zambezi basin cannot be adequately described with the standard equation used for the other basins (see Equation 3.33), because the outflow from Lake Kariba is regulated in order to optimize hydropower generation. The main objective of Lake Kariba management is to maintain a certain fixed level of energy production. The second objective is to avoid high discharges through the floodgates (Gandolfi and Salewicz, 1991). The first implies that the outflows are not allowed to fall below a certain level, while the second means that, once a certain maximum storage has been reached, the outflow will be increased in order to avoid excessive filling of the reservoir. To simulate this type of outflow management, the following outflow formulation has been used, modified from Gandolfi and Salewicz (1991):

$$R_{riv}(t) = \begin{cases} 0 & \text{if } S_{fsw}(t) \leq S_{fsw,min} \\ R_{firm} & \text{if } S_{fsw,min} < S_{fsw}(t) \leq S_{fsw,max} \\ R_{firm} + S_{fsw}(t) - S_{fsw,max} & \text{if } S_{fsw}(t) > S_{fsw,max} \end{cases} \quad \text{[kg/yr]} \quad (8.2)$$

where R_{firm} represents the minimum outflow needed to attain a certain energy production, $S_{fsw,min}$ the dead reservoir storage and $S_{fsw,max}$ the maximum reservoir storage capacity. The lowest average monthly outflow from Lake Kariba has been taken as a measure of R_{firm}. In the period 1962-1988 this was 2.2×10^{12} kg/month, occurring in September (SADC-ELMS and DANIDA, 1994). The dead reservoir storage has been assumed at 116×10^{12} kg and the maximum storage capacity at 185×10^{12} kg (Vörösmarty and Moore, 1991), which implies an active or live storage of about 70×10^{12} kg.[1] The land cover factors for the Middle Zambezi have been calibrated by correlating simulated and observed annual runoff from the basin. The resulting factors are: 1.38 for forests, 1.0 for grassland and 1.25 for cropland. Due to the stabilizing effect of Lake Kariba, it is difficult to calibrate the ratios of direct runoff to net precipitation. For this reason, averages of the values found for the other sub-basins have been used, i.e. a ratio of 0.64 for forests and 0.70 for grassland and cropland. Simulated river runoff from the Middle Zambezi basin resulting from the above assumptions is shown in Figure 8.12. It can be seen that the simulated hydrograph resembles the observed one reasonably well, the Willmott index being 0.83.

The Lake Malawi - Shire basin
For simulating river runoff from the Lake Malawi - Shire basin, the following equation has been used, adapted from Calder *et al.* (1995):

$$R_{riv}(t) = \begin{cases} \left(\dfrac{S_{fsw}(t - T_{fsw}) - S_{fsw,min}}{\kappa_{fsw}} \right)^x & \text{if } \kappa_{fsw} > 0 \\ R_{riv0}(t - T_{fsw}) & \text{if } \kappa_{fsw} = 0 \end{cases} \quad \text{[kg/yr]} \quad (8.3)$$

where κ_{fsw} represents a particular response factor and x an exponent determining the extent of linearity between active storage and outflow. This equation can be regarded as a generalization of Equation 3.33, where the exponent was assumed to be equal to one. As shown by Calder *et al.* (1995), Lake Malawi cannot adequately be described as a linear store. In such a case, the lag time of the store varies as a function of its storage:

$$k_{fsw}(t) = \kappa_{fsw}^x \times \left(S_{fsw}(t) - S_{fsw,min} \right)^{1-x} \quad \text{[yr]} \quad (8.4)$$

If $x = 1$, lag time k_{fsw} is equal to response factor κ_{fsw} and Equation 8.3 corresponds to Equation 3.33. If x is larger than one, as suggested by Calder *et al.* (1995) in the case of Lake Malawi, the lag time increases if the storage decreases. In contrast to Calder

---

[1] See also Gandolfi and Salewicz (1991). SADCC (1990) gives a slightly lower live storage of 65×10^{12} kg.

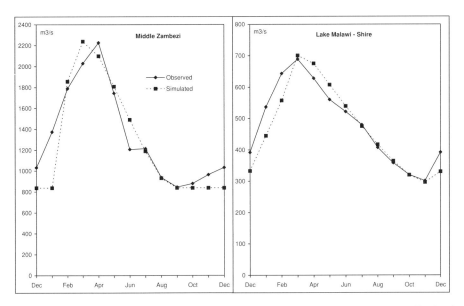

Figure 8.12. Simulated and observed hydrographs for an average year, for the Middle Zambezi basin and the Lake Malawi - Shire basin.

et al. (1995), a constant response factor κ_{fsw} has been used, instead of a different value for each month of the year. The dead storage of Lake Malawi, the storage below which there is no outflow, has been assumed at 7600×10^{12} kg.[1] The ratios of direct runoff to net precipitation could not be calibrated, because the outflow from Lake Malawi is very insensitive to these parameters due to the stabilizing effect of the lake. Therefore, averages of the values found for the other sub-basins have been taken, i.e. a ratio of 0.64 for forests and 0.70 for grassland and cropland. The following parameters have been calibrated: κ_{fsw}, T_{fsw}, x and the land cover factors lcf. The calibration resulted in $\kappa_{fsw} = 1.12 \times 10^{10}$, $T_{fsw} = 0$, $x = 3.03$ and the following land cover factors: 1.75 for forests, 1.27 for grassland and 1.59 for cropland. The residence time k_{fsw} of Lake Malawi appears to vary from five and a half years in the wet period to ten years in the dry period if measured relative to the *active* storage (as in Equation 8.4), which is equal to about 350 to 800 years if measured relative to the *total* water storage in Lake Malawi. Simulated and observed runoff are shown in Figure 8.12. The Willmott index is 0.86.

---

[1] Assumptions: the no-outflow level lies 4-5 metres below the average lake level during the past fifty years (Calder et al., 1995), the lake area is about 30×10^3 km² and the average storage amounts to 7720×10^{12} kg (Korzun et. al., 1978). In fact, the dead storage is not a crucial parameter, as this part of the storage does not take part in the outflow dynamics of the lake.

The Lower Zambezi basin

It seems plausible that the river runoff from this basin is dominated by Lake Cahora Bassa, although not all outflow from the basin, including the part of the flow originating in the Lake Malawi - Shire basin, passes through this reservoir. It is unclear to what extent the outflow has in fact been regulated during the past two decades, because the power plant of Cahora Bassa has operated during a few years only, due to the civil war in Mozambique. Unfortunately, adequate river runoff data for the Lower Zambezi basin are lacking, so that runoff parameters for the Lower Zambezi basin could not be calibrated. Provisionally, the parameter values have been assumed to equal the averages of the values for the other basins.

8.4 Discussion of results

For an average year, the model simulates a total runoff from the Zambezi basin of 110×10^{12} kg/yr. The largest contributions come from the Upper Zambezi, Lower Zambezi and Luangwa basins, each of which generates roughly 20 per cent of the total runoff. These basins produce the highest runoff not only in an absolute sense, but also in terms of specific runoff, i.e. the runoff per square kilometre. The Cuando-Chobe and Middle Zambezi basins produce least runoff. These outcomes correspond to the analysis made by SADC-ELMS and DANIDA (1994). The current water quality in the basin is found to be generally adequate to good. The fraction of surface waters of inadequate or bad quality (classes C and D) varies between 1 and 4 per cent. The best quality is found in the Upper Zambezi, Barotse, Cuando-Chobe and Luangwa basins. These estimates are based on rough appraisals of pollution by nutrients. Not accounted for in the model is the pollution by metals, due to the mining activities in the Copperbelt area, in the Kafue basin.

On the basis of the model results, the conclusion can be drawn that more than half of present water use in the Zambezi basin consists of water supply for irrigation (Table 8.12). Domestic water use contributes about a quarter of total water use. The largest part of domestic water use, about two thirds, is in the urban areas. Industrial water use accounts for about 14 per cent and livestock water use for 6 per cent. Zambia and Zimbabwe are the largest water users, together responsible for about 60 per cent of the total water use in the Zambezi basin. About half of the total water use is consumptive use, which means that half of the total water withdrawal does not return to rivers or ground water, but is lost through evaporation. Under current assumptions, this anthropogenic evaporation does not return as precipitation, but leaves the basin, resulting in a reduction in the original runoff from the basin of about 1 per cent.

Table 8.12. Water supply and scarcity in the Zambezi basin in the year 1990.

	Ang.	Nam.[3]	Bots.[3]	Zamb.	Zimb.	Tanz.[3]	Mal.	Moz.	Total
Water supply (10^9 kg/yr)									
- domestic	7.1	1.8	0.2	164	144	10	127	32	486
- irrigation	60	1.6	0.7	248	251	0	200	234	994
- livestock	1.4	1.7	0.3	31	54	3.9	11	2.4	106
- industrial	6.5	1.9	0.4	102	114	0.6	20	1.7	247
- total	75	7	2	545	562	14	359	269	1833
- consumptive use	50	3	1	258	274	5	191	190	972
Potential water supply[1] (10^9 kg/yr)									
- hierarchist world-view	3,100	16,000	22	11,000	29,000	1,100	4,300	49,000	49,000
- egalitarian world-view	3,100	16,000	22	11,000	29,000	1,100	4,300	49,000	49,000
- individualist world-view	20,000	38,000	470	54,000	54,000	2,700	11,000	100,000	110,000
Water scarcity[2] (%)									
- hierarchist world-view	1.6	0.0	3.6	2.4	1.0	0.4	4.4	0.4	2.0
- egalitarian world-view	2.4	0.0	6.7	5.2	1.9	1.3	8.3	0.5	3.7
- individualist world-view	0.2	0.0	0.2	0.4	0.5	0.2	1.6	0.2	0.9

[1] Potential water supply is defined as the stable runoff (hierarchist and egalitarian world-view) or as the total runoff (individualist world-view).
[2] Water scarcity is defined as the ratio of consumptive water use to potential water supply (hierarchist and individualist world-view) or as the ratio of total water supply to potential water supply (egalitarian world-view).
[3] Namibia, Botswana and Tanzania have very small territories within the Zambezi basin, which implies that the model output for these areas is relatively sensitive to the input data which were estimated on the basis of the assumption of homogeneity within the basin countries. See Section 8.2.

The model outcomes show that, at present, water is not a scarce commodity in the Zambezi basin, in any of its regions. The greatest scarcity is found in Malawi, with 2 to 8 per cent (on a scale of 0 to 100 per cent), depending on the exact definition of water scarcity applied. This is well within the 'low risk' category, which ranges from 0 to 25 per cent, and it means that only general management problems occur (see Section 7.6). Malawi is the only area where the 'water competition level' exceeds 600 persons per 1 million m^3/yr, a criterion often used as a first sign of 'water stress' (see Section 2.4). The relatively low water availability found for Malawi equates with the fact that this is the region with the highest population density.

Relatively low values for water scarcity are found for the basin areas of Namibia, Zimbabwe and Mozambique due to the fact that these areas have a comparatively high potential water supply from external sources. Namibia can draw on the

Zambezi river flow at the north-eastern border of the Caprivi Strip. Zimbabwe can draw on the Zambezi river flow from the Victoria Falls to a few kilometres upstream of Cahora Bassa. Mozambique can use the entire downstream Zambezi river flow. If both internal and external water sources are considered Zimbabwe has a water scarcity of 0.5 to 2 per cent, but if only internal sources were looked at current scarcity would be 3-6 per cent. For Mozambique, a scarcity of 0.2-0.5 per cent is found if both internal and external sources are considered, but 1-9 per cent if only internal sources were taken into account. At present there is no reason not to count on the external water sources, as upstream water withdrawals do not significantly affect downstream river runoff anywhere in the basin. As mentioned above, total water use in the basin has reduced total runoff by only about 1 per cent.

The above results suggest that the present situation can be characterized as water abundance, with a high potential for further water resources development. Many regional studies offer a similar picture and propose ambitious development strategies, particularly with regard to irrigation and hydropower development. A few observations should be made. First, the above results refer to an *average* climatic year. In a relatively dry year, as in the recent hydrological years 1991-92 or 1994-95, water scarcity is much greater than suggested by the average figures. In general, the basin is more sensitive to dry than to wet years. This is illustrated in Figure 8.13, where the results are shown of a series of model experiments with relatively dry and wet years. The most vulnerable areas appear to be the upstream areas without external water sources, such as the basin areas of Angola, Zambia and Malawi. Downstream areas such as the basin area of Mozambique are less vulnerable, due to the presence of the Zambezi river flow, which is less vulnerable to drought than its tributaries, mainly because of the flow regulation by Lake Kariba and Cahora Bassa. The basin appears to be most sensitive to increased water scarcity during droughts of medium intensity (with annual precipitation of about 85-90 per cent of the average value). During more severe droughts water scarcity continues to increase, but to a lesser extent, because the basin's water stores still deliver water. If drought conditions were to persist this effect would diminish.[1] Furthermore, as can be seen from Figure 8.13, the lag in delivery does not apply to Malawi, due to the fact that the Lake Malawi basin does not generate any runoff below a certain lake level (see also Calder *et al.*, 1995).

---

[1] In the drought experiments discussed here, a drought period of five years has been assumed.

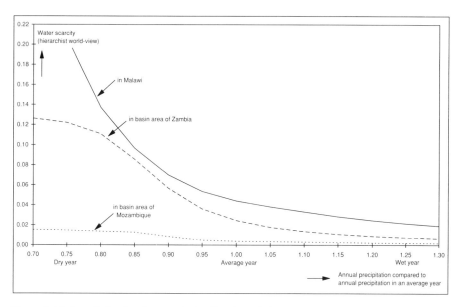

Figure 8.13. Sensitivity of water scarcity to climatic variation.

A second rider to the optimistic picture is that water scarcity has been considered here in physical terms. In reality, water may be abundant in a physical sense, but people's actual access to water may be inadequate, due to socio-economic reasons, a lack of proper management or political obstacles. In Angola and Mozambique for instance, many irrigation schemes are probably not in operation as a result of the recent civil wars in these countries. But it is likely that in other regions of the basin too a significant part of the water supply systems is not properly operated and maintained, resulting in water shortages under water abundance conditions. A third critical observation is that the low level of present water use does not mean that there are no significant environmental consequences. SARDC *et al.* (1994) and SADC *et al.* (1996) observe for instance an increasing pressure on wetland habitats, a significant loss of water by evaporation from artificial lakes, and siltation of dams. According to the model, evaporation from Lake Kariba is about 7×10^{12} kg/yr (=1300 mm/yr[1]) and from Cahora Bassa 4×10^{12} kg/yr (=1600 mm/yr). Together, this amounts to approximately twelve times current consumptive water use in the entire

---

[1] This is smaller, but in the same order of magnitude as the value of 1610 mm/yr given by Calder *et al.* (1995). SADCC and UNEP (1991) give a value of 4660 mm/yr, which seems to be very unrealistic.

basin. A fourth shadow over the optimistic picture is that the current rate of water use is expected to increase rapidly. At present there is not much reason for conflict among different users or countries, but this situation is likely to change in the coming decades. This will be the subject of the next chapter.

9 A water assessment for the Zambezi basin

Several studies with regard to water resources development in the Zambezi basin, both national and basin-wide, are currently being carried out. Some of these studies aim to develop short-term objectives, while others are intended to support planning for the next five to ten years. However, despite the variety of studies, a long-term and integrated view on development is lacking. In this chapter, I explore some possible long-term futures for the Zambezi basin, contemplating the future up to the year 2050. The AQUA Zambezi Model is used as a supportive tool. The international character of the basin gives an extra dimension to the kind of problems which can be expected. Three utopias and a number of dystopias are considered. The chapter concludes with an overview of the main risks and a proposal for policy priorities to reduce these risks.

9.1 Introduction

Several authors have stated that water resources development in the upstream parts of the Zambezi basin will reduce the possibilities for downstream development to a considerable extent, and that the increasing demand for water throughout the basin will certainly affect its current ecological functioning. It is recognized that most of the problems which might be expected will not emerge immediately as the result of one particular project, but rather that they will develop as the long-term net result of a combination of activities.[1] However, most water studies of the region are intended to explore the feasibility of short-term objectives, such as meeting the increased water demand of a particular city or constructing a new dam or hydropower plant to stimulate economic development. This type of study is always linked to one specific development project. A smaller number of studies aims to support water resources planning for the next five to ten years. Two examples of such medium-term studies are the *Regional Irrigation Development Strategy* by SADCC (1992) and the *Sub Saharan Africa Hydrological Assessment* by the World Bank and UNDP (1990). These studies do not mention the development period they examine in any explicit way, but if one reads them carefully it is obvious that they are useful for planning over a timespan of at most a decade. They lack projections for population or economic growth and do not consider mechanisms which become visible only over a longer period, such as for instance the effects of increasing prosperity and urbanization on domestic water demand, the possible long-term effects of changes in

---

[1] See for instance: Howard (1994), Matiza *et al.* (1995), SARDC *et al.* (1994) and SADC *et al.* (1996).

the water tariff system, and the effects of attempts to increase water-use efficiency. As a result, several fundamental issues remain unexplored, for example water pricing, technological development, water use education, and trade-offs between different types of water use. There are many individual plans for new irrigation schemes and hydropower plants, and there is even a plan for large-scale water transport from the Zambezi basin to South Africa, but it is open to question whether these plans can go ahead collectively. The failure of medium-term studies is that they do not consider long-term goals or possibilities, on a scale of several decades.

Since 1987 the Southern African Development Community (SADC) has run a programme called ZACPLAN (Zambezi Action Plan). The aim of this programme is to stimulate mutual co-operation among the Zambezi basin states and to achieve coherent and environmentally-sound management of the basin's water. The programme includes nineteen projects, four of which have now been (partly) funded. These are the listing and evaluation of completed, ongoing and planned projects and the initiation of a basin-wide exchange of information (ZACPRO 1), the development of legislation for common management of the basin (ZACPRO 2), the development of a basin-wide water quality and quantity monitoring system (ZACPRO 5), and the development of an integrated water resources management plan (ZACPRO 6). Despite all efforts undertaken, progress in the Zambezi Action Plan is behind original expectations, partly because of lack of funding, but also due to failures within the funded projects (SADC-ELMS, 1994). As mentioned in the introduction to the previous chapter, a rapid integrated water assessment project was established in 1994 as a joint effort of SADC and UNEP. This rapid assessment was intended to provide a focus for the ZACPLAN activities, by signalling the most urgent long-term future water supply issues in the basin. In this way, SADC could be in a more informed position to set priorities for the ZACPLAN-studies yet to be funded. The results of the rapid assessment study are reported in Bannink (1998). During the assessment study two workshops with regional experts were organized, one in January 1995 in Lilongwe, Malawi, and one in November 1996 in Harare, Zimbabwe. The aim of the first workshop was to identify and analyse the most important water policy issues in the Zambezi basin and to gather basic information on these issues (UNEP, 1995b). The second workshop aimed to explore possible water futures and develop promising water policy strategies for the region as a whole (UNEP, 1996). As I was involved in organizing these workshops and carrying out the rapid assessment, I decided to use this work as an illustration of the concept of 'integrated water assessment' at river-basin level within the context of this thesis. The results of this chapter differ from the results reported in Bannink (1998), because I have advanced a few steps. First, as stated in the previous chapter, I have made some improvements to the AQUA Zambezi Model since the workshop in Harare. Second, I use different context scenarios for population and economic growth to those in the rapid assessment study, following suggestions made at the Harare workshop.[1] Third, possible futures according to different perspectives are described, while the rapid assessment study was based on just one perspective (the hierarchist). Finally, I use a time horizon of 2050 instead of 2025.[2]

This chapter is composed as follows. In the next section, I give a short description of the Zambezi basin, considering its present state as well as current plans for development. In the third section, three water utopias are presented, according to the hierarchist, egalitarian and individualist perspectives. As described in Chapter 5, a utopia is characterized by a coherent set of assumptions with regard to exogenous developments, world-view and water management style. The fourth section discusses a number of different dystopian futures, which will evolve if a fatalist management style is applied, if there are restrictions in respect of the level of expenditure in the water sector or if the hierarchist or egalitarian utopia is confronted with high population growth and rapid economic development. These dystopias give some insight into the risks associated with each of the utopias. The effect droughts can have in each utopia is also shown. The fifth section reflects on the water policy priorities proposed by the participants of the workshop in Harare, discussing the kinds of risk which will emerge if these policies are put into practice. The last section of the chapter provides some suggestions for policy priorities which may reduce the various types of risk.

9.2 The Zambezi basin: current issues and prospects

The catchment area of the Zambezi lies in the tropics, between 9 and 20 degrees south of the equator, and encompasses humid, semi-arid and arid regions. The rainy season is from November to April, in the southern summer. Annual rainfall varies between 600 mm/yr in the southern part of the basin and 1200 mm/yr in the

---

[1] The main difference is that low economic growth is now primarily related to low population growth and high economic growth to high population growth (see Section 9.3). In the rapid assessment study, we used low economic with high population growth and high economic with low population growth, which seemed unrealistic to several participants in the Harare workshop.

[2] In the rapid assessment study the time horizon of 2025 was chosen because developments beyond that year were considered 'too uncertain'. However, in an integrated assessment, the presence of great uncertainties beyond a certain range is not a valid reason for limiting the scope to what is relatively better known (see Chapter 3). Instead, it is better to widen the scope and use an appropriate methodology to account for the uncertainties. For this purpose I use the perspective approach (Chapter 5).

northern part. The lowest mean monthly temperature ranges from 16 to 18 °C and the highest monthly temperature from 22 to 24 °C (Leemans and Cramer, 1991). The Zambezi river rises at an altitude of 1500 to 1600 metres above sea level on the Central African Plateau in the north-western part of Zambia, near the Kalene Hills, close to the borders with Angola and Zaire (see Figure 8.1). From its source, it flows into Angola and then, just before the Chavuma Falls, back into Zambia. After the Chavuma Falls, at an altitude of about 1100 metres, the Zambezi slowly descends in a southerly direction, passes through the Barotse Flood Plain, and then turns eastwards. For about 110 kilometres the river forms the border between Zambia and Namibia, and then, for a much greater distance, between Zambia and Zimbabwe. From the Victoria Falls, where the Zambezi falls about a hundred metres[1], the river regime changes dramatically. The river rapidly descends through the Batoka Gorge and the Devil's Gorge, to enter Lake Kariba, an artificial reservoir primarily created to generate hydropower. From there the Zambezi river continues to form the border between Zambia and Zimbabwe, until it enters Mozambique and flows into a second man-made reservoir, Lake Cahora Bassa, also used for hydropower generation. From here the river begins its descent from the Central African Plateau to the coastal plain, taking nearly 600 km before finally draining into the Indian Ocean. The delta of the river is characterized by vast marshes and sandbars. The total length of the main river is 2500 to 3000 km.[2]

One of the major tributaries of the Zambezi river is the Cuando-Chobe, which flows into the Zambezi near Kasane, a little upstream of the Victoria Falls, where the borders of Zambia, Namibia, Botswana and Zimbabwe converge. Although the Cuando-Chobe basin forms a considerable part of the Zambezi basin, it barely contributes to the Zambezi river flow. In wet years, part of the water in the Zambezi can even flow into the Chobe, to be lost through evaporation in the Chobe Swamp, the Linyanti Swamp or one of the other swampy areas in that region (see Chapter 8). Real contributions to the Zambezi river runoff come from more downstream tributaries: the Kafue and Luangwa rivers, both draining into the Zambezi between Lake Kariba and Lake Cahora Bassa, and the Shire river, entering the Zambezi about 200 km before it reaches the Indian Ocean.

For a further description of the Zambezi basin, a schematization into eight sub-basins will be used, as shown in the previous chapter. The Upper Zambezi basin, defined as the area upstream of Lukulu, is partly in Angola, partly in Zambia. The

---

[1] Showers (1989): 92 m; Van der Leeden *et. al.* (1990): 108 m.
[2] Howard (1994): 2524 km; Korzun *et al.* (1978) and Czaya (1981): 2660 km; Masundire and Matiza (1995): 3000 km; Van der Leeden *et al.* (1990): 3500 km. After having had a close look on a large-scale map, the last estimate seems very implausible to me.

> **The Ku-omboka ceremony**
> Every year towards the end of the rainy season, when the water in the Barotse Flood Plain rises, the Lozi people ceremonially move to higher grounds. Ku-omboka means 'to get out of the water onto dry ground'. When the chief of the village decides that it is time to leave, which might be any time from February to May, the drums sound the signal to the people. They pack their belongings into canoes and the whole tribe departs, the chief and his family in their barge with a troop of traditionally dressed paddlers in the lead. It takes about six hours to cover the distance between the dry season capital Lealui and the wet season capital Limulunga. In the latter village, a successful move is celebrated with traditional singing and dancing. This ceremony dates back more than 300 years, when the Lozi people broke away from the great Lunda Empire to settle in the upper regions of the Zambezi. The vast plains with abundant fish were ideal for settlement, but the annual floods could not be controlled, so every year the people move to higher ground until the floods subside. Source: internet site http://www.africa-insites.com/zambia.

basin is sparsely populated and there are no major cities or industrial activities. At present, neither Angola nor Zambia have plans for large-scale irrigation development in the Upper Zambezi basin (World bank and UNDP, 1990b, 1990d).

The Barotse basin, the area upstream of the Chobe confluence, receives the runoff from the Upper Zambezi basin and is largely situated in the Western Province of Zambia, with a small part in Angola. The area is thinly populated by shepherds, farmers and fishermen. The annual flooding in the Zambezi valley has given rise to the so-called Ku-omboka Ceremony, the sign for thousands of inhabitants to move to higher ground when the Zambezi water overflows onto the low lying plains (see text box). At present there is little irrigation in the Barotse basin, nor are there plans for large-scale irrigation in the near future (World Bank and UNDP, 1990b, 1990d). However, it has been recognized that it could be profitable to withdraw water from the Zambezi river near Katima Mulilo, at the lower end of the Barotse basin, for export to South Africa to supply the water needs of Johannesburg, Pretoria, and surrounding agricultural areas. Because South Africa is not naturally entitled to use Zambezi water, agreement would have to be reached among the states affected by any reduction in flow due to this inter-basin water transfer (Basson, 1995). It is difficult to say whether such a transfer will materialize, because the export plan has not passed beyond the conceptual stage.

The Cuando-Chobe basin forms, together with the Upper Zambezi and Barotse basins, the least densely populated area of the Zambezi basin. Here too, there is little irrigation or industrial activity. However, both Angola and Namibia have identified areas in the basin where large-scale irrigation could be developed (World Bank and

UNDP, 1990b; Heyns, 1995). The schemes in Angola would use water from the downstream Cuando; the Namibian schemes would rely on water from the Zambezi.

The largest part of the Middle Zambezi basin is in Zimbabwe; a smaller part lies within Zambia. The Zambezi river and Lake Kariba form the border between the two countries. The Kariba dam was build in the late 1950s, when both Zambia and Zimbabwe were still under British colonial rule. When the dam was completed in 1960, it was the largest ever built. More than fifty thousand people were displaced; some animals were transported to other areas, but many were killed by the rising water. A hydropower generation capacity of 1266 MW was installed, partly on Zambia's and partly on Zimbabwe's side of the lake. Nowadays, Lake Kariba not only provides hydroelectricity, but also supports a thriving commercial fishing industry.

The Kafue basin has an area of 158×10^3 km^2 and is entirely in Zambia, covering about 20 per cent of Zambia's total area. The average annual outflow from the basin is only 6 per cent of the mean annual rainfall, due to high open water losses in areas with impeded drainage and high evaporation losses in groundwater seepage zones (Burke *et al.*, 1994). All major industrial towns and centres in Zambia are located along the Kafue, which makes the Kafue basin the most important and most utilized river basin in Zambia (Kasimona and Makwaya, 1995). One of the main industries is mining in the Copperbelt area, in the northern, upstream part of the basin. The mining activities, which started in the 1930s, have resulted in widespread deforestation, intensive water use and pollution. The principal development of the Kafue river has been the construction of a dam and power plant at Kafue Gorge, completed in 1971, and the subsequent construction of another dam at Itezhitezhi, completed in 1978. At present there is relatively little irrigation in the Kafue basin, but there is great potential for further development. According to Kasimona and Makwaya (1995), the limiting factor will be shortage of water, not of suitable land. From an ecological point of view, the most interesting areas in the basin are probably the Kafue Flats, a swampy area upstream of the Kafue reservoir, and the Lukanga and Busanga Swamps upstream of Itezhitezhi.

The Luangwa basin is about the same size as the Kafue basin and is also in Zambia. The population density is however less than in the Kafue basin. The Luangwa river flows into the Zambezi about 170 kilometres downstream of the Kafue confluence. Dry season flows are so low that they are already needed in their entirety for dilution of wastewater. According to the World Bank and UNDP (1990d), further development of the basin's resources would require storage provisions. However, the Luangwa, with high sediment loads, has limited potential for damming (Kasimona and Makwaya, 1995).

Approximately 70 per cent of the Lake Malawi - Shire basin is in Malawi. Smaller parts of the basin lie in Tanzania and Mozambique. About one fifth of the basin area is formed by Lake Malawi (formerly Lake Nyasa). The Shire river is the only outlet of Lake Malawi and drains into the Zambezi about 200 kilometres before it reaches the Indian Ocean. The land area is the most densely populated in the Zambezi basin, with about 75 people per square kilometre. There is relatively little industrial activity and most people depend on agriculture. Malawi has an irrigated area of about 200 km^2, while a further 1000 km^2 has been identified for potential irrigation development (World Bank and UNDP, 1990g). However, it is recognized that the actual development of this potential would reduce the hydropower generation capacity of the Shire river considerably. The basin is very sensitive to droughts. If a succession of relatively dry years occurs the water level in Lake Malawi can fall so much that there is no outflow at all. In the period 1915-1937 for instance, the outflow from the lake ceased completely.

The Lower Zambezi basin receives water from the more upstream sub-basins. For Mozambique, which forms the main part of the Lower Zambezi basin, this means that there is a large potential supply from external water sources compared to the potential supply from internal sources. As a result the area is less sensitive to droughts than for instance the Lake Malawi - Shire basin. At present only a small fraction of the available surface water resources are being used. According to the World Bank and UNDP (1990h), Mozambique has an irrigated area in the Lower Zambezi basin of approximately 100 km^2, to be extended in the near future to about 2400 km^2. Runoff from the Lower Zambezi basin is greatly influenced by Lake Cahora Bassa. The Cahora Bassa dam, constructed in the 1970s, was primarily intended to produce hydroelectric power for export to South Africa. This export of electricity was never fully realized, however, due to the destruction of power lines during the civil war in Mozambique. Today plans exist to increase the installed hydropower generation capacity at Lake Cahora Bassa from the present 2075 MW to 2625 MW (Dale, 1995). World Bank and UNDP (1990h) cite a study showing that the installed capacity at Lake Cahora Bassa could even be increased up to 3600 MW. According to this study, the Zambezi river offers an additional potential for hydropower development of 3700 MW between Lake Cahora Bassa and the outflow into the Indian Ocean.

From the above it is clear that there will be an increasing number of claims on the water within the Zambezi basin, particularly for irrigation and hydropower, but domestic and industrial water demand are also expected to rise and a plan for export of water to South Africa is still being considered. It is unclear to what extent different claims will impede each other, either directly or through increasing risks.

This is the subject of the next sections, which analyse what kind of futures will evolve under different assumptions and development strategies.

9.3 Three water utopias

To get a first impression of possible changes within the Zambezi basin, Table 9.1 shows some recent growth rates in population, gross national product, total cropland area and irrigated area. It appears that the population in the Zambezi basin has increased about 1.5 times faster than gross basin product, which implies that the average income per capita has decreased considerably. The total cropland area has grown at a rate of about 10 per cent of population growth, while irrigated cropland increased by nearly as much as the population. In the past fifteen years the population in the Zambezi basin grew by 60 to 70 per cent and the total irrigated cropland area by 50 to 60 per cent.

For exploring possible futures in respect of water, three different 'contexts' have been formulated: a hierarchist, an egalitarian and an individualist context.[1] The hierarchist context is largely an extrapolation of the recent trends shown in Table 9.1. The egalitarian context is characterized by more modest growth rates. The individualist context represents a future of rapid economic growth. For each context a water utopia has been formulated. The hierarchist utopia will evolve within the hierarchist context, assuming that the world functions according to the hierarchist world-view and that a hierarchist management style is adopted. The egalitarian and individualist utopias have been formulated in a similar way. For a detailed description and explanation of the different world-views and management styles the reader is referred to Chapter 5, but a concise overview is given in Table 9.2.

The hierarchist water utopia
An important assumption behind the hierarchist water utopia is that the economies of the Zambezi countries will show moderate growth during the 21st century, slightly higher than during the past fifteen years. The population will continue to increase, but growth rates will decline according to the medium population scenario of the United Nations. Average gross basin product per capita will grow steadily, from about 340 US$/yr in 1990 to 450 US$/yr in 2050 (Figure 9.1). The annual growth rates of total and irrigated cropland are assumed to equal the average growth rates of the past fifteen years, i.e. 0.4 and 3 per cent respectively. In accordance with

---

[1] A 'context' refers to exogenous developments which are not part of this study and for which assumptions have to be made.

Table 9.1. Average annual growth rates (%).

	Ang.	Nam.	Bots.	Zamb.	Zimb.	Tanz.	Mal.	Moz.	Zambezi basin[3]
Population[1]									
1980-1994	3.1	2.7	3.4	3.4	3.2	3.2	4.1	1.8	3.4
Gross national product[2]									
1980-1991	n.a.	1.6	9.3	0.7	3.6	2.0	3.5	-1.1	2.3
Total cropland area[1]									
1980-1994	0.21	0.05	0.35	0.23	0.83	1.5	1.8	0.23	0.36
Irrigated cropland area[1]									
1980-1994	0.0	2.9	0.0	6.5	2.8	1.6	3.2	3.6	3.0

[1] Calculated on the basis of data from FAO (1996).
[2] WRI (1994).
[3] Data for the Zambezi basin have been calculated on the basis of the national data, weighted according to the relative contribution of a nation to the total population (gross national product, total cropland area or irrigated cropland area) in the Zambezi basin.

current plans, a new dam and hydropower plant will be build at Batoka Gorge (with an installed capacity of 1600 MW) and the present hydropower generation capacity at Cahora Bassa will be extended from 2075 to 2625 MW. Starting in the year 2015, water from the Zambezi river will be exported to South Africa. The diverting point will be at Katima Mulilo in Namibia and the volume of export will be 1.5×10^{12} kg/yr, half of what can be withdrawn without having to provide storage in the Zambezi river (SARDC *et al.*, 1994; Basson, 1995).

Under these conditions, total water supply in the Zambezi basin will grow by a factor of about 7.5 in the period 1990-2050 (Figure 9.2). In 2050, water export will constitute 11 per cent of the total water withdrawal in the basin and 21 per cent of the consumptive water use. The irrigation sector will remain the largest water user, both in terms of total withdrawal and in terms of consumptive water use. Water supply for irrigation will increase slightly less than the area of irrigated land, because the average water application factor per hectare will decrease by about 6 per cent, due to the introduction of more efficient irrigation techniques. The domestic sector will remain the second largest water user. The relative growth of water use in this sector will be greater than in the irrigation, livestock and industrial sectors, because not only will the population grow, but so will average water demand per capita, as a result of increasing prosperity, urbanization and public water supply coverage. The relative importance of the different factors can be seen in Figure 9.3.

Table 9.2. Major characteristics of the three water utopias in the Zambezi basin

	Hierarchist utopia	Egalitarian utopia	Individualist utopia
Socio-economic context			
Gross national product[1]	Growth rate 3.0% yr^{-1}	Growth rate 2.0% yr^{-1}	Growth rate 4.0% yr^{-1}
Population[2]	UN medium scenario	UN low scenario	UN high scenario
Total cropland area[1]	Growth rate 0.4% yr^{-1}	Growth rate 0.3% yr^{-1}	Growth rate 0.6% yr^{-1}
Irrigated cropland area[1]	Growth rate 3.0% yr^{-1}	Growth rate 2.0% yr^{-1}	Growth rate 4.5% yr^{-1}
New hydropower plants	Extension Cahora Bassa, Batoka	No new large dams	Extension Cahora Bassa, Batoka
World-view			
Measure of potential water supply	Stable runoff	Stable runoff	Total runoff
Measure of water scarcity	Consumptive water use / stable runoff	Total water supply / stable runoff	Consumptive water use / total runoff
Growth elasticities water demand[3]	Medium	Low	High
Price elasticities water demand[3]	Zero	Low	High
Fractions consumptive water use[3]	Medium	High	Low
Water supply costs[4]	Increase moderately	Increase rapidly	Increase slowly
Management style			
Water export to South Africa	1.5×10^{12} kg/yr, starting in 2015	No export	3×10^{12} kg/yr, starting in 2015
Technological diffusion rate[5]	High	Low	Zero
Technological development rate[5]	High	Low	Zero
Percentage water price / actual cost	Growing towards 75% in 2025	Growing towards 110% in 2025	Growing towards 100% in 2025
Public water supply coverage	Growing towards 100% in 2050	Growing towards 100% in 2025	Depending on economic growth
Sanitation coverage	Growing towards 100% in 2050	Growing towards 100% in 2025	Depending on economic growth
Fraction of wastewater treated	Growing towards 100% in 2050	Growing towards 100% in 2050	Depending on economic growth

[1] Growth rates have been assumed equal for all countries within the basin and for the entire simulation period. In reality, it is likely that growth rates will fluctuate over the years and differ in the countries. However, it is assumed here that the average growth rates in the long term will be distributed quite uniformly over the basin states.

[2] The UN scenarios provide country specific data up to 2025 (UN, 1993); for the period 2025-2050, the 2025 growth rates have been assumed.

[3] The exact figures correspond to the values used in the AQUA World Model (see Section 6.3).

[4] Cost curves differ per water-use sector, quality of the intake water and region. Figure 8.6 gives an example of cost curves for public water supply and water quality class A in the part of the Zambezi basin in Zimbabwe.

[5] These parameters refer to (non-price driven) improvements in water-use efficiency. The exact figures correspond to the values used in the AQUA World Model (see Section 6.3).

The most dominant factor after population growth appears to be urbanization, resulting in an increase in domestic water use of about 40 per cent.

Within the hierarchist world-view, water scarcity is defined as the ratio of consumptive water use to potential water supply. The latter is assumed to be equal to stable runoff. Applying these definitions, water scarcity in the Zambezi basin will grow from 2 per cent in 1990 to 15 per cent in 2050 (Figure 9.4). Water scarcity will be highest in Malawi, largely because of the high population density, which will reach 440 people per km^2 of land in 2050, greater than the present densities in countries such as India or Japan and nearly as high as the current density in the Netherlands (WRI, 1994). Water scarcity in Malawi will grow from 4.4 per cent in 1990 to 27 per cent in 2050 (Figure 9.5). Although the Zambezi basin as a whole will remain within the 'low risk' zone, the Malawian part will enter the 'moderate risk' zone.[1] This means that serious water supply problems could occur in several parts of the country, probably mainly in urban areas such as Lilongwe, Blantyre, Mzuzu and Zomba. From a hierarchist point of view, the ultimate solution to Malawi's water scarcity problems of the future is to rely on water supply from Lake Malawi. After Malawi, greatest water scarcity will occur in the parts of the Zambezi basin in Zambia, Zimbabwe and Mozambique. As an example, Figure 9.5 shows total water supply, consumptive water use and potential water supply in Zimbabwe's section of the basin. There are two important reasons why water scarcity in Zimbabwe's part of the basin will remain much lower than in Malawi. The first is that the population density will continue to be lower, with about 110 people per km^2 in 2050, the second is the availability of external water sources. As discussed in the previous chapter, the external water sources of Zimbabwe are formed by the runoff from the Barotse and Kafue basins. However, it has to be realized that, as can be seen in the figure, the potential supply from external sources will decrease as a consequence of increased water use upstream, the most severe effect coming from the withdrawal of water at Katima Mulilo for export to South Africa.

In the hierarchist utopia water supply costs in the Zambezi basin will grow by a factor of nearly 2 (Figure 9.6). The average price of water will increase much more, because the fraction of the total costs charged to the consumer will go up from about 25 to 75 per cent. The increase in costs is the net result of greater water scarcity and improved water quality. Expenditure in the water sector, expressed as a fraction of gross basin product, is shown in Figure 9.7. Despite large investment in wastewater treatment, resulting in extended treatment coverage, water quality will decrease

---

[1] Following the approach in Chapter 7, the designation 'low risk' is used if water scarcity ranges between 0 and 25 per cent, 'moderate risk' if scarcity ranges between 25 and 50 per cent and 'high risk' if water scarcity exceeds 50 per cent.

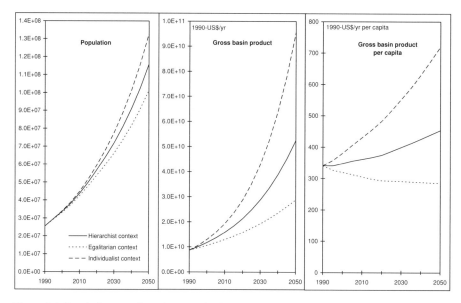

Figure 9.1. Population growth and economic development in the Zambezi basin.

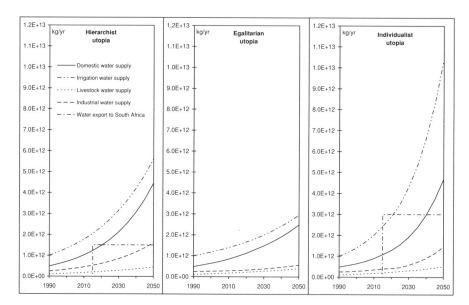

Figure 9.2. Sector water supplies in the Zambezi basin in each of the three utopias. Water export from the Zambezi basin to South Africa is regarded as a separate sector.

slightly during the first quarter of the 21st century, due to the increase in total wastewater production. However, during the second quarter of the century the effect of growing treatment coverage will become greater than the effect of increased wastewater production, resulting in an improvement in water quality. This is shown for the Lake Malawi - Shire basin in Figure 9.8.

The hierarchist utopia can best be characterized as a world balanced between the desirable (high growth) and the possible (limited availability of resources). The demand for water will increase rapidly, as in the individualist utopia, but water availability is clearly limited, as it is in the egalitarian utopia. As a result, water will become scarcer in the hierarchist utopia than in the other two utopias. High-tech infrastructure is needed to supply the water requirements of each sector of society. In the year 2050, urban water supply in Bulawayo, Zimbabwe, will for instance depend largely on water from Lake Kariba, about 350 kilometres to the north of the city. Lake Kariba might in fact be regarded as the ultimate source of water for a large part of Zimbabwe (see also SARDC *et al.*, 1994). The same will apply to Lake Malawi for Malawi.

The egalitarian water utopia
The egalitarian utopia is a future with a relatively slow increase in population, accompanied by relatively low economic growth. The most basic difference with the hierarchist utopia is that egalitarians prefer not to balance on the edge of the maximum possible, but rather to stay on a comfortable level below this maximum. This means that priority is given to water demand policy over water supply policy. Instead of building new large dams, governments will stimulate more efficient water-use. With regard to water supply, the main concern in the egalitarian utopia is to increase the number of people with proper water supply and sanitation, because an estimated 10 million people in the Zambezi basin presently lack access to such facilities, i.e. about 40 per cent of the entire population of the basin. Improving water supply and sanitation conditions is expected to raise living standards and reduce the number of people affected by waterborne diseases. In the egalitarian utopia, high investment in public water supply and sanitation aims to attain full coverage by the year 2025.

Total water supply in the Zambezi basin will grow much less than in the hierarchist utopia (Figure 9.2), partly because of the absence of water export to South Africa. Contrary to the hierarchist utopia, people generally obtain their water from nearby sources, which is possible because water demand is much smaller and a relatively high level of water self-sufficiency can be maintained. Within the egalitarian world-view, water scarcity is defined as the ratio of total to potential water

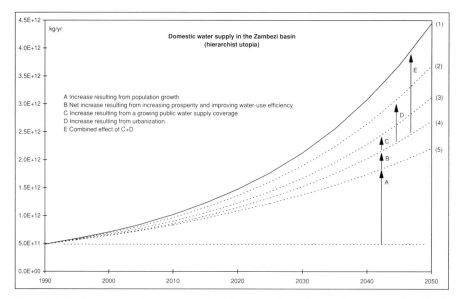

Figure 9.3. Driving forces behind the growth of the domestic water supply in the Zambezi basin. (1) Hierarchist utopia; (2) Hierarchist utopia without increasing public water supply coverage; (3) Hierarchist utopia without urbanization; (4) Hierarchist utopia without urbanization and without increasing public water supply coverage; (5) Hierarchist utopia with constant demand per capita.

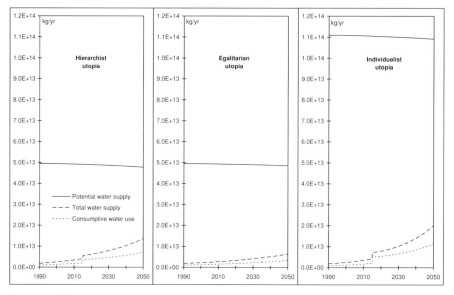

Figure 9.4. Total water supply and consumptive water use compared to potential water supply in the Zambezi basin, in each of the three utopias. Water export to South Africa in the hierarchist and individualist utopias starts in the year 2015, which induces a sudden increase in total water supply and consumptive water use in the basin.

supply, the latter being equal to stable runoff. Using this definition, water scarcity in the Zambezi basin will grow from 4 per cent in 1990 to 13 per cent in 2050 (Figure 9.4). As a result of the increasing scarcity, water supply costs will increase, but less so than in the hierarchist utopia (Figure 9.6). However, water prices will increase much more than in the hierarchist utopia, due to the strong pricing policy introduced, including a water tax of 10 per cent of actual costs. This policy is partly responsible also for the only modest increase in total water demand. Expenditure in the water sector will be much less than in the hierarchist utopia, even if expressed as a fraction of gross basin product (Figure 9.7).

The individualist water utopia
The individualist utopia is a future of rapid growth. An annual economic growth of 4 per cent is assumed for the entire Zambezi basin and for the whole period under consideration. This scenario presumes a strong revival of the national economies, especially the war-stricken economies of Mozambique and Angola. Population is supposed to grow according to the high population scenario of the United Nations. In the period 1990-2050, gross basin product per capita more than doubles, resulting in improved water supply and sanitation conditions. Due to the favourable economic conditions, there is room for high investment in advanced water supply infrastructure, water re-use techniques and wastewater treatment, all more than in the hierarchist and egalitarian utopias. However, the need for high investment in the individualist utopia is also greater than in the other utopias, as a result of relatively high water demand and wastewater production. Especially water supply for irrigation will be much more extensive than in the other two utopias (Figure 9.2). Furthermore, it is assumed that the SADC countries will achieve agreement on exporting water from the Zambezi river to South Africa. As in the hierarchist utopia, the water will be diverted at Katima Mulilo, starting in 2015, but the volume of export is asssumed to be 3×10^{12} kg/yr, twice as large as in the hierarchist utopia. This is the quantity which can be diverted without having to provide storage in the Zambezi river.

Within the individualist world-view water scarcity is defined as the ratio of consumptive water use to potential water supply. Potential water supply is assumed to equal total runoff. Using these definitions, water scarcity in the Zambezi basin will grow from 1 per cent in 1990 to 10 per cent in 2050 (Figure 9.4). The costs of water supply will grow at the same rate as in the hierarchist utopia (Figure 9.6). Water quality improvements will suppress the growth in water supply costs in the period 2030-2050 to some extent, due to decreasing costs of water purification. In the period 2000-2025 a market-pricing policy will be introduced, which means that

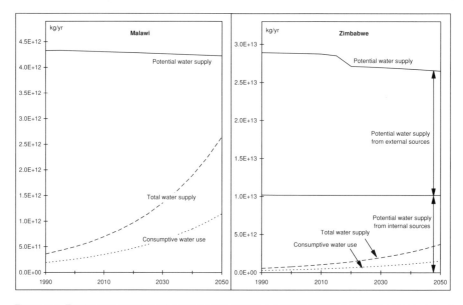

Figure 9.5. Total water supply and consumptive water use compared to potential water supply in two specific regions of the Zambezi basin, in the hierarchist utopia. In the case of Zimbabwe, potential water supply consists of an internal and an external component. In the case of Malawi, potential water supply depends entirely on internal sources.

water tariffs will increase towards 100 per cent of actual costs, including both depreciation costs and operational and maintenance costs. In absolute terms, expenditure in the water sector in the individualist utopia will be much larger than in the hierarchist or egalitarian utopias, but they will be lower if expressed as a fraction of gross basin product (Figure 9.7).

The effects of increasing water use throughout the basin will be most noticeable in the downstream parts of the basin, where all effects accumulate. Figure 9.9 shows how upstream water consumption affects river runoff from the Middle and Lower Zambezi basins. Because the outflow from Lake Kariba is regulated by man, minimum river runoff from the Middle Zambezi basin will not change, but the effects will become visible in the size of peak flows. In the case of the Lower Zambezi basin, however, there will also be significant effects on the minimum river runoff (a reduction of 12 per cent in the period 1990-2050). The individual effect of water export is shown by presenting the resulting hydrographs if there were no water export. Increased water consumption in the Zambezi basin will influence hydropower generation at both the existing hydropower plants and the plants yet to be constructed. The annual outflow from Lake Kariba in the year 2050 will be 17 per cent lower than the current outflow. If there is no spillage, as occurred in the

A water assessment for the Zambezi basin 275

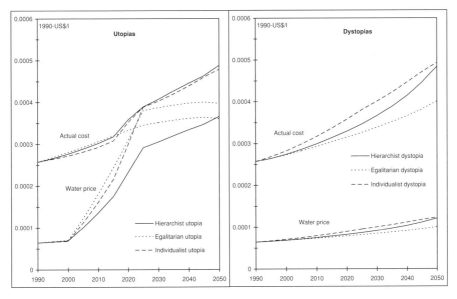

Figure 9.6. Average water costs and prices in the Zambezi basin. In all possible futures, water costs will increase as a result of increasing water scarcity. In the hierarchist and individualist utopias the effect of water export can be seen in the sharp increase in costs in the year 2015. In the three utopias, water prices will increase most strongly during the first quarter of the 21st century, due to active pricing policies. In the three dystopias, costs not only rise as a result of increased scarcity but also as a result of decreased water quality. In the dystopias active pricing policies are lacking, which keeps prices low but results in less efficient water use and higher water demand.

1980s (SADC-ELMS and DANIDA, 1994), any reduction in outflow will lead to a comparable reduction in electrical output. In relatively wet years, when spillage is not nil, intelligent operation of the reservoir can diminish the effect of upstream water consumption on hydropower generation to some extent (by reducing the spill flow), but in dry years any reduction in reservoir outflow can be linearly translated into a reduction in electricity generation. The generation potential of the existing hydropower plants at Lake Cahora Bassa will barely if at all be affected because there is an installed capacity of only 2075 MW, which is far below the potential of the lake. According to SADCC (1990), the discharge at maximum electrical output of the existing plants is about 5.9×10^{17} kg/month. As can be seen in Figure 9.9, this flow is available throughout the year, not only now but also in the year 2050. This means that, in an average year, increased water consumption will not affect hydropower generation at the existing plants of Cahora Bassa. Only in dry years and in the case of ineffective operation of the reservoir, might upstream water consumption reduce hydropower generation at the existing plants. However, in the

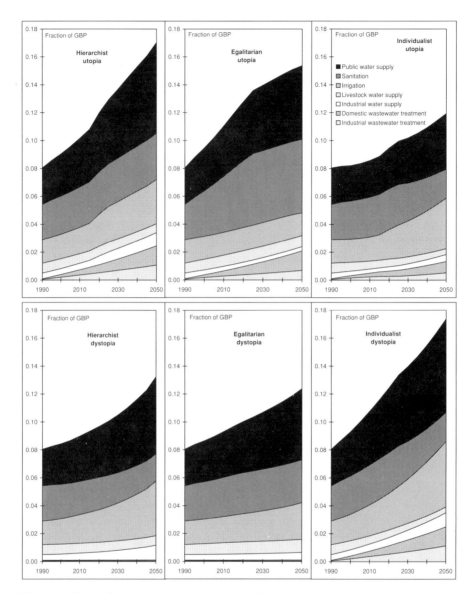

Figure 9.7. Expenditure in the water sector in the Zambezi basin, expressed as a fraction of gross basin product. In the hierarchist and egalitarian dystopias, total expenditure in the water sector is lower than in the respective utopias, due to the lack of investment in public water supply, sanitation and wastewater treatment. Nevertheless, expenditure still grows, mainly because of inefficient water use and the increasing costs of water purification. In the individualist case, the same is true, but to such an extent that total expenditure in the dystopia exceeds expenditure in the utopia.

individualist utopia the total installed hydropower generation capacity at Cahora Bassa will be extended by another 550 MW, requiring an extra flow of about 1.6×10^{12} kg/month, which is not available in every month. As a result of increased water consumption in the period 1990-2050, the utilization fraction will be 3.5 per cent less in an average year; the effect will be smaller in wet years and greater in dry years. Another plan to be carried out in the individualist utopia is the construction of a dam and hydropower plant at Batoka Gorge, just downstream of the Victoria Falls. The capacity of this plant will be 1600 MW, to be shared between Zambia and Zimbabwe (Tapfuma, 1995). The Batoka installation will be a run-of-river plant without monthly storage, thus not influencing regional evaporation or river runoff patterns. Complete use of the installed hydropower generation capacity requires a flow of about 1120 m^3/s, which is far from available during dry months. The reason for installing a capacity of 1600 MW is the benefit which can be obtained from conjunctive use of the Batoka and Kariba plants: the Batoka plant can make full use of the high natural flows during the wet months while Kariba can run at reduced capacity and store the incoming water (SADCC, 1990). Due to the instream character of the Batoka plant, a reduction in river runoff as a result of upstream water consumption will also reduce hydropower generation. For this reason hydropower generation at Batoka Gorge in the individualist utopia will in 2050 be about 7 per cent lower than it could be today (utilization fraction of 77 per cent instead of 83 per cent).

9.4 Dystopias and risks

Each of the three water utopias discussed in the previous section can be regarded as a 'best possible future' according to a particular perspective.[1] It has been my intention to show that the current situation in the Zambezi basin can evolve along quite different development paths, none of which can be said to be superior or fundamentally more attractive. Each utopia is preferable from a particular point of view, but none of the utopias can be called more desirable from an objective standpoint. In this section I will discuss several types of future which are not preferable from any particular perspective, but which will emerge if disparate elements from the three utopias are combined or if the fatalist management style is applied. These dystopian futures show what risks are attached to the three utopias.

---

[1] As illustrated in the previous section, these 'best possible futures' are not ideal worlds, in which no trade-offs would take place between different sectors or between upstream and downstream development. Such trade-offs are unavoidable in any possible future.

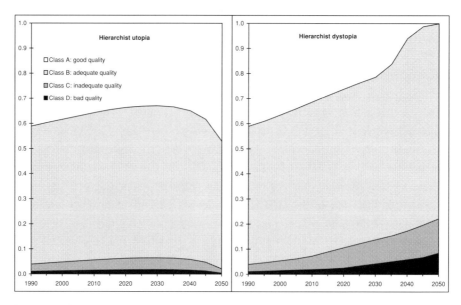

Figure 9.8. Surface water quality in the Lake Malawi - Shire basin in the hierarchist utopia and dystopia. In the dystopia, an increasing volume of untreated wastewater will cause severe pollution problems. Such problems are prevented in the utopia by investing heavily in wastewater treatment, eventually leading to better water quality than today.

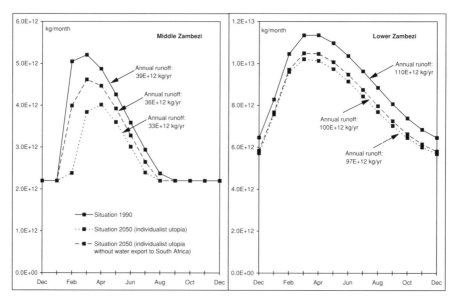

Figure 9.9. Changes in the hydrographs for the Middle and Lower Zambezi basins in the individualist utopia. The effect of water export to South Africa is shown by presenting hydrographs for both the case with export and the case without export.

Below I will first examine what happens in each of the three utopias if the fatalist management style is applied instead of the utopian style. Secondly, I will show what happens in the utopias if there are external constraints on the amount of investment in the water sector, which may be the case if no development aid is provided to support the investment needed or if there is lack of political support within the region. Thirdly I will discuss a type of dystopian future which is worth considering because it is the most catastrophic of all possible futures: the hierarchist or egalitarian utopia confronted by rapid growth. Finally, I will consider the effects of drought in each of the three utopias.

The first risk is that policy does not develop according to the utopian management style. In general, the fatalist management style appears to work least well in all utopias. The principal characteristic of the fatalist management style is that no new water policy measures are implemented and that current practice remains more or less unchanged (see Chapter 5). Applying the fatalist management style in the hierarchist or the egalitarian utopia is most catastrophic in the following fields: public water supply, sanitation, and water quality. In addition, there will be no improvement in water-use efficiency, which means that more water is withdrawn for the same purposes. However, in the hierarchist dystopia total water withdrawal by the year 2050 will be less than in the hierarchist utopia, due to the absence of water export to South Africa. The water scarcity situation in the hierarchist and egalitarian dystopias will not differ greatly from that in the respective utopias. The increase in water costs will be of the same order of magnitude, but prices will be kept low (Figure 9.6). Expenditure for irrigation, livestock water supply and industrial water supply will be higher than in the utopias (Figure 9.7). However, expenditure in public water supply, sanitation and wastewater treatment will be lower, resulting in a sharp decline in public water supply, sanitation and wastewater treatment coverage. Applying the fatalist management style in the individualist utopia has rather different consequences to applying it in the hierarchist or egalitarian utopias. Within the individualist world-view improvements in water supply and sanitation conditions depend on economic growth, and not the other way round as in the hierarchist and egalitarian world-views. As a result, the fatalist management style will not lessen the increase in public water supply and sanitation coverage. Also, water quality will improve as a result of increasing wastewater treatment coverage. The greatest problem in the individualist dystopia is inefficient water use, resulting in a total water withdrawal which is nearly 25 per cent larger than in the individualist utopia and total expenditure for water supply which is nearly 50 per cent higher (Figure 9.7).

A second risk is a lack of investment capacity. In each of the three utopias, total expenditure in the water sector increases considerably, even if expressed as a fraction of gross basin product. It is questionable whether expenditure exceeding 10 per cent of gross basin product is still realistic. In the formulation of the utopias, it was assumed that possible bottlenecks in the financing of future development projects would be solved by external support from donor countries. This is not necessarily unrealistic: development assistance to the Zambezi basin states in 1991 varied between 5 and 70 per cent of the gross national products of these states (Chandiwana and Snellen, 1994). However, what would happen if a constraint is put on investment through either a lack of development aid or a lack of political support within the region? I will only look at constraints on the expenditure for public water supply, sanitation and irrigation, because the other items of expenditure never exceed 1.5 per cent of gross basin product and in most cases are much less (see Figure 9.7). Of the three utopias, the hierarchist is most vulnerable if constraints are applied to the expenditure for public water supply. This can be understood by the fact that public water supply is relatively expensive in the hierarchist utopia: population growth is moderate, but investment capacity is moderate also and improvements in water-use efficiency are relatively small (compared to the egalitarian or individualist utopias). If in the hierarchist utopia expenditure for public water supply in each basin country is limited to 5 per cent of gross national product, the public water supply coverage in the basin would be only 61 per cent in 2050, instead of 100 per cent. The egalitarian utopia is most vulnerable if constraints are put on sanitation expenditure. This is caused by the fact that the egalitarian utopia has an ambitious target in respect of sanitation improvements, combined with a low investment capacity. Application of a '5 per cent of GNP' constraint on sanitation expenditure in all basin countries results in sanitation coverage in the basin of 69 instead of 100 per cent. The individualist utopia is most vulnerable to constraints on irrigation expenditure. This can be explained by the fact that the individualist utopia has by far the most ambitious programme of irrigation development, requiring relatively high investment. If irrigation expenditure in each country is limited to 5 per cent of gross national product, the irrigated cropland area in the Zambezi basin in 2050 will be 22 per cent smaller than without this constraint. The main reduction will be in Mozambique, where *current* irrigation expenditure already exceeds 5 per cent of gross national product.

A third risk to be considered here is the confrontation of the egalitarian or the hierarchist utopia with high population growth and rapid economic development (the individualist context). In both utopias, rapid growth will be disastrous. In the egalitarian utopia total water withdrawal in the Zambezi basin would increase towards a level of 21×10^{12} kg/yr by the year 2050, entirely for use within the basin (there is no water export). This is more than three times the level in the egalitarian

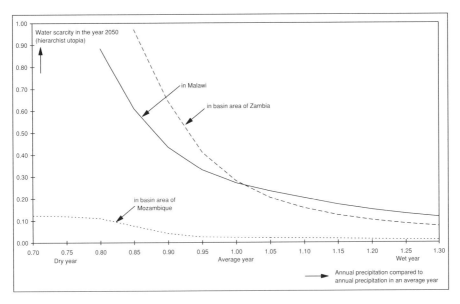

Figure 9.10. Sensitivity of water scarcity to climatic variation in the year 2050.

utopia. Water scarcity in the basin as a whole will then grow to 44 per cent in 2050 (compared to 13 per cent in the utopia) and the costs of water supply will increase to an average of nearly 1 US$ per cubic metre. Expenditure in the water sector will reach nearly a quarter of gross basin product. The situation will be worst in the parts of the basin in Malawi (scarcity at 92 per cent) and Zambia (scarcity at 56 per cent). In the hierarchist utopia rapid growth will have an effect which is even worse than in the egalitarian utopia, due to the lack of a water-pricing system which would discourage inefficient water use.

The last but not the least important risk which will be considered here is that of droughts. Given the growing demand for water, it is clear that the effects of droughts will become increasingly severe. At the end of the previous chapter an analysis was made of the current vulnerability to drought, showing that the regions most in danger are the upstream areas which lack external water sources. This pattern will not change, but the effects of droughts will become much more serious. This is illustrated in Figure 9.10, which shows the sensitivity of water scarcity to climatic variation in three different regions in 2050. The data refer to the hierarchist utopia, which is more affected in this respect than the other utopias. Figure 9.10 has been based on experiments with drought periods of five years. The effect of droughts on water scarcity will be more severe if the drought periods were longer. In the context of global climate change, such a scenario should not be ignored.

Of all risks discussed above, droughts are in a special category, for two reasons. First, droughts will occur in all possible futures. Whereas the fatalist management style, a lack of investment capacity or rapid growth are risks which can possibly be avoided, droughts will certainly have to be dealt with. Second, droughts are not merely a risk in themselves, they also constitute an opportunity for other risks to become manifest. In this sense, droughts are a kind of early warning system. Growing competition between water users can become really fierce during a succession of dry years. If a number of relatively wet years follows, problems may seem to be solved, but they will probably return in a more severe form during the next series of dry years.

9.5 Reflection on the 'Harare priorities'

As mentioned in the introduction to this chapter, a workshop on the Zambezi basin was held in Harare in November 1996 with regional water experts. The aim of the workshop was to explore possible water futures and to develop promising water policy strategies for the region (UNEP, 1996). The participants were confronted with the same kind of analytical results as have been presented in the previous sections.[1] At the end of the workshop, the participants were asked to translate their insights into policy priorities on empty 'priority sheets'. For this purpose they were grouped according to their home country. Table 9.3 presents the policy priorities proposed by the participants. Here I will interpret the results of this exercise and discuss what risks will typically emerge if the policy priorities proposed by the participants of the workshop were to be put into practice.

The most striking result is that all participants were very explicit in giving first priority to water supply policy and second priority to water demand policy. Apart from the fact that the participants disapproved of the idea of water export to South Africa, their approach typically fits within the hierarchist perspective, which might be an indication that the hierarchist view on water is dominant in the basin at present. Although the results of the workshop are far from sufficient to draw a final conclusion on this subject, suppose that the future management of the Zambezi basin will resemble the hierarchist management style, although excluding water export to South Africa. The way in which this type of management will work in the next few decades depends strongly on external factors such as population growth and economic development and on the rules according to which matters within the

---

[1] At the workshop, results were presented which were based on an earlier version of the AQUA Zambezi Model. The terminology of 'utopias' and 'dystopias' was not used at the workshop, but basically the same types of more and less desirable futures were analysed and discussed.

Table 9.3. Policy priorities proposed by the participants of the workshop in Harare (UNEP, 1996).

Policy	Nam.[1]	Zamb.[1,2]	Zimb.[1,2]	Tanz.[1]	Mal.[1]	Moz.[1]
General						
- water demand policy	25	20 → 40	20 → 30	2	+	+
- water supply policy	75	80 → 60	80 → 70	1	++	++
Water demand policy						
- water pricing (removing subsidies)	33	60 → 25	10 → 10	2	++++	++++
- water-use efficiency	33	25 → 40	5 → 10	1	+++	+++
- water education	33	15 → 35	5 → 10	3	++	++
- water export	0	0 → 0	0 → 0	4	0	----
Water supply policy						
- infrastructure policy	60	60 → 50	70 → 50	1	++++	++++
- public water supply		40 → 20	30 → 20	1	+++	+++
- sanitation		10 → 20	10 → 10	1	+	+
- irrigation		10 → 10	30 → 20	1	+++	+++
- water quality policy	20	20 → 30	5 → 10	3	++	++
- land and soil policy	20	15 → 15	5 → 10	2	+++	++
- climate policy	0	5 → 5	0 → 0	4	0	0

[1] Each country used its own method of presenting the priorities. Namibia, Zambia and Zimbabwe used a priority scale from 0 (low) to 100 (high); Tanzania ranked the priorities from 1 (high!) to 4 (low!); Malawi and Mozambique used a scale from 0 (low) to ++++ (high). Figures for Angola and Botswana are not available because there were no participants from these countries.

[2] Whereas the other country participants did not respond to the request to distinguish between past or present and future priorities, the participants from Zambia and Zimbabwe did, resulting in the trends as indicated by arrows.

Zambezi basin will proceed (i.e. according to which world-view). By varying the conditions, I arrive at the nine different scenarios shown in Table 9.4. Looking at the criteria of water scarcity, water costs and vulnerability to drought, the Harare priorities will work most favourably if operated under *low growth* conditions in a world which functions according to the *hierarchist world-view*. Under *medium growth* conditions the Harare priorities can also support socio-economic development effectively, although the trade-off now is a rather high vulnerability to drought (comparable to the vulnerability to drought in the hierarchist utopia; see previous section and Figure 9.10). Apart from the vulnerability to drought, application of the Harare priorities will carry two further risks. One of these is that water use will be inefficient, resulting in higher water demand and greater pressure on the water system than is necessary. This occurs if the world does not function

Table 9.4. What will happen - under different conditions - if the 'Harare priorities' are put into practice.

	Hierarchist context (medium growth)	Egalitarian context (low growth)	Individualist context (high growth)
Hierarchist world-view	Water demand increases by a factor of 6 to 7 in the period 1990-2050. Total water demand per capita grows by about 50%. Water scarcity grows from 2% to 12%. Average water costs per litre increase by about 60%. Water policy adequately supports continued economic growth, but the trade-off is a rather high vulnerability to drought.	The increase in water demand is relatively modest and can be satisfied without major problems. Water scarcity and water costs increase less than in the case of medium growth. Vulnerability to drought is relatively low. Water problems do not impede socio-economic development. Water policy appears to be effective.	Water demand increases relatively fast. The emphasis on supply policy appears to be inadequate. Water supply will fall short. Investment to further extend supply infrastructure cannot be afforded. Vulnerability to drought becomes very high. Strong demand policy is needed.
Egalitarian world-view	Water demand, water scarcity and water costs increase by about the same percentages as in the scenario above. Water policy adequately supports continued economic growth, but the trade-off is a rather high vulnerability to drought. Furthermore, a lack of an appropriate pricing policy results in waste of water.	Lack of an appropriate pricing policy, resulting in waste of water. However, water scarcity and water costs remain relatively low, due to a relatively small increase in water demand. Water demand can be supplied without major problems. Vulnerability to drought is relatively low.	Water demand increases relatively fast. The emphasis on supply policy appears to be inadequate. Water supply will fall short. Investment to further extend supply infrastructure cannot be affroded. Vulnerability to drought becomes very high. Strong demand policy is needed.
Individualist world-view	Lack of an appropriate pricing policy, resulting in inefficient water use. Despite the fact that water demand increases by a factor of 5, water scarcity and water costs remain relatively low, due to a high supply potential.	Lack of an appropriate pricing policy, resulting in inefficient water use. However, water scarcity and water costs remain relatively low, due to a relatively low increase in water demand and a high supply potential.	Lack of an appropriate pricing policy, resulting in inefficient water use. Despite the rapid increase in water demand, water scarcity and water costs do not increase greatly, due to a high supply potential.

Note: The 'Harare priorities' have been translated into the model by assuming the hierarchist management style, but excluding water export to South Africa. Relative values in the table are in reference to the case of medium growth in a world which functions according to the hierarchist world-view.

A water assessment for the Zambezi basin 285

according to the hierarchist, but according to the egalitarian or the individualist world-view. The other risk is that water supply will fall short while investment capacity is not large enough to further extend supply infrastructure. This happens if the world functions according to the hierarchist or the egalitarian world-view under high growth conditions. In these two scenarios, water expenditure would have to grow to about 25 per cent of gross basin product in order to supply the water demanded. As can be seen from Table 9.4, these two scenarios are the worst developments which could occur within the Zambezi basin if the Harare priorities were to be put into practice.

In conclusion, I will attempt a tentative estimate of the risk of putting the Harare priorities into practice. Let us presuppose that future growth in the Zambezi basin might be low or medium, but could also be high, and that the three world-views are equally sound. In this case, one can say that proper application of the Harare priorities roughly carries the following odds. There is a chance of 1 in 9 (about 10 per cent) that socio-economic development will be supported effectively without the necessity for trade-offs; a chance of 6 in 9 (about 70 per cent) that socio-economic development will be supported effectively with relatively inefficient water use or a rather high vulnerability to drought, or both, as trade-off; and a chance of 2 in 9 (about 20 per cent) that application of the Harare priorities will be ineffective. In the last case, water supply will fall short and expenditure in the water sector will become extraordinarily high. This might occur under high growth conditions, and would be a reason to be alert if application of the Harare or similar priorities were combined with rapid population growth and economic development.

9.6 Conclusion

The previous sections have shown that the set of possible water futures in the Zambezi basin includes a wide variety of developments. A few are desirable and a few catastrophic, two extreme types of development which have received most attention. A lot of possible futures, however, will be neither desirable from any particular point of view nor catastrophic, but lie somewhere in between. As a result, one might easily get the impression that 'anything can happen'. Although this is not the case, it is true that the analytical results presented diverge to such an extent that they are quite useless as predictions. This is not a surprise in itself: it would be impertinent to suggest that one could predict what will happen during the next fifty years. However, if not in terms of predicting, what can we learn from the results of this study? To begin with, different mechanisms have been identified which could become important in the basin's future: the mechanism of rapid growth, the mechanism of balancing the desirable and the possible, and the mechanism of stabilization. Each mechanism is preferable from a particular point of view. Next, it has been shown that these mechanisms can interfere with each other, resulting in less desirable futures. These less desirable futures can be regarded as risks attached to the more desirable futures. This has been analysed by regarding 'dystopias' as derivations of the three 'utopias'.

A major risk in both the hierarchist and the egalitarian utopia is that public water supply, sanitation and wastewater treatment fail to improve, as a result of either mismanagement or a lack of investment capacity. Furthermore, both utopias are vulnerable to high growth conditions, which can lead to absolute water scarcity conditions in several parts of the basin. The main risk to the individualist utopia is inefficient water-use and extraordinarily high expenditure for water supply, which can result from a price policy which does not conform to the market.

As concluded in Section 9.4, a major risk in *all* utopias is vulnerability to drought. Although future droughts can to a large extent be regarded as unavoidable, their effects will depend on the type of development path that is followed. Independent of the world-view applied, the effects will be greater under high growth than under low growth conditions. Minimizing the risks of droughts will therefore require low growth. In this respect, the egalitarian socio-economic context is less perilous than the hierarchist context, which, in turn is less risky than the individualist context. With regard to management style, the question of which type of management increases the effects of future droughts and which reduces the effects is slightly more complex. The only advantage of the fatalist management style in this respect is the absence of water export to South Africa, because water export heightens the vulnerability to droughts considerably. On the other hand however, the fatalist management style lacks elements which improve water-use efficiency and it thus increases demand and vulnerability to droughts. In addition, the fatalist management style introduces other types of risk, related for instance to public health and water quality, which makes this kind of management not at all desirable. The hierarchist management style focuses strongly on increasing water supply rather than on reducing demand, which will result in relatively high specific water demands and an increase in society's vulnerability to drought. In this respect, the egalitarian and individualist management styles have the advantage of changing pricing structures more radically, thus reducing demands, which makes these types of management preferable to the hierarchist management style. With regard to the effects of droughts, the main disadvantage of the individualist management style is the export of water from the basin to South Africa. From the above, it can be concluded that the egalitarian context in combination with the egalitarian management style will reduce the risk of droughts most sharply. This would mean: low growth, no water export to South Africa and strong efforts to increase water-use efficiency through improving water-pricing structures and 'water education'.

10 Discussion

In the first chapter of this book, it has been argued that studies about water and sustainable development require an integrated and explorative approach, in which water problems are studied in relation to long-term socio-economic development and environmental change. The challenge of this study has been to search for a methodology to effectively adopt such an approach. In this chapter, I will evaluate the methodology which has been developed and reported in the previous chapters. First, I reflect on the tool developed. The validity of the tool is supposed to depend on three factors: can it serve its intended purpose, does it have a sound scientific basis and how generic is it. Second, I evaluate the perspective-based scenario approach as a method for handling uncertainties. The chapter concludes with a discussion of interesting research lines for the future.

10.1 Validity of the AQUA tool

The main objective of this study has been to improve the methodology of integrated water assessment through a better integration of instruments. For this purpose, elements from different approaches and sciences have been brought together. In respect of knowledge of water, concepts have been taken from hydrology, water management, economics and the environmental sciences. With regard to the method of research, concepts have been used from systems analysis, system dynamics, indicator research, uncertainty analysis and cultural theory. The research has resulted in a generic tool for integrated water assessment, the AQUA tool, which has been expanded and used on two different spatial levels: the global and the river-basin. The purpose of this section is to examine the validity of the tool developed. It is assumed here that the validity of a tool is not an attribute in itself, but something which depends entirely on what is expected from the tool. For this reason, I repeat here the main aims of the AQUA tool. As described in Chapter 3, the tool should be useful to:

- improve insight into the interrelations between water and long-term socio-economic development,
- show how these interrelations can be perceived according to different world-views,
- analyse the effects and risks of different water policy strategies,
- support agenda-setting and the formulation of water policy priorities.

In order to link scientific analysis to the interests of policy makers, the tool should be able to translate scientific model results into policy-relevant information (indicators). Furthermore, the tool should be reliable - well-founded from a scientific point of view - and generic - applicable at both global and river-basin level. A final design criterion was user-friendliness, including a short running time, so that interactive use would be possible. The user-friendliness of the tool will not be evaluated here, because this was not a subject of this study. However, the AQUA tool will be examined below regarding each of the other requirements.

Let me start with the question of usefulness. Has it been sufficiently demonstrated that the tool can be helpful in improving insight into the interaction between water and development, in showing the role of world-views, in analysing different water policy strategies and in formulating water policy priorities? In order to answer this question I will consider the fresh insights which have been obtained while developing and using the AQUA tool. For this purpose, I will now summarize the main findings in respect of water in the 21st century, both at global level (Chapters 6-7) and for the Zambezi basin (Chapters 8-9).

A surprising result of the global water study is that loss of ground water through groundwater exploitation may be an important driving force behind sea-level rise. This contradicts the current belief in the global change research community that climate change is the only significant influence in past and future sea-level rise. Mechanisms of long-term water loss on land have been recognized, but it is generally argued that possible water losses in some places are gains in others (Warrick et al., 1996). In this study, where the possible contribution from land has not been excluded in advance, this water loss on land is estimated to have contributed between 30 and 40 per cent to sea-level rise in the 20th century, depending on the world-view adopted (Figure 6.3). In the three utopias for the 21st century, the contribution of water loss on land to sea-level rise varies from about 25 per cent in the egalitarian utopia (where the climate-related contributions are relatively large) to about 45 per cent in the hierarchist and individualist utopias (Table 7.1). The results have been obtained by regarding ground water as a dynamic store with two inflows (natural and artificial groundwater recharge) and three outflows (groundwater withdrawal, delayed surface runoff and subsurface runoff). The natural groundwater outflow (i.e. delayed surface runoff plus subsurface runoff) has been assumed to be a function of the groundwater storage.[1] Because our present knowledge on the behaviour of ground water on a large scale contains great

---

[1] Natural groundwater outflow has been modelled according to Equation 3.29. As explained in Section 6.3 the behaviour of the groundwater system depends greatly on the value of exponent p. The conclusion that groundwater losses may be an important driving force behind sea-level rise is based on various experiments with different values for p. Assumptions with regard to future groundwater use and artificial groundwater recharge have been varied too.

uncertainties, a conclusive answer on the role of groundwater losses cannot be given on the basis of this or any other research, but at least one thing may be concluded: excluding this factor from studies of past or future sea-level rise is not justified. The results of this study clearly suggest that further investigation is needed into the importance of groundwater losses. As sea-level rise as such has already been recognized as a political issue, new findings concerning the driving mechanisms behind this phenomenon may lead directly to adjustments in political agendas.[1]

A second insight obtained is that water scarcity could easily become a global issue if there is no rapid and structural change in the way people use water. Such a change could be achieved by a worldwide modification of water tariff systems, by promoting water-saving technology and by starting programmes of 'water education'. The observation that water demand policy will become increasingly important is not an understanding arising from this study alone. Over the last few years, an increasing number of authors have reached this conclusion (Young et al., 1994). However, most integrated studies so far have been qualitative in nature. This study provides the first integrated *quantitative* assessment, by showing in quantitative terms the need for and effectiveness of different strategies of water demand policy under a number of scenarios. In both the global and the Zambezi study, it has been shown which policy strategies are needed within three different utopias and what happens if these strategies are not brought into effect. Additionally, the risks of different policy strategies have been analysed by considering what happens if the world does not behave as expected. In these respects the approach in this study offers more insight than earlier qualitative studies. There have been a number of quantitative assessments of future water demand, but these studies did not meet the requirements of an integrated approach. Recent scenarios of global water demand in the 21st century developed by other authors are based on trend extrapolation rather than on a study of the actual driving mechanisms of demand.[2] As a result, different processes and assumptions are implicit only and the role of policy remains unclear. It is for example not clear which types of pricing measures are implicitly assumed in these scenarios and how sectors will respond to rising prices. Another drawback of these studies is that they focus on so-called 'best estimates', thus ignoring fundamental uncertainties. As a result, each of the studies provides either just one scenario or a few scenarios which differ in some of the quantitative assumptions but are in fact all of the same kind. Because basic

---

[1] For the sake of clarity: the results of this study do *not* suggest that the contribution of climate change to sea-level rise is probably smaller than suggested by the IPCC. The results suggest that there is a further influence, which has probably been underestimated.

[2] See for instance: Margat (1994), Shiklomanov (1997), and Raskin et al. (1997).

assumptions are not varied and policy remains implicit, this type of scenario cannot be used to assess the risks of different policy strategies.

Both the global and the Zambezi water assessment show that, although the type of future water policy will strongly influence the extent of future water scarcity, there are factors which are much more important determinants of this scarcity. These factors are population and economic growth. Although this may seem trivial, it in fact implies that water can no longer be treated as a separate policy domain, to be addressed within a given context of development and exclusively by traditional establishments such as ministries of water, water boards and similar institutions. During the 21st century, the fast growing regions in the world in particular will find that water availability is a serious constraint on development, and demand management will become an essential part of development planning. Development policy in such regions should explicitly include a water paragraph, as an integral part of the policy as a whole. The Zambezi study illustrates this well: any development policy in this region is doomed to failure if water resources are not carefully planned and if demand policy is not formulated at an early stage. It has been shown that it would be wise if the Zambezi basin states take possible future water scarcity into account when formulating national or basin-wide development strategies, because the vulnerability to water scarcity can vary greatly between the different kinds of strategy.

Fresh insight has also been obtained regarding the dynamics between water and development. From model experiments, both at global and at river-basin level, it has been shown that the character of these dynamics changes when water gets scarcer. In order to describe the changing interaction between water and development, the hypothesis of a 'water transition' has been formulated. According to this hypothesis, the changing relationship between water and development can be divided into three phases. In the first phase, new water resources are continuously developed, to supply an increasing number of people, to improve water supply and sanitation conditions and to support irrigation development. This phase is self-reinforcing: the development of water resources supports socio-economic development and this in turn drives a more intensive exploitation of the water system. The first phase can be distinguished by increasing rates of water withdrawal and pollution. In the second phase, growing water scarcity and water supply costs induce competition between water users. Water supplies may become less reliable, supplies may temporarily fall short, yields in irrigated agriculture may be affected, pollution begins to kill aquatic life and fishing industries are threatened. Growth rates will be moderated. In the third phase water scarcity has reached a level at which people are forced to increase water-use efficiency and reduce pollution. Under favourable socio-economic conditions (stabilizing population, reasonable living standards), the third phase can conclude with a stabilization of water demand and the restoration of the quality of the natural water system. Each phase brings with it its own type of policy. From the first to the third phase, priorities gradually move from supply towards demand policy. A few recent policy documents have adopted the water-transition concept to refer to a priority shift which needs to take place from 'water exploitation' to 'efficient water use' (UN, 1997b; Raskin et al., 1997).

Another understanding reached in the previous chapters is that each of the three phases of water transition can be said to be dominated by one particular perspective. The individualist perspective clearly prevails in the first phase of exponential growth. In the second phase, balancing the desirable and the possible, the hierarchist perspective becomes dominant. Finally, in the third phase of stabilization, the egalitarian perspective dominates. The third phase is not necessarily the end of the line, which will be reached sooner or later. If, at some time during the stabilization phase, it becomes apparent that more water is available than was thought or if it appears that water can be used much more efficiently, it is likely that the third phase of a transition will move into the first phase of a next transition, which will then take place under the new conditions. In this case, a shift will occur from the egalitarian to the individualist perspective. From the controversies in the literature on the type of management strategy to be adopted (see Chapter 2), we can learn that in many regions of the world consensus is lacking on the current phase of water transition. The model experiments in the previous chapters show that there is indeed room for all three perspectives, the individualist, the hierarchist and the egalitarian. According to the individualist perspective, continued growth can take place, because there are still great possibilities for more efficient water use and a large volume of water remains unexploited. The hierarchist clearly sees limitations and is more prudent in assessing future possibilities. According to the egalitarian perspective, reducing water demand is a sheer necessity, to prevent future catastrophes such as droughts and absolute water scarcity. In Chapters 7 and 9, I have shown that the egalitarian management style carries the lowest risks. As a consequence, if one prefers to avoid high-risk developments, one would have to adopt many elements of the egalitarian perspective.

To obtain a clearer picture of the AQUA tool, I should also discuss its weaknesses. Whereas the previous chapters have shown that AQUA can be useful as a tool for learning, they have also illustrated that it *cannot* be used to predict. As Oreskes et al. (1994) argue, this feature is inherent in any open-system model. Being primarily a heuristic tool, intended to guide further study, the model itself is not susceptible to proof. It is a structure of 'if then' statements rather than a straightforward representation of how the world works. As a consequence it is more difficult to evaluate the scientific reliability of the model than is the case with a predictive model, something which will be discussed further below. The weakest part of the AQUA simulation model is probably the response sub-model. This is closely related to the modelling method followed. A fundamental characteristic of the simulation approach is that the condition of a system at a certain point in time depends essentially on its condition at a previous point in time. The method thus ignores the fact that tomorrow's situation might also partly depend on an expectation of the conditions the day after tomorrow. Especially in systems where humans are involved, who think ahead and act accordingly, it would be useful to be able to feed the possible future system's behaviour back as a determinant of the system's behaviour of today. The fundamental difficulty lies in the fact that people in the real system will respond to possible futures of the system (having some insight into the dynamics of the system), whereas people in the modelled system only respond to the actual state or history of the system. Within the simulation modelling approach, this problem cannot be solved without changing the basis of the method.[1] A practical solution is to exogenously feed the human response into the system. Such a human response could be found via workshops where people 'play' with the model, learn about possible futures and develop response strategies themselves. This is the way in which the model has been used in the context of the Harare workshop, as mentioned in Chapters 8-9. As a result, the weakness of the approach is at the same time a strength.

Another weak point, which can also equally be called a strength, is the lack of much detail, particularly in the AQUA World Model. On the one hand phenomena are placed within a broad context, which makes it possible to understand them better, but on the other hand they are described on such an abstract level that it might be questionable whether any concrete conclusion can be reached. One could call this paradox the 'restriction of comprehensiveness'. In the case of the AQUA tool, this paradox could be solved to some extent by increasing the spatial resolution (especially in the World Model), distinguishing more water-demanding sectors (e.g. distinction between different types of industrial water demand, and, in the case of the World Model, between urban and rural supply), refining the cost calculations (e.g. explicit distinction between investment costs and operational and maintenance costs) or introducing additional modules (e.g. sources of pollution other than untreated wastewater, effects of erosion, calculation of energy demand for different types of

---

[1] It would be possible to explicitly distinguish a variable 'expectation of future's system behaviour', but this variable would have to be based on an extrapolation of historical trends and will not be dynamic in itself. As a result, the variable will differ from the real future's system behaviour, which *is* dynamic.

water supply). However, the primary approach will remain comprehensive, which implies that it will never be possible to be as detailed with regard to specific subjects as one can be in studies with a more bottom-up approach.

Scientific reliability

There are several methods of evaluating the reliability of a model. As will be discussed below, each method is incomplete and cannot be used as *the* method to assess scientific reliability. For this reason, more than one method has been used. The first method is empirical validation, which means that simulation results are compared to observations. This has been done in Chapter 6 for the AQUA World Model and in Chapter 8 for the AQUA Zambezi Model. In general, one can say that the historical simulations of both models do not contradict observations. However, this requires two critical notes. The first is that real observations are rarely available, particularly on a global level. The AQUA tool necessarily works with aggregate variables which are never observed as such. At best, they can be calculated from more detailed data, but often they can only be estimated in an indirect way. For instance, if we speak of the growth in global water demand or of sea-level rise during the 20th century, we have to rely on assessments rather than direct empirical data. As a result, the concept of 'empirical validity' has to be taken with due caution. The second critical comment concerns the independence of empirical data. Empirical data are generally used when calibrating a model, which makes it impossible to use them for validation later on.[1] This was also the case with the AQUA World Model and the AQUA Zambezi Model. After calibration, only a few data were left for independent validation. In Section 6.3 for instance, the simulated contributions of global groundwater loss, deforestation, drainage of wetlands and artificial surface reservoirs to sea-level rise have been compared with independent estimates in the literature. In Section 6.4, the simulated global expenditure on water supply and sanitation has been compared with some independent World Bank estimates. An indirect way of empirical validation is to compare parameter values obtained by calibration to independent estimates by others. In the case of the AQUA World Model for example, the growth elasticities obtained through calibration were compared to the elasticities found in a number of national studies. In the case of the AQUA Zambezi Model, it was possible in some cases to evaluate the land cover factors and river flow delays obtained through calibration on the basis of

---

[1] It would be nonsense to use the fact that predictions are close to observations as a confirmation of model validity if predictions have previously been correlated with observations through calibration (Oreskes *et al.*, 1994). Confirmation of model outcomes can only take place for variables which have not been correlated in the calibration phase.

independent values in the literature. On the whole, one can say that the AQUA World and Zambezi Models pass the empirical validity test, but it needs to be admitted at the same time that this outcome should be appreciated in a proper way, i.e. modestly. As Oreskes *et al.* (1994) argue, this is inherent in models of open systems, which can have only a limited predictive value. As a result, empirical validation can play no more than a limited role in evaluating the scientific reliability of open-system models.

A second method of testing the reliability of a model is to carry out sensitivity analyses. Varying the input values of a model can be very helpful, not in order to reject the model if sensitivities appear to be rather high[1], but as an instrument to learn about model behaviour. Within this context, validation means comparing the behaviour of the model to the behaviour of the modelled system. If - within acceptable parameter ranges - a model behaves in a way which is clearly impossible in reality, the reliability of the model should be questioned. In the case of AQUA, most parameter changes cause 'ripples' throughout the model, but these become weaker rather than being reinforced. In this sense, the AQUA model is fairly robust. However, if water scarcity increases, say to above 75 per cent, model outcomes can become highly vulnerable to perturbations. Under these circumstances, water pricing appears to be very effective. Changes in the supply-cost curves and price elasticities of water demand in particular have a considerable effect on model behaviour. If water scarcity rapidly approaches 100 per cent, the model can even produce cyclic behaviour, due to the feedback between demand and prices. The general picture which emerges is that the model world is more vulnerable to perturbations if water use becomes more intensive, particularly if water scarcity has reached a relatively high level, and most prominently if the *rate* of scarcity increase is also relatively high. This increasing vulnerability in the model world can be interpreted in real-world terms as follows. As long as water is abundant, it is reasonably cheap, and the disturbance of the hydrological cycle through water withdrawals is relatively small. A variation of, for instance, water-pricing structures or hydrological parameters has minor effects: water will continue to be quite cheap and disturbances will remain small. However, if water is being used intensively and has become a scarce commodity, any change regarding the way of management or the vulnerability of the hydrological system can have a great effect. A small shift from surface water to groundwater use will for instance affect the extent of groundwater-level decline

---

[1] As Lewandowski (1982) observes, there is a school of researchers holding the view that a model has to be rejected if a sensitivity analysis shows that model outcomes respond strongly to parameter changes. This could make sense for a model which describes the behaviour of a well-known isolated system, but not for models of open systems with unpredictable and possibly complex behaviour.

more strongly if total water demand is high than if it is low, and a 10 per cent price increase will have a considerably greater effect if prices are already high than if they are relatively low. In the regions of the world where water is most scarce at present, as in the Middle East, the high vulnerability of the system can already be seen in the form of political tension over water.

A third method of testing model reliability is to use alternative conditions, for instance a different period, area or spatial level. In this sense, the application of the generic AQUA tool to the Zambezi basin can be regarded as an additional validity test of the global application and *vice versa*. A fourth method, particularly useful in evaluating the scientific reliability of an integrated model, is to select parts of the model and compare these to other, more detailed models. In the case of the AQUA World Model, such a comparison has been made for instance with the ice-sheet model (Warrick *et al*., 1996), the glacier model (Wigley and Raper, 1995) and the sea-level rise impact model (Hoozemans *et al*., 1993; RA, 1994). For the AQUA Zambezi Model the part selected was the complete hydrological sub-model, for which the grid-based runoff model of Vörösmarty and Moore (1991) has been used as a reference. More specifically, the equation for the outflow from Lake Malawi has been adapted according to insights gained from the model of Calder *et al*. (1995). Furthermore, in an early phase of this research the AQUA model was applied to the Ganges-Brahmaputra basin (Van Rijswijk, 1995) and compared to a grid-based runoff model of this basin (Van Deursen and Kwadijk, 1994; Van Deursen, 1995). This comparison showed that, despite its relative simplicity, AQUA can reproduce observed river runoff at two downstream outflow points as accurately as the more detailed model (Hoekstra, 1995). In general, most parts of the AQUA simulation model reflect insights obtained from more detailed water models. However, the parts of the model which describe water demand and supply dynamics are original, although the concepts used have already been applied in other research fields, such as energy resources.

Generic
The term 'generic' is understood here as the possibility to apply a model at different spatial levels without changing the model structure. In this study it has been shown that the AQUA tool can be applied at both global and river-basin level without changing the model structure or equations. The tool can probably be used on a continental level or for river basins other than the Zambezi basin without major modifications.

It is not self-evident that a model developed at global level is applicable at river-basin level or *vice versa*. For instance, when modelling runoff in a river basin, the basin is preferably schematized in such a way that the effects of individual large dams can be taken into account. When modelling runoff on a global scale, one will have to be less detailed, so the effects of dams should be accounted for in some lumped way. As a result, one might choose different model formulations on the two different spatial levels. In this study, the AQUA World Model and the AQUA Zambezi Model have been developed parallel to each other, so experiences gained in one case could be used in the other and model formulations could be tuned. Let me give three examples. In the global water study, it became clear that artificial groundwater recharge is a variable to be taken into account, because large-scale groundwater depletion and sea-level rise could become an actual issue. In the Zambezi basin, artificial groundwater recharge is not a theme at all at present, so that - without the experience in the global study - this variable would probably have been left out of the AQUA Zambezi Model. Now, however, this variable has been included, strengthening the integrated character of this model. A second example deals with the simulation of water demand. In the global study, a sufficient amount of data could be collected to formulate a water demand module and to find reliable ranges for the water demand parameters (see Section 6.3). This was impossible for the Zambezi basin, where we had to rely on the results obtained in the global study. A third example deals with the simulation of seasonal variations in runoff. In the AQUA World Model it was difficult to calibrate and validate the simulation of these variations. In the river-basin application, most of the calibration and validation effort has therefore been focused on the hydrological sub-model, to test the adequacy of the runoff equations (see Section 8.3). Within this context, it is interesting to refer to the research paradigm of 'strategic cyclical scaling' as described by Root and Schneider (1995), a concept used to refer to a continuous cycling between large- and small-scale studies.

The global and the river-basin application of AQUA differ only with regard to the schematization and input values used. However, there are also a few structural differences: the Zambezi Model does not include the modules which describe atmospheric, oceanic and ice-sheet processes, nor the module which simulates impacts of sea-level rise. This means that climate and sea-level rise are not endogenous as in the World Model, but exogenous. As a consequence, the Zambezi Model does not include the possibility of feedback through the climate system. Land use changes and consumptive water use within the basin do not influence the model climate, but they might influence the real climate, as Savenije (1995) for instance has shown with regard to land use changes in the Sahel region. In the Zambezi study the impacts of possible changes in climate have been studied separately, by carrying out experiments with different inputs for precipitation. The possible impacts of sea-level rise have not been considered in the Zambezi study.

10.2 Perspectives as a rationale for composing scenarios

As explained in the first chapter, one aim of this study has been to examine the use of perspectives as a rationale for scenario development. It was chosen to use four perspectives from the cultural theory of Thompson *et al.* (1990), namely the hierarchist, egalitarian, individualist and fatalist. In Chapters 5 to 9 the approach has been developed by implementing perspectives in the generic AQUA model and applying the model to long-term water policy analyses both for the globe as a whole and for the Zambezi basin. In this section, I evaluate this approach and give an overview of my experiences, and the questions and obstacles which arose.

A fundamental assumption of the perspective approach is that there are inherent uncertainties in our knowledge of the world and that, as a consequence, there is not one truth, nor one best perspective (or at least no one can claim to have the only truth or the best perspective). It is therefore misleading to speak about 'best' or 'central' estimates with regard to future developments, because these estimates generally refer to a *medium* estimate, and not necessarily to *the most probable* estimate. In this research, I have shown for the field of water that distinctly possible futures can differ so much that presenting just one medium estimate is not useful. The plausibility of a certain future does not depend on its 'distance' from a particular medium projection, but on the extent to which this future can be logically explained from a specific perspective or combination of perspectives. In this study I have presented several fundamentally different scenarios, each of which can be explained through some combination of the perspectives of the cultural theory. In my opinion, this approach provides a better understanding of the future than the single-scenario approach followed by for instance Shiklomanov (1997) or the multiple, but *ad hoc*, scenario approach of Raskin *et al.* (1997). I also think that this approach can be more useful to policy makers, because the political debate is provided with visions of the future which are explicitly characterized by different sets of assumptions rather than with *ad hoc* projections. On the basis of scientifically-based utopias and dystopias, political controversies can be made more explicit, understandable and easy to debate, which might narrow the gap between what scientists produce and what policy makers would like to have.

A critical response I often receive is that this approach means that politicians are provided with a scientific basis for their own opinion. It is indeed possible that politicians could find scientific arguments for their own position, but they are in fact provided with a broader framework, in which they will also find the rationale for others' positions.[1] This can structure and simplify the debate, so that it can focus on the fundamental decisions which have to be made. However, it should be realized that the perspective approach is still in the first phase of development and has not been used in any actual policy-making process yet, so that many aspects remain which deserve further consideration.

The distinction between world-view and management style
One such an aspect is the distinction between world-view and management style. The purpose of this distinction is to contrast the world 'as it is' with the world 'in as far as people (can) actively influence it'. A sharp distinction cannot be made, which can be traced back to the old philosophical question of determinacy and indeterminacy. There still is no final answer showing which part of human behaviour comes under the heading 'functioning of the world' and thus under world-view, and which under 'free action' and thus under management style. The perception of what people can actively do is in fact subject to different perspectives. Hierarchists for example believe more strongly in the functioning of legislation, egalitarians in education and common endeavour, and individualists in free entrepreneurship. The problem of a perspective-dependent distinction between world-view and management style has been solved in this study by making a number of perspective-specific assumptions about what to classify as world-view and what as management style. For the hierarchist and egalitarian, for example, improvement in water-use efficiency has been assumed to be part of the management style, while for the individualist this is part of the world-view. The individualist actively stimulates a proper functioning of the market mechanism (management style), and efficiency improvements are supposed to be induced by price rises (world-view). This solution does not, however, solve the problem that some processes of change are part of *both* the world 'as it is' and the world 'as actively managed'. Theoretically a solution could be found by distinguishing an autonomous and a human-driven component in these processes, but this has not been done in this study, which might lead to some misinterpretation of the results. For example, categorizing water-use efficiency improvements as part of the hierarchist management style, as done in this study, might erroneously suggest that these improvements are fully directed by human beings. However, considering these improvements part of the world-view would suggest that they are completely autonomous and unsteerable, which is also not correct.

---

[1] As mentioned earlier, ideas which contradict empirical evidence are excluded.

Another problem in making a distinction between world-view and management style is that it is important to know what is meant when one speaks of 'people' who actively influence the world. This can refer to *all* people, but policy makers are more interested in what *they* can do (in their duty as policy maker) and have to consider the behaviour of the rest of society part of the world 'as it is'. Policy makers can certainly achieve less than all people together, but how much influence they have depends on one's perspective. Hierarchists believe more in powerful governments than egalitarians and individualists; egalitarians have more faith in the power of 'organized people with a mission' (such as non-governmental organizations, and possibly but not necessarily also government departments); and individualists believe in the power of the individual. In this study, I have used the concept of management style in a broad sense as referring to what people including governments can do. In interpreting the results for policy-making, one should therefore be aware of the limitations to the role of government policy. Strangely enough, most policy analysis studies pay very little attention to the uncertainty about what governments can or cannot achieve. By not emphasizing or even mentioning the limited influence of policy intervention in development as a whole, one implicitly adheres to the hierarchist view.

A further issue which deserves attention is the difference between world-view and management style in respect of historical simulation. To explore the future I have used different world-views and management styles, but in explaining the past I have only applied different world-views.[1] An important reason for simulating historical developments according to different world-views is that if history cannot be explained in terms of different world-views, there are no grounds for applying these world-views to the exploration of the future. Using different world-views to explain history is therefore an essential element of the perspective-based scenario approach. This is not so with management style. A management style used to explore future developments is not necessarily applicable when explaining historical developments, because a future management style may constitute a break with the past. In the case of world-views, one could say: if the world may have worked in the past according to three alternative views, it might work in the future according to one of these views too. In the case of management styles, this is different: historical action is already fixed (although possibly uncertain), but future action is still open.

---

[1] For the historical simulations, I used the data of the realized management path. This could be done only for these manageable parameters for which historical values were available or could be estimated. The remaining manageable parameters were assumed according to a particular management style, corresponding to the world-view under consideration.

Some people question the applicability of perspectives to handling uncertainties which pertain to physical processes and parameters. According to Risbey *et al.* (1996), uncertainties in these sorts of parameter are at best very weak functions of world-view, and have more to do with epistemological ignorance about complex properties of the earth system than with cultural characteristics. In fact, I fully agree with this observation, if formulated in this way. To give an example, let us consider the sensitivities of the Greenland and Antarctica ice sheets. Is it not nonsense to say that if Greenland has a high sensitivity to climate change, Antarctica also has a high sensitivity, and *vice versa*? Indeed, this is absurd, because the two ice sheets lie at opposite poles and have their own unique and independent characteristics. But is this not what I do say? Yes and no! Yes indeed, I have assumed high sensitivities (i.e. high contributions to sea-level rise) in the egalitarian world-view, medium sensitivities in the hierarchist world-view and low sensitivities in the individualist world-view (see Tables 5.3 and 6.5). But this does not mean that I suppose that the sensitivities of the Greenland and Antarctica ice sheets correlate in any physical sense! I do not want to suggest more than that a few particular combinations of sensitivities exist which fit well within certain world-views. I think it is useful to distinguish these few combinations (instead of looking at all combinations at once), because people act in correspondence to particular world-views. As a consequence, although the combinations may have no meaning in a physical sense, they have in a societal sense. The possible combination of high contributions to sea-level rise from both Greenland and Antarctica (plus a high contribution from glaciers, etc.) drives the egalitarians to call for immediate action. The possible overall robustness of the climate system is a sufficient reason for individualists to wait until there is clear evidence of undesirable disturbances. In my opinion, the value of applying cultural perspectives in the earth sciences is not to 'uncover' interdependencies between uncertain processes or parameters, but to show how the various uncertainties can be interpreted according to different world-views and how they can give rise to alternative response strategies.

Interaction between perspectives
An issue which equally deserves further attention is the interaction between perspectives. According to Thompson *et al.* (1990), the perspectives are not viable on their own but need each other in order to exist. As they put it, each myth of nature is a partial representation of reality. If nature everywhere and always corresponded to one myth, this myth would become dominant and the others would rapidly fade into extinction. They further argue that a persistent pattern of surprises, i.e. discrepancies between the actual and the expected, forces individuals to alter their

perspective (micro-changes). Several individual changes together might result in a shift in the cultural bias on a higher level of organization (macro-changes). These reflections might give rise to the question: why then have the perspectives been implemented in the AQUA model in such a static way? The futures which have been explored in this study are all based on a fixed set of context, world-view and management style (see Chapters 7 and 9), with no allowance for changes in the set during the period of simulation. As Risbey *et al.* (1996) observe, a static view of cultural theory fails to capture the dynamic nature of human perspectives, which change over time, partly as a result of earlier decisions made according to previous preferences. I agree with this kind of criticism, but would like to point out that a static perspective approach is still a step forward compared to a non-perspective approach. In this study, one step has been achieved, not two steps at once. A dynamic perspective approach is an interesting direction for further research. The utopias which have been presented here are the possible futures without surprises. These futures will not change, because there is no reason for it. However, in dystopian futures, which are in fact 'surprise futures', individuals' perceptions will change and as a result so will the overall cultural bias. This means that the dystopian futures as presented in this thesis will not continue to evolve along the same lines, but will change direction, in response to the new dominant cultural bias. I think that it is very useful to elaborate on this idea by making dystopias dynamic. Each time, cultural macro-changes will tend to push a particular dystopia in the direction of a utopia, but new surprises will mean constant direction changes. Interesting experiments with a dynamic implementation of perspectives in a simple simulation model have been carried out by Janssen (1996), who speaks of 'the battle of the perspectives'. I think this method deserves further expansion, at least from a methodological point of view. However, as a first step in the actual use of the perspective-based method of scenario development, I would recommend the simpler static approach which has been followed in this study.

Biases
In implementing the perspectives in the model, I found that I had to overcome three types of bias: scientific, methodological and personal. The scientific bias is the predisposition of the scientific community to overestimate mainstream thinking. Dissenting from the majority opinion often means being classed as a scientific outcast. A relevant question in this study has been where to draw the line between the scientific and the unscientific. There are always positions which are called scientific by a minority of researchers but unscientific by the majority. This in itself is not a reason to exclude these positions ('the truth cannot be decided upon

democratically'), but where to draw the line? Who are the experts? Should one take *all* scientists seriously? I have heard respectable hydrologists say that someone like Malin Falkenmark - although a distinguished Swedish hydrologist - cannot be taken very seriously, with her continual warnings of a future world water crisis. Should I disregard Falkenmark's opinion for this reason? In this specific case, I think that Falkenmark takes a representative and valuable position in the debate about possible water futures (and is of course attacked by her opponents). The same can be said for her opponents, for example someone like the Russian hydrologist Mark L'vovich, one of the pioneers in the study of global hydrology, who wrote: "The views of certain authors who believe that population growth and economic development will be limited by the shortage of fresh water can serve as the most vivid manifestation of pessimism." (L'vovich, 1979). A practical aspect of the same problem is: which uncertainty ranges should be assumed for poorly-known parameters? There is a tension between the aim of being honest in admitting the uncertainties in future projections and the aim of being taken seriously by the scientific community, which crowds around some central best estimates. The dominant view is still: the further away from these central estimates, the more exceptional, improbable, and therefore unscientific. In this study I have tried to take into account as broad a range of views as possible, but I have only considered views represented in scientific literature.

The second bias which had to be overcome - in which I was less successful - is a methodological bias. As noted by Geels (1996), analytical methods in the environmental sciences are often biased in a particular direction. The simulation modelling approach fits better within the hierarchist world-view than within the individualist or egalitarian world-views. It is for instance difficult to translate the individualist concepts of learning and adaptive behaviour and evolution into the mechanistic structure of a simulation model. A simulation model can equally not easily represent the change in social structures advocated by egalitarians, or the capricious, discontinuous and unpredictable behaviour of ecosystems. Furthermore, a simulation model cannot address risks of ignorance and indeterminacy of knowledge, major concerns of the egalitarian. On the other hand, the causal and mechanistic approach of simulation modelling corresponds well to the hierarchist perception of the world as a manageable, cybernetic machine. The problem of the methodological bias is that, metaphorically speaking, if you have a hammer, everything looks like a nail. In the same way, a hierarchist method of analysis will make the world look like a hierarchist world. According to Risbey *et al.* (1996), who speak about *model bias*, any single model will always exclude certain facets and arbitrarily include others when a particular model structure is chosen. In my research, the development and use of the AQUA model takes central place, which definitely gives a somewhat one-sided view of reality, despite my efforts to include different perspectives within the model framework. Here, I cannot do other than observe this and emphasize that this study should be regarded as just one element in an ongoing diverse process, in which others contribute other necessary approaches.

Apart from the scientific and methodological biases, there is a personal bias to overcome. I think it is not necessary to argue over whether all of us have our own preferences, I certainly have. A factor which can reinforce this bias is the tension which exists between honesty and the ambition to influence the actual political process. Honesty should force scientists to admit that the uncertainties in long-term scenarios are actually *very* great. However, it is easier to get a message across and make it clear to other people if one highlights and overstates one particular outcome of a study and omits the nuances. Another factor which might strengthen the personal bias is the tension between honesty and the political expectations of others (Geels, 1996). Because most politicians dislike research results which have too much of a flavour of 'it depends on', it is tempting for scientists to put particular results forward with much more certainty and less nuance than they would have done without this social pressure, especially as scientists depend on these same politicians for financial support. In this study, I have tried to put equal value on each of the different perspectives, and not give way under the pressure by overstating things or giving a relatively high value to my own view and preferences.

Beyond the perspectives
Once one is familiar with the use of different perspectives, an interesting question emerges: is it possible to go beyond the perspectives and make some general statements which are valid independent of the perspectives? As shown in the last sections of Chapters 7 and 9, it is indeed possible to exceed the perspectives, provided one is able to accept that none of the perspectives has a higher value than any of the others. The approach has been to study whether there are policies which work well (or quite well) under a wider range of circumstances than do others, although these other policies might work best in some specific circumstances. Let me summarize the approach by simplifying. Suppose there are two perspectives: the egalitarian and the individualist. The individualist wants growth and advocates a number of measures. The egalitarian does not particularly want growth but equal distribution, and proposes a different set of measures. If the individualist measures are put into practice, the individualist rates the situation which will develop at ten points. The egalitarian, who sees that the pressure on the environment is reaching an unacceptable level and that differences between poor and rich are growing, rates the situation at three points. If the egalitarian measures are put into practice, the

egalitarian rates the resulting situation at ten points, but the individualist's rating is seven points, because the egalitarian measures will limit economic growth. By adding up the points given by the individualist and the egalitarian, the two situations are rated at thirteen and seventeen points respectively. Overall, one can then conclude that the egalitarian type of measures are more desirable. In my opinion, this can be a useful approach and it has been applied in calculating the 'total perceived risks' for different types of development.[1] This has led to some general conclusions, such as that the individualist context of high growth carries relatively high risks if one wishes to safeguard water supply for future generations, or that, of all management styles, the fatalist style is least preferable.

I would like to make two critical comments regarding the approach followed. First, it is crucial that all perspectives are treated and valued equally. This seems reasonable, but it will often be difficult in practice. Second, the evaluation criterion of 'total perceived risk' is debatable. A strategy of avoiding high-risk developments corresponds to the 'precautionary principle', which might be regarded as typically egalitarian. This would make 'total perceived risk' unacceptable as a perspective-exceeding evaluation criterion. However, it can also be maintained that the precautionary principle is rather a principle of survival and can therefore bridge the contradictions between different morals (Van de Poel, 1991). In my opinion, the criterion of 'total perceived risk' is at least an attempt to arrive at some perspective-exceeding evaluation criterion, which makes it worthwhile to study its consequences, but other types of criteria could be suggested and applied as well. However, I think it will remain difficult to find criteria which are at the same time perspective-exceeding and clear enough to be of practical use in policy evaluation.

The problem of ignorance

Perspectives naturally refer to what people are used to experience, and thus exclude everything that falls outside human perception. To say that the existing perspectives together embody all possibilities would be a denial of ignorance. Applying widely diverging perspectives in an assessment does not mean that one cannot be mistaken. As Silver and DeFries (1990) observe, our vulnerability to error is greatest not from the things we include in the model, but from the prophesies we leave out entirely. I think that this is true not only in traditional single-perspective models but also for multiple-perspective models. Although perspectives have been implemented throughout the model, it is impossible to claim that I have taken into account every

---

[1] In the terminology introduced in Section 7.6: the total (second-level) risk of a certain development path is calculated as the sum of the (first-level) risks as perceived within the different world-views.

uncertain issue. It may well be that I have left out issues and phenomena which will emerge in the future, but cannot be foreseen at the moment. This risk is inevitable, even when the perspective approach is used. One should be extremely careful in overestimating humanity's ability to determine the borders between the possible and the impossible or the plausible and the implausible. The 'danger' in using the perspective approach is that one is inclined to think that we now - unlike when we do not take different perspectives into account - have incorporated the whole range of possible futures. A limited number of alternative formulations has been implemented in the AQUA model, in spots where it seemed to be necessary. But there will certainly be some areas where realistic possibilities have been overlooked.

As a final remark, I would say that - despite all the questions and problems I have just put forward - the perspective-based scenario approach looks very interesting from a methodological point of view. However, the approach needs to be refined and further research will have to be carried out before it can be useful in the actual process of policy-making. I would therefore recommend that the method, including the questions and obstacles mentioned above, is further investigated.

10.3 Future research

Integrated water assessment is still a very new field of research. Hopefully, this thesis makes a substantial contribution to the field, but it is clear that many questions remain unanswered. My research has probably even raised more new questions than it has given conclusive answers. However, this is typical for this kind of exploratory research. The challenge is to take the new questions on and continue in the chosen direction. Just as in this study, most future research is likely to have two components: a specific water component and a more general methodological component. A water-related question which could be studied in further detail is what the contribution of groundwater losses to sea-level rise might be. The integrated approach followed in this study has shown that worldwide groundwater exploitation could be a relevant factor in addition to climate change, but more research is needed to arrive at firmer conclusions. Another interesting research subject would be 'hidden water use', the indirect use of water in a particular region through the import of water-intensive products from outside the region, a topic closely related to the issue of water dependency or water self-sufficiency. In this study, it has been shown that a worldwide trade in 'virtual water' may be undesirable from a certain perspective (the egalitarian), but necessary and unavoidable from another (the individualist). A further issue which needs elaboration is the concept of water recycling on different scales. At first sight this seems to be an old issue, because the concept of the 'hydrological cycle' has already existed for many years. However, no one has ever asked the fundamental question on what scale could or should the phenomenon of water recycling be exploited, so as to be most beneficial? In this study I have shown that one approach could be to regard water consumption as a mechanism for enhancing the global hydrological cycle, which is in principle profitable to both man and environment. It has been shown that an alternative approach would be to minimize a global disturbance of the hydrological cycle, because it might increase the unequal distribution of water, and to strive for water recycling on a *small* scale, through small-scale water re-use. Further study of this issue could improve insight into the different approaches and support a well-considered strategy, which is completely lacking today.

These are typical questions in the research field of integrated water assessment. Whether the AQUA tool can be helpful, either in its current or a modified form, whether other computer tools will have to be developed or whether such tools are not needed at all, will vary from case to case. However, because the development of AQUA has taken central place in this work, I will briefly discuss the kind of improvements and extensions which could be useful with regard to this tool. A difficulty well-known among modellers is that 'a model is never finished'. As a result, I can make as long a list of possible improvements and extensions as I like. The interests of future users of the AQUA tool will strongly determine which efforts are most desirable. Here I will list only a few types of improvements and extensions which to my mind deserve priority. For a more refined global water assessment, it would be most useful to regionalize the AQUA World Model. Whether one would schematize the world into a few subcontinents or into smaller country clusters and river basins, should depend on the requirements of the assessment and the time available. From an economic point of view, it could be useful to distinguish - in both the World and the Zambezi Model - between investment and depreciation costs on the one hand and operational and maintenance costs on the other. This would make it possible to study the dynamics of water supply and expenditure if investments fall short over a long period (e.g. ten or twenty years). From an environmental point of view, it might be desirable to improve the part of the model which simulates water quality, because the present quality module could be too simple for specific purposes. Furthermore, one could add a module to calculate the effects of erosion on hydrological processes and water quality. Another interesting addition could be to calculate energy demand as a function of different types of water supply (e.g. energy demand for desalination or pumping ground water) and to discount energy prices in water supply costs. If linked to an energy model, as within the framework of TARGETS, this could provide insights in the interrelation between water and energy demand. If one wishes to apply the AQUA tool to a river basin with large snow storage in winter, a snow storage and melt module should be added. Finally, an interesting step could be to embed the AQUA Zambezi Model, or a different regional application, in the AQUA World Model. This would make it possible to do exploratory research at river-basin level within a dynamic global context. An even more advanced possibility would be to develop several regional applications of AQUA and combine them under the umbrella of the AQUA World Model.

My research has not only led to questions specifically relating to water, but also to a number of methodological questions. In this study, a number of 'water indicators' have been proposed, to translate scientific information for the public and policy makers, but it has not been studied whether they can indeed be effective. Within the context of this research there was no room for an effectiveness study, because this is a separate area, but also because this would take much more time. At the Harare workshop we in fact used pressure, state, impact and response indicators to communicate model results, but I did not systematically observe and analyse the feedback from participants. At present, there are a lot of studies introducing water indicators, but the next step should be one of practical use and evaluation. Another field which still needs attention is uncertainty research. As concluded in the previous section, perspective-based exploration of possible futures is an interesting approach which offers the possibility of addressing structural uncertainties in a systematic way, but the approach requires much more refining. As part of a further investigation of the method, one could choose an alternative for the cultural theory, in order to evaluate the usefulness of different social theories for the purpose of scenario development. One could also apply analytical techniques other than simulation to meet the specific requirements of different perspectives.

Finally, it is interesting to note that progress in methodological thinking in the field of water and sustainable development is behind that made in several other fields. Whereas concepts such as markets, prices, efficiency, demand management and scenarios have become commonly used in energy discussions for instance, these concepts often have yet to be introduced in debates on water. For example, it is striking that, in assessing future global water demand, the Russian State Hydrological Institute recently developed just *one* scenario for the year 2025 (Shiklomanov, 1997). It is also striking that the Harare workshop on freshwater use in the Zambezi basin, with experts from the region (see Chapters 8-9), did not result in much discussion about the necessity, desirability and feasibility of water-pricing policy and efficiency improvements, because most participants still had to be acquainted with the whole *idea* of water demand management. The obvious advantage of this lagging behind in the field of water policy and research is that one can benefit from lessons learnt in other disciplines. It is at least my personal experience that working as a 'water researcher' in an interdisciplinary team strongly stimulates creative thinking in my own field. I would therefore promote the further exploration of the integrated approach not only because it is essential for understanding the interaction between water and mankind in a broad context, but also because of the beneficial side effects it may bring.

References

Achterberg, W. (1986) *Partners in de natuur: Een onderzoek naar de aard en de fundamenten van een ecologische ethiek* Uitgeverij Jan van Arkel, Utrecht, the Netherlands.

Adriaanse, A. (1993) *Environmental policy performance indicators: A study on the development of indicators for environmental policy in the Netherlands* SDU Publishers, The Hague, the Netherlands.

Ambroggi, R.P. (1977) 'Underground reservoirs to control the water cycle' *Scientific American* 236(5): 21-27.

Ambroggi, R.P. (1980) 'Water' *Scientific American* 243(3): 90-105.

Anderson, T.L. (1995) 'Water options for the blue planet' In: R. Bailey (ed.) *The true state of the planet*, 267-294, The Free Press, New York, USA.

Arnoldus, H.M.J. (1978) 'An approximation of the rainfall factor in the universal soil loss equation' In: M. de Boodt and D. Gabriels (eds.) *Assessment of erosion*, John Wiley & Son, New York, USA.

Bakkes, J.A., G.J. van den Born, J.C. Helder, R.J. Swart, C.W. Hope and J.D.E. Parker (1994) 'An overview of environmental indicators: State of the art and perspectives', Environment Assessment Technical Reports 94-01, United Nations Environment Programme, Nairobi, Kenya.

Bakkes, J.A. and J. van Woerden (eds.) (1997) 'The future of the global environment: A model-based analysis supporting UNEP's first global environmental outlook', DEIA/TR.97-1, United Nations Environment Programme, Nairobi, Kenya.

Balon, E.K. and A.G. Coche (1974) *Lake Kariba: A man-made tropical ecosystem in Central Africa* Monographiae Biologicae 24, W. Junk, The Hague, the Netherlands.

Bankes, S. (1993) 'Exploratory modeling for policy analysis' *Operations Research* 41(3): 435-449.

Bannink, B.A. (ed.) (1998) *The Zambezi: Reliable water for all, forever? A case-study of integrated sustainability assessment, Draft report* United Nations Environment Programme, Nairobi, Kenya.

Barbour, I.G. (1980) *Technology, environment, and human values* Praeger, New York, USA.

Barney, G.O. (ed.) (1980) *The global 2000 report to the president: Entering the twenty-first century* US Council on Environmental Quality and the Department of State, Washington, D.C., USA.

Basson, M.S. (1995) 'South African water transfer schemes and their impact on the southern African region' In: T. Matiza, S. Crafter and P. Dale (eds.) *Water resource use in the Zambezi basin: Proceedings of a workshop held at Kasane, Botswana*, 41-48, IUCN, Gland, Switzerland.

Baumgartner, A. and E. Reichel (1975) *The world water balance: Mean annual global, continental and maritime precipitation, evaporation, and runoff* R. Oldenbourg Verlag, München, Germany.

Bazilevich, N.I., L.Y. Rodin and N.N. Rozov (1971) 'Geographical aspects of biological productivity' *Soviet Geography* 12: 293-317.

Beck, U. (1997) *De wereld als risicomaatschappij: Essays over de ecologische crisis en de politiek van de vooruitgang* De Balie, Amsterdam, the Netherlands.

Bendix, R. (1977) *Max Weber: An intellectual portrait* University of California Press, Berkeley, USA.

Bergsma, E. (1981) 'Indices of rain erosivity: a review' *ITC Journal* (4). 460-484.

Berry, B.J.L. (1990) 'Urbanization' In: B.L. Turner II, W.C. Clarck, R.W. Kates, J.F. Richards, J.T. Mathews and W.B. Meyer (eds.) *The earth as transformed by human action: global and regional changes in the biosphere over the past 300 years*, 103-119, Cambridge University Press, Cambridge, UK.

Binns, J.A. (1990) 'Is desertification a myth?' *Geography* 75: 106-113.

Bolton, P. (1983) *The regulation of the Zambezi in Mozambique: A study of the origins and impacts of the Cabora Bassa Project* Ph.D. thesis, University of Edinburgh, UK.

Bossel, H. (1996) '20/20 vision: Explorations of sustainable futures', Center for Environmental Systems Research, University of Kassel, Germany.

Bower, B.T., J. Kindler, C.S. Russell and W.R.D. Sewell (1984) 'Water demand' In: J. Kindler and C.S. Russell (eds.) *Modeling water demands,* 1-22, Academic Press, London, UK.

Brown, L.R., H. Kane and E. Ayres (1993) *Vital signs 1993-1994: The trends that are shaping our future* Earthscan Publications, London, UK.

Brown, L.R., J. Abramovitz, C. Bright, C. Flavin, G. Gardner, H. Kane, A. Platt, S. Postel, D. Roodman, A. Sachs and L. Starke (1996) *State of the world 1996: A Worldwatch Institute report on progress toward a sustainable society* W.W. Norton & Company, London, UK.

Budyko, M.I. (1982) *The earth's climate: past and future* International Geophysics Series Vol. 29, Academic Press, New York, USA.

Budyko, M.I. (1986) *The evolution of the biosphere* D. Reidel Publishing Company, Dordrecht, the Netherlands.

Buijs, P.H.L. and J. Dogterom (1995) 'Concepts for indicator application in river basin management', ICWS-report 95.01, International Centre for Water Studies, Amsterdam, the Netherlands.

Burke, J.J., M.J. Jones and V. Kasimona (1994) 'Approaches to integrated water resource development and management of the Kafue basin, Zambia' In: C. Kirby and W.R. White (eds.) *Integrated river basin development,* 407-423, John Wiley & Sons, Chichester, UK.

Calder, I.R., R.J. Harding and P.T.W. Rosier (1983) 'An objective assessment of soil-moisture deficit models' *Journal of Hydrology* 60: 329-355.

Calder, I.R., R.L. Hall, H.G. Bastable, H.M. Gunston, O. Shela, A. Chirwa and R. Kafundu (1995) 'The impact of land use change on water resources in sub-Saharan Africa: A modelling study of Lake Malawi' *Journal of Hydrology* 170: 123-135.

Carson, R. (1962) *Silent spring* Houghton Mifflin, Boston, USA.

Chandiwana, S.K. and W.B. Snellen (1994) 'Incorporating a human health component into the integrated development and management of the Zambezi River Basin', PEEM River Basin Series No. 2, World Health Organization, Geneva, Switzerland.

Christmas, J. and C. de Rooy (1991) 'The decade and beyond: At a glance' *Water International* 16(3): 127-134.

Clark, W.C. and R.E. Munn (eds.) (1986) *Sustainable development of the biosphere* Cambridge University Press, Cambridge, UK.

Clarke, R. (1991) *Water: The international crisis* Earthscan Publications, London, UK.

Colby, M.E. (1991) 'Environmental management in development: The evolution of paradigms' *Ecological Economics* 3: 193-213.

Conway, D., M. Krol, J. Alcamo and M. Hulme (1996) 'Future availability of water in Egypt: The interaction of global, regional, and basin scale driving forces in the Nile basin' *Ambio* 25(5): 336-342.

Couillard, D. and Y. Lefebvre (1985) 'Analysis of water quality indices' *Journal of Environmental Management* 21: 161-179.

Czaya, E. (1981) *Rivers of the world* Van Nostrand Reinhold, New York, USA.

Dale, A.P. (1995) 'An energy sector overview of Zambezi basin developments' In: T. Matiza, S. Crafter and P. Dale (eds.) *Water resource use in the Zambezi basin: Proceedings of a workshop held at Kasane, Botswana,* 147-152, IUCN, Gland, Switzerland.

Dávid, L.J., G.N. Golubev and M. Nakayama (1988) 'The environmental management of large international basins: The EMINWA programme of UNEP' *International Journal of Water Resources Development* 4(2): 103-107.

Dean, R.B. and E. Lund (1981) *Water reuse: Problems and solutions* Academic Press, London, UK.

Deichman, U. (1994) 'A medium resolution population database for Africa', National Center for Geographic Information and Analysis, University of California, Santa Barbara, USA.

De Vries, H.J.M. and M.A. Janssen (1997) 'The energy submodel: TIME' In: J. Rotmans and H.J.M. de Vries (eds.) *Perspectives on global change: The TARGETS approach,* 83-106, Cambridge University Press, Cambridge, UK.

Donkers, H. (1994) *De witte olie: Water, vrede en duurzame ontwikkeling in het Midden-Oosten,* Novib, The Hague; Uitgeverij Jan van Arkel, Utrecht, the Netherlands.

Douglas, M. (1970) *Natural symbols: Explorations in cosmology* Barrie and Rockcliff, London, UK.

Douglas, M. and A. Wildavsky (1982) *Risk and culture: An essay on the selection of technical and environmental dangers* University of California Press, Berkeley, USA.

Dunn, W.N. (1981) *Public policy analysis: An introduction* Prentice-Hall, Englewood Cliffs, New Jersey, USA.

Durham, D.S. (1995) 'The use of Zambezi water for urban supplies in Zimbabwe: reflections on current demands' In: T. Matiza, S. Crafter and P. Dale (eds.) *Water resource use in the Zambezi basin: Proceedings of a workshop held at Kasane, Botswana,* 69-73, IUCN, Gland, Switzerland.

Edgerton, L.T. (1991) *The rising tide: Global warming and world sea levels* Natural Resources Defense Council, Island Press, Washington, D.C., USA.

Edmonds, J. and J.M. Reilly (1985) *Global energy: Assessing the future* Oxford University Press, Oxford, UK.

Ehrlich, P.R. and A.H. Ehrlich (1991) *Healing the planet: Strategies for resolving the environmental crisis* Addison-Westley Publishing Company, Reading, Massachusetts, USA.

ESRI (1992) 'ArcWorld 1:3 M Database', Environmental Systems Research Centre, Redlands, USA.

Euroconsult (1989) *Agricultural compendium: For rural development in the tropics and subtropics,* Third revised edition, Elsevier Science Publishers, Amsterdam, the Netherlands.

Eurostat (1995) *Europe's environment: Statistical Compendium for the Dobrís assessment* Office for Official Publications of the European Communities, Luxembourg.

Fairbanks, R.G. (1989) 'A 17,000-year glacio-eustatic sea level record: Influence of glacial melting rates on the Younger Dryas event and deep-ocean circulation' *Nature* 342: 637-642.

Falkenmark, M. (1989) 'The massive water scarcity now threatening Africa: Why isn't it being addressed?' *Ambio* 18(2): 112-118.

Falkenmark, M. and G. Lindh (1974) 'How can we cope with the water resources situation by the year 2015?' *Ambio* 3(3-4): 114-122.

Falkenmark, M. and G. Lindh (1976) *Water for a starving world* Westview Press, Boulder, Colorado, USA.

Falkenmark, M., L. da Cunha and L. David (1987) 'New water management strategies needed for the 21st century' *Water International* 12(3): 94-101.

Falkenmark, M., J. Lundqvist and C. Widstrand (1989) 'Macro-scale water scarcity requires micro-scale approaches: Aspects of vulnerability in semi-arid development' *Natural Resources Forum*: 258-267.

FAO (1995a) *Reforming water resources policy: A guide to methods, processes and practices* Food and Agriculture Organization, Rome, Italy.

FAO (1995b) *Irrigation in Africa in figures* Food and Agriculture Organization, Rome, Italy.

FAO (1996) 'FAOSTAT, version 1996', Food and Agriculture Organization, Rome, Italy.

Forrester, J.W. (1961) *Industrial Dynamics* Wright-Allen Press, Cambridge, USA.

Forrester, J.W. (1968) *Principles of systems* Wright-Allen Press, Cambridge, USA.

Forse, B. (1989) 'The myth of the marching desert' *New Scientist* 1650: 31-32.

Furon, R. (1963) *Le problème de l'eau dans le monde* Payot, Paris, France.

Gandolfi, C. and K.A. Salewicz (1991) 'Water resources management in the Zambezi valley: Analysis of the Kariba operation' In: F.H.M. van de Ven, D. Gutknecht, D.P. Loucks and K.A. Salewicz (eds.) *Hydrology for the water management of large river basins,* IAHS Publication No. 201, 13-24, IAHS Press, Wallingford, UK.

Gates, W.L., J.F.B. Mitchell, G.J. Boer, U. Cubasch and V. Meleshko (1992) 'Climate modelling, climate prediction and model validation' In: J.T. Houghton, B.A. Callander and S.K. Varney (eds.) *Climate change 1992: The supplementary report to the IPCC scientific assessment* Cambridge University Press, Cambridge, UK.

Gates, W.L., A. Henderson-Sellers, G.J. Boer, C.K. Folland, A. Kitoh, B.J. McAvaney, F. Semazzi, N. Smith, A.J. Weaver and Q.-C. Zeng (1996) 'Climate models - Evaluation' In: J.T. Houghton, L.G. Meira Filho, B.A. Callander, N. Harris, A. Kattenberg and K. Maskell (eds.) *Climate change 1995: The science of climate change* Cambridge University Press, Cambridge, UK.

Geels, F. (1996) 'De wonderbaarlijke terugkeer van de onzekerheid en het einde van de moderniteit: Een onderzoek naar natuurbeelden, concepten in de ecologie, en verwachtingen omtrent wetenschappelijke kennis en beheersing inzake milieuproblemen' M.Sc. thesis, University of Twente, Enschede, the Netherlands.

Gibbons, D.C. (1986) *The economic value of water* Resources for the Future, Washington, D.C., USA.

Gleick, P.H. (1987a) 'The development and testing of a water balance model for climate impact assessment: Modeling the Sacramento basin' *Water Resources Research* 23(6): 1049-1061.

Gleick, P.H. (1987b) 'Regional hydrological consequences of increases in atmospheric CO_2 and other trace gases' *Climatic Change* 10: 137-160.

Gleick, P.H. (ed.) (1993a) *Water in crisis: A guide to the world's fresh water resources* Oxford University Press, New York, USA.

Gleick, P.H. (1993b) 'Water and conflict: Fresh water resources and international security' *International Security* 18(1): 79-112.

Gleick, P.H. (1995) 'Global water resources in the 21st century: Where should we go and how should we get there?' Proceedings Fifth Stockholm Water Symposium, Stockholm Water Company, Stockholm, Sweden.

Gleick, P.H. (1996) 'Basic water requirements for human activities: Meeting basic needs' *Water International* 21: 83-92.

Golubev, G. and O. Vasiliev (1978) 'Interregional water transfers as an interdisciplinary problem' *Water Supply & Management* 2: 67-77.

Gornitz, V., S. Lebedeff and J. Hansen (1982) 'Global sea level trend in the past century' *Science* 215: 1611-1614.

Gornitz, V. and S. Lebedeff (1987) 'Global sea-level changes during the past century' In: D. Nummedal, O.H. Pilkey and J.D. Howard (eds.) *Sea-level fluctuation and coastal evolution: Based on a symposium in honor of William Armstrong Price,* 3-16, Society of Economic Paleontologists and Mineralogists, Tulsa, Oklahoma, USA.

Gouzee, N., B. Mazijn and S. Billharz (eds.) (1995) 'Indicators of sustainable development', Report of the Workshop of Ghent, Belgium, 9-11 January 1995, submitted to the UN CSD, Federal Planning Office of Belgium, Bruxelles, Belgium.

Groenen, W., E. Pommer, M. Ras and J. Blank (1993) 'Milieuheffingen en consument', Social and Cultural Planning Office, Rijswijk, the Netherlands.

Groenendijk, H. (1989) 'Estimation of the waterholding-capacity of soils in Europe: The compilation of a soil dataset', Simulation Report CABO-TT 19, Centre for Agrobiological Research, Agricultural University, Wageningen, the Netherlands.

Harjono, M., F. Hoefnagels, V. de Lange, S. van Bennekom, C. Besselink and A. Ellenbroek (1996) 'Nederlands ruimtebeslag in het buitenland', Environmental Strategies Series No. 1996/9, Ministry of Housing, Planning and Environment, The Hague, the Netherlands.

Harris, J.M. (1990) *World agriculture and the environment* Garland Publishing, New York, USA.

Heilig, G.K. (1994) 'How many people can be fed on earth?' In: W. Lutz (ed.) *The future population of the world: What can we assume today?,* 207-261, Earthscan Publications, London, UK.

Helldén, U. (1991) 'Desertification: Time for an assessment?' *Ambio* 20(8): 372-383.

Heyns, P. (1995) 'Irrigation needs in the eastern Caprivi region of Namibia' In: T. Matiza, S. Crafter and P. Dale (eds.) *Water resource use in the Zambezi basin: Proceedings of a workshop held at Kasane, Botswana,* 101-110, IUCN, Gland, Switzerland.

Hoekstra, A.Y. (1995) 'AQUA: A framework for integrated water policy analysis', Report No. 461502006, National Institute of Public Health and the Environment, Bilthoven, the Netherlands.

Hoekstra, A.Y. (1996) 'Water policy and sustainable development: An integrated approach' In: W. Fischer, C.R. Karger and F. Wendland (eds.) *Wasser: Nachhaltige Gewinnung und Verwendung eines lebenswichtigen Rohstoffs,* Konferenzen des Forschungszentrums Jülich, Band 16, 231-255, Forschungszentrum Jülich, Germany.

Hoekstra, A.Y. (1997) 'The water submodel: AQUA' In: J. Rotmans and H.J.M. de Vries (eds.) *Perspectives on global change: The TARGETS approach,* 107-134, Cambridge University Press, Cambridge, UK.

Hoekstra, A.Y. and J.W.D. Vis (1996) 'The AQUA Zambezi Model' Workshop on the Sustainability of Freshwater Resources in the Zambezi Basin, November 25-28, 1996, Harare, Zimbabwe.

Hoekstra, A.Y., A.H.W. Beusen, H.B.M. Hilderink and M.B.A. Van Asselt (1997) 'Water in crisis?' In: J. Rotmans and H.J.M. de Vries (eds.) *Perspectives on global change: The TARGETS approach,* 291-317, Cambridge University Press, Cambridge, UK.

Hofstede, G. (1991) *Cultures and organizations: Software of the mind* McGraw-Hill, London, UK.

Hoozemans, F.M.J., M. Marchand and H.A. Pennekamp (eds.) (1993) *Sea level rise: A global vulnerability assessment,* Second revised edition, Delft Hydraulics, Delft, the Netherlands.

Houghton, J.T., L.G. Meira Filho, B.A. Callander, N. Harris, A. Kattenberg and K. Maskell (1996) *Climate change 1995: The science of climate change, Contribution of WGI to the Second Assessment Report of the Intergovernmental Panel on Climate Change* Cambridge University Press, Cambridge, UK.

Houghton, R.A., J.E. Hobbie, J.M. Melillo, B. Moore, B.J. Peterson, G.R. Shaver and G.M. Woodwell (1983) 'Changes in the carbon content of terrestrial biota and soils between 1860 and 1980: A net release of CO_2 to the atmosphere' *Ecological Monographs* 53(3): 235-262.

House, M.A. (1990) 'Water quality indices as indicators of ecosystem change' *Environmental Monitoring and Assessment* 15(3): 255-263.

Howard, G.W. (1994) 'The Zambezi river basin and its management' Proceedings of the Seminar on Water Resources Management, September 12-16, 1994, Economic Development Institute, Tanga, Tanzania.

ICWE (1992a) 'Report of the conference', International Conference on Water and the Environment, Dublin, Ireland.

ICWE (1992b) 'The Dublin statement on water and sustainable development', International Conference on Water and the Environment, Dublin, Ireland.

IUCN, UNEP and WWF (1980) *World conservation strategy: Living resource conservation for sustainable development* International Union for the Conservation of Nature and Natural Resources, Gland, Switzerland.

IWPDC (1991) *International water power and dam construction handbook* Reed Enterprise, Sutton, UK.

Janssen, M.A. (1996) 'Meeting targets: Tools to support integrated assessment modelling of global change' Ph.D. thesis, University of Maastricht, the Netherlands.

Janssen, P.H.M., P.S.C. Heuberger and R. Sanders (1994) 'UNCSAM: A tool for automating sensitivity and uncertainty analysis' *Environmental Software* 9: 1-11.

JICA (1995) 'The study on the national water resources master plan in the Republic of Zambia', Japan International Cooperation Agency, Japan.

Jones, P.D., T.M.L. Wigley and K.R. Briffa (1994) 'Global and hemispheric temperature anomalies - land and marine instrumental records' In: T.A. Boden, D.P. Kaiser, R.J. Sepanski and F.W. Stoss (eds.) *Trends '93: A compendium of data on global change,* ORNL/CDIAC-65, 603-608, Carbon Dioxide Information Analysis Center, Oak Ridge National Laboratory, Oak Ridge, Tennessee, USA.

Kahn, H. (1982) *The coming boom* Simon and Schuster, New York, USA.

Kaltenbrunner, H.F. (1992) 'Summary of the data used for the prototype Zambezi EIA-model', Resources Planning Consultants, Delft, the Netherlands.

Kasimona, V.N. and J.J. Makwaya (1995) 'Present planning in Zambia for the future use of Zambezi river waters' In: T. Matiza, S. Crafter and P. Dale (eds.) *Water resource use in the Zambezi basin: Proceedings of a workshop held at Kasane, Botswana,* 49-56, IUCN, Gland, Switzerland.

Kattenberg, A., F. Giorgi, H. Grassl, G.A. Meehl, J.F.B. Mitchell, R.J. Stouffer, T. Tokioka, A.J. Weaver and T.M.L. Wigley (1996) 'Climate models: Projections of future climate' In: J.T. Houghton, L.G. Meira Filho, B.A. Callander, N. Harris, A. Kattenberg and K. Maskell (eds.) *Climate change 1995: The science of climate change, Contribution of WGI to the Second Assessment Report of the Intergovernmental Panel on Climate Change,* Cambridge University Press, Cambridge, UK.

Keilani, W.M., P.H. Peters and P.J. Reynolds (1974) 'A water quality economic index' 9th Canadian Symposium on Water Pollution Research, University of Western Ontario, Canada.

Keller, W.J. and J. van Driel (1985) 'Differential consumer demand systems' *European Economic Review* 27: 375-390.

Kindler, J. and C.S. Russell (eds.) (1984) *Modeling water demands* Academic Press, London, UK.

Kivugo, M.M. and A.B. Nnunduma (1994) 'Current status in Lake Nyasa basin in Tanzania', Workshop on the development of an integrated water resources management plan for the Zambezi river basin, Livingstone, Zambia, 2-6 May 1994, The United Republic of Tanzania, Ministry of Water, Energy and Minerals, Tanzania.

Klein Goldewijk, C.G.M. and J.J. Battjes (1997) 'A hundred year (1890-1990) database for integrated environmental assessments (HYDE, version 1.1)', Report No. 422514002, National Institute of Public Health and the Environment, Bilthoven, the Netherlands.

Knoppers, R. and W. van Hulst (1995) *De keerzijde van de dam* Novib, The Hague; Uitgeverij Jan van Arkel, Utrecht, the Netherlands.

Kolars, J. (1994) 'Problems of international river management: The case of the Euphrates' In: A.K. Biswas (ed.) *International waters of the Middle East: From Euphrates-Tigris to Nile*, Water Resources Management Series No. 2, Oxford University Press, Oxford, UK.

Kooreman, P. (1993) 'De prijsgevoeligheid van huishoudelijk waterverbruik' *Economisch Statistische Berichten* 78: 181-183.

Korzun, V.I., A.A. Sokolov, M.I. Budyko, K.P. Voskresensky, G.P. Kalinin, A.A. Konoplyantsev, E.S. Korotkevich and M.I. Lvovich (eds.) (1977) *Atlas of world water balance* UNESCO, Paris, France.

Korzun, V.I., A.A. Sokolov, M.I. Budyko, K.P. Voskresensky, G.P. Kalinin, A.A. Konoplyantsev, E.S. Korotkevich and M.I. Lvovich (eds.) (1978) *World water balance and water resources of the earth* Studies and Reports in Hydrology 25, UNESCO, Paris, France.

Kosko, B. (1993) *Fuzzy thinking: The new science of fuzzy logic* HarperCollins Publishers, London, UK.

Kuik, O.J. and H. Verbruggen (eds.) (1991) *In search of indicators of sustainable development* Kluwer Academic Publishers, Dordrecht, the Netherlands.

Kulshreshtha, S.N. (1993) 'World water resources and regional vulnerability: Impact of future changes', Research Report RR-93-10, International Institute for Applied Systems Analysis, Laxenburg, Austria.

Kwadijk, J.C.J. (1993) 'The impact of climate change on the discharge of the River Rhine' Ph.D. thesis, University of Utrecht, the Netherlands.

La Rivière, J.W.M. (1989) 'Threats to the world's water' *Scientific American* 261(3): 48-55.

Leemans, R. and W.P. Cramer (1991) 'The IIASA database for mean monthly values of temperature, precipitation and cloudiness on a global terrestrial grid', Research Report RR-91-18, International Institute for Applied Systems Analysis, Laxenburg, Austria.

Leemans, R. and G.J. van den Born (1994) 'Determining the potential distribution of vegetation, crops and agricultural productivity' *Water, Air, and Soil Pollution* 76: 133-161.

Leggett, J., W.J. Pepper and R.J. Swart (1992) 'Emissions scenarios for IPCC: An update' In: J.T. Houghton, B.A. Callander and S.K. Varney (eds.) *Climate change 1992: The supplementary report to the IPCC scientific assessment*, Cambridge University Press, Cambridge, UK.

Lélé, S.M. (1991) 'Sustainable development: A critical review' *World Development* 19(6): 607-621.

Lewandowski, A. (1982) 'Issues in model validation', Research Report RR-82-37, International Institute for Applied Systems Analysis, Laxenburg, Austria.

Lindblom, C.E. (1977) *Politics and markets* Basic Books, New York, USA.

Liverman, D.M., M.E. Hanson, B.J. Brown and R.W. Merideth (1988) 'Global sustainability: Toward measurement' *Environmental Management* 12(2): 133-143.

Lowi, M. and J. Rothman (1993) 'Arabs and Israelis: The Jordan river' In: G.O. Faure and J.Z. Rubin (eds.) *Culture and negotiation: The resolution of water disputes*, 156-175, Sage Publications, Newbury Park, California, USA.

Luyten, J.C. (1995) 'Sustainable world food production and environment', Report 37, Research Institute for Agrobiology and Soil Fertility, Wageningen, the Netherlands.

L'vovich, M.I. (1973a) 'The water balance of the world's continents and a balance estimate of the world's freshwater resources' *Soviet Geography*: 135-152.

L'vovich, M.I. (1973b) 'The global water balance' *US IHD Bulletin*(23): 28-42.

L'vovich, M.I. (1977) 'World water resources: Present and future' *Ambio* 6(1): 13-21.

L'vovich, M.I. (1979) *World water resources and their future* LithoCrafters, Chelsea, Michigan, USA.

L'vovich, M.I. and G.F. White (1990) 'Use and transformation of terrestrial water systems' In: B.L. Turner II, W.C. Clarck, R.W. Kates, J.F. Richards, J.T. Mathews and W.B. Meyer (eds.) *The earth as transformed by human action: global and regional changes in the biosphere over the past 300 years,* 235-252, Cambridge University Press, Cambridge, UK.

Maltby, E. and R.E. Turner (1983) 'Wetlands of the world' *The Geographical Magazine* 55: 12-17.

Malthus, T.R. (1798) *An essay on the principle of population* World's Classics Edition 1993, Oxford University Press, Oxford, UK.

Margat, J. (1994) 'Les utilisations d'eau dans le monde, état présent et essai de prospective, Contribution au Projet M-1-3 du Programme Hydrologique International, PHI-IV', UNESCO, Paris, France.

Masundire, H.M. and T. Matiza (1995) 'Some environmental aspects of developments in the Zambezi basin ecosystem' In: T. Matiza, S. Crafter and P. Dale (eds.) *Water resource use in the Zambezi basin: Proceedings of a workshop held at Kasane, Botswana,* 137-146, IUCN, Gland, Switzerland.

Matiza, T., S. Crafter and P. Dale (1995) *Water resource use in the Zambezi basin: Proceedings of a workshop held at Kasane, Botswana* IUCN, Gland, Switzerland.

Matthews, E. and I. Fung (1987) 'Methane emissions from natural wetlands: global distribution, area, and environmental characteristics of sources' *Global Biogeochemical Cycles* 1: 61-86.

McCaffrey, S.C. (1993) 'Water, politics, and international law' In: P.H. Gleick (ed.) *Water in crisis: A guide to the world's fresh water resources,* 92-104, Oxford University Press, New York, USA.

Meadows, D.H., D.L. Meadows, J. Randers and W.W. Behrens (1972) *The limits to growth* Universe Books, New York, USA.

Meadows, D.H., D.L. Meadows and J. Randers (1991) *Beyond the limits: Confronting global collapse, envisioning a sustainable future* Earthscan Publications, London, UK.

Meybeck, M. (1976) 'Total mineral dissolved transport by world major rivers' *Hydrological Sciences Bulletin* XXI(2): 265-284.

Meybeck, M. (1982) 'Carbon, nitrogen, and phosphorus transport by world rivers' *American Journal of Science* 282: 401-450.

Meybeck, M. (1988) 'How to establish and use world budgets of riverine materials' In: A. Lerman and M. Meybeck (eds.) *Physical and chemical weathering in geochemical cycles,* 247-272, Kluwer Academic Publishers, Dordrecht, the Netherlands.

Meybeck, M. (1992) 'C, N, P and S in rivers: From sources to global inputs' In: R. Wollast, F.T. Mackenzie and L. Chou (eds.) *Interactions of C, N, P and S biogeochemical cycles and global change,* NATO ASI Series I Vol. 4, 163-193, Springer Verlag, Berlin, Germany.

Meybeck, M. and R. Helmer (1989) 'The quality of rivers: From pristine stage to global pollution' *Palaeogeography, Palaeoclimatology, Palaeoecology* 75: 283-309.

Miller, J.R. and G.L. Russell (1992) 'The impact of global warming on river runoff' *Journal of Geophysical Research* 97(D3): 2757-2764.

Milliman, J.D. and R.H. Meade (1983) 'World-wide delivery of river sediment to the oceans' *The Journal of Geology* 91(1): 1-21.

Mitchell, B. (1990) 'Integrated water management' In: B. Mitchell (ed.) *Integrated water management: International experiences and perspectives,* 1-21, Belhaven Press, London, UK.

Mitsch, W.J. and J.G. Gosselink (1993) *Wetlands,* Second edition, Van Nostrand Reinhold, New York, USA.

Morgan, M.G. and M. Henrion (1990) *Uncertainty: A guide to dealing with uncertainty in quantitative risk and policy analysis* Cambridge University Press, Cambridge, UK.

Morgan, M.G. and H. Dowlatabadi (1996) 'Learning from integrated assessment of climate change' *Climatic Change* 34(3-4): 337-368.

Mukosa, C., G. Pitchen and C. Cadou (1995) 'Recent hydrological trends in the Upper Zambezi and Kafue basins' In: T. Matiza, S. Crafter and P. Dale (eds.) *Water resource use in the Zambezi basin: Proceedings of a workshop held at Kasane, Botswana,* 85-98, IUCN, Gland, Switzerland.

Murakami, M. and K. Musiake (1994) 'The Jordan river and the Litani' In: A.K. Biswas (ed.) *International waters of the Middle East: From Euphrates-Tigris to Nile,* Oxford University Press, New Delhi, India.

Nace, R.L. (1967) 'Water resources: A global problem with local roots' *Environmental Science and Technology* 1(7): 550-560.

Nace, R.L. (1969) 'World water inventory and control' In: R.J. Chorley (ed.) *Water, earth, and man: A synthesis of hydrology, geomorphology, and socio-economic geography,* Methuen & Co., London, UK.

Nemec, J. (1993) 'Comparison and selection of existing hydrological models for the simulation of the dynamic water balance processes in basins of different sizes and on different scales', Report No. II-7, International Commission for the Hydrology of the Rhine basin (CHR/KHR), Lelystad, the Netherlands.

Newman, W.S. and R.W. Fairbridge (1986) 'The management of sea-level rise' *Nature* 320: 319-321.

Nicholls, N., G.V. Gruza, J. Jouzel, T.R. Karl, L.A. Ogallo and D.E. Parker (1996) 'Observed climate variability and change' In: J.T. Houghton, L.G. Meira Filho, B.A. Callander, N. Harris, A. Kattenberg and K. Maskell (eds.) *Climate change 1995: The science of climate change, Contribution of WGI to the Second Assessment Report of the Intergovernmental Panel on Climate Change,* Cambridge University Press, Cambridge, UK.

Nieswiadomy, M.L. (1992) 'Estimating urban residual water demand: Effects of price structure, conservation and education' *Water Resources Research* 28(3): 609-615.

OECD (1976) 'Measuring social well-being: A progress report on the development of social indicators', Organisation for Economic Co-operation and Development, Paris, France.

OECD (1982) 'The OECD list of social indicators', Organisation for Economic Co-operation and Development, Paris, France.

OECD (1993) 'OECD core set of indicators for environmental performance reviews', Environment Monographs No. 83, Organisation for Economic Co-operation and Development, Paris, France.

Oerlemans, J. (1989) 'A projection of future sea level' *Climatic Change* 15: 151-174.

Oldeman, L.R. and H.T. van Velthuyzen (1991) 'Aspects and criteria of the agro-ecological zoning approach of FAO', International Soil Reference and Information Centre (ISRIC), Wageningen, the Netherlands.

Olson, J., J.A. Watts and L.J. Allison (1985) 'Major world ecosystem complexes ranked by carbon in live vegetation: A database,', NDP-017, Oak Ridge National Laboratory, Oak Ridge, Tennessee, USA.

Öquist, M.G. and B.H. Svensson (1996) 'Non-tidal wetlands' In: R.T. Watson, M.C. Zinyowera and R.H. Moss (eds.) *Climate change 1995: Impacts, adaptations and mitigation of climate change: Scientific-technical analyses,* Cambridge University Press, Cambridge, UK.

Oreskes, N., K. Shrader-Frechette and K. Belitz (1994) 'Verification, validation, and confirmation of numerical models in the earth sciences' *Science* 263: 641-646.

Ott, W.R. (1978) *Environmental indices: Theory and practice* Ann Arbor Science Publishers, Ann Arbor, Michigan, USA.

Passmore, J. (1980) *Man's responsibility for nature: Ecological problems and western traditions* Second edition, Duckworth, London, UK.

Pearce, F. (1992) *The dammed: Rivers, dams, and the coming world water crisis* The Bodley Head, London, UK.

Peixoto, J.P. and M.A. Kettani (1973) 'The control of the water cycle' *Scientific American* 228(4): 46-61.

Penman, H.L. (1948) 'Natural evaporation from open water, bare soil and grass' *Proc. Roy. Soc. A.* 193: 120-146.

Penman, H.L. (1956) 'Evaporation: An introductory survey' *Netherlands Journal of Agricultural Science* 4: 9-29.

Pereira, A.R. and A. Paes de Camargo (1989) 'An analysis of the criticism of Thornthwaite's equation for estimating potential evapotranspiration' *Agricultural and Forest Meteorology* 46: 149-157.

Peterson, D.F. (1987) 'Some issues in irrigation macro strategies' *Water International* 12(3): 102-109.

Pinay, G. (1988) 'Hydrobiological assessment of the Zambezi river system: A review', Working Paper WP-88-089, International Institute for Applied Systems Analysis, Laxenburg, Austria.

Postel, S. (1992) *Last oasis: Facing water scarcity* W.W. Norton & Company, New York, USA.

Postel, S. (1993) 'Water and agriculture' In: P.H. Gleick (ed.) *Water in crisis: A guide to the world's fresh water resources,* 56-66, Oxford University Press, New York, USA.

Postel, S. (1996) 'Dividing the waters: Food security, ecosystem health, and the new politics of scarcity', Worldwatch Paper 132, Worldwatch Institute, Washington, D.C., USA.

Postel, S.L., G.C. Daily and P.R. Ehrlich (1996) 'Human appropriation of renewable fresh water' *Science* 271: 785-788.

Prentice, I.C., W. Cramer, S.P. Harrison, R. Leemans, R.A. Monserud and A.M. Solomon (1992) 'A global biome model based on plant physiology and dominance, soil properties and climate' *Journal of Biogeography* 19: 117-134.

Prentice, I.C., M.T. Sykes and W. Cramer (1993) 'A simulation model for the transient effects of climate change on forest landscapes' *Ecological Modelling* 65: 51-70.

Press, W.H., B.P. Flannery, S.A. Teukolsky and W.T. Vetterling (1987) *Numerical recipes: The art of scientific computing* Cambridge University Press, Cambridge, UK.

Priestley, C.H.B. and R.J. Taylor (1972) 'On the assessment of surface heat flux and evaporation using large scale parameters' *Mon. Weather Rev.* 100: 81-92.

Probst, J.L. and Y. Tardy (1987) 'Long range streamflow and world continental runoff fluctuations since the beginning of this century' *Journal of Hydrology* 94: 289-311.

Pulles, J.W. (1985) *Beleidsanalyse voor de waterhuishouding in Nederland / Policy analysis of water management for the Netherlands (PAWN)* Rijkswaterstaat, The Hague, the Netherlands.

RA (1994) 'Socio-economic impacts study of international environmental problems, Phase 3, Part I: software implementation activities, Part II: data gathering for the world version, Part III: data collection for coastal defense model', RA/94-150, 151 & 159, Resource Analysis, Delft, the Netherlands.

Raskin, P., E. Hansen and R. Margolis (1995) 'Water and sustainability: A global outlook', Polestar Report Series No. 4, Stockholm Environment Institute, Boston, USA.

Raskin, P.D., E. Hansen and R.M. Margolis (1996) 'Water and sustainability: global patterns and long-range problems' *Natural Resources Forum* 20(1): 1-5.

Raskin, P., P. Gleick, P. Kirshen, R.G. Pontius and K. Strzepek (1997) 'Water futures: Assessment of long-range patterns and problems, Background document for Chapter 3 of the comprehensive assessment of the freshwater resources of the world', Stockholm Environment Institute, Boston, USA.

Renzetti, S. (1992) 'Evaluating the welfare effects of reforming municipal water prices' *Journal of Environmental Economics and Management* 22(2): 147-163.

Richards, J.F. (1990) 'Land transformation' In: B.L. Turner II, W.C. Clarck, R.W. Kates, J.F. Richards, J.T. Mathews and W.B. Meyer (eds.) *The earth as transformed by human action: global and regional changes in the biosphere over the past 300 years,* 163-178, Cambridge University Press, Cambridge, UK.

Rind, D. (1988) 'The doubled CO_2 climate and the sensitivity of the modeled hydrologic cycle' *Journal of Geophysical Research* 93(D5): 5385-5412.

Rind, D., R. Goldberg, J. Hansen, C. Rosenzweig and R. Ruedy (1990) 'Potential evapotranspiration and the likelihood of future drought' *Journal of Geophysical Research* 95(D7): 9983-10004.

Risbey, J., M. Kandlikar and A. Patwardhan (1996) 'Assessing integrated assessments' *Climatic Change* 34(3-4): 369-395.

Rogers, P.P. (1985) 'Fresh water' In: R. Repetto (ed.) *The global possible: Resources, development, and the new century,* 255-298, Yale University Press, New Haven, USA.

Root, T.L. and S.H. Schneider (1995) 'Ecology and climate: Research strategies and implications' *Science* 269: 334-341.

Rosegrant, M.W. (1997) 'Water resources in the twenty-first century: Challenges and implications for action', Food, Agriculture, and the Environment Discussion Paper 20, International Food Policy Research Institute, Washington, D.C., USA.

Rotmans, J. (1990) *IMAGE: An integrated model to assess the greenhouse effect* Kluwer Academic Publishers, Dordrecht, the Netherlands.

Rotmans, J. (1997) 'Indicators for sustainable development' In: J. Rotmans and H.J.M. de Vries (eds.) *Perspectives on global change: The TARGETS approach,* 187-204, Cambridge University Press, Cambridge, UK.

Rotmans, J. and H.J.M. de Vries (eds.) (1997) *Perspectives on global change: The TARGETS approach* Cambridge University Press, Cambridge, UK.

Rozanov, B.G., V. Targulian and D.S. Orlov (1990) 'Soils' In: B.L. Turner II, W.C. Clarck, R.W. Kates, J.F. Richards, J.T. Mathews and W.B. Meyer (eds.) *The earth as transformed by human action: global and regional changes in the biosphere over the past 300 years,* 203-214, Cambridge University Press, Cambridge, UK.

SADC, IUCN and SARDC (1996) *Water in Southern Africa* Southern African Development Community, Maseru, Lesotho; The World Conservation Union, Harare, Zimbabwe; Southern African Research and Documentation Centre, Harare, Zimbabwe.

SADC-ELMS (1994) 'Zambezi river system action plan - ZACPLAN: Progress review and reflections on its programme of action', Workshop on the development of an integrated water resources management plan for the Zambezi river basin, Livingstone, Zambia, 2-6 May 1994, Southern African Development Community, Environment and Land Management Sector, Maseru, Lesotho.

SADC-ELMS and DANIDA (1994) 'Water resources assessment study, ZACPRO 6, Sector Studies Activity 2.1, Project Activity 12150', Denconsult International Development Consultants, Denmark.

SADCC (1990) 'Simulation of the joint operation of major power plants in the Zambezi basin, Final report', Southern African Development Coordination Conference, Maseru, Lesotho.

SADCC (1992) 'Regional irrigation development strategy, Final report', Southern African Development Coordination Conference, Maseru, Lesotho.

SADCC and UNEP (1991) 'ZACPRO 5: Development of a basin-wide unified environmental monitoring system related to water quality and quantity', Southern African Development Coordination Conference, Maseru, Lesotho, and United Nations Environmental Programme, Nairobi, Kenya.

Saeijs, H.F.L. and M.J. van Berkel (1995) 'Global water crisis: The major issue of the 21st century, a growing and explosive problem' *European Water Pollution Control* 5(4): 26-40.

Sahagian, D.L., F.W. Schwartz and D.K. Jacobs (1994) 'Direct anthropogenic contributions to sea level rise in the twentieth century' *Nature* 367: 54-57.

Salati, E. and P.B. Vose (1984) 'Amazon basin: A system in equilibrium' *Science* 225: 129-138.

SARDC, IUCN and SADC (1994) *State of the environment in Southern Africa* Southern African Research and Documentation Centre, Harare, Zimbabwe, The World Conservation Union, Harare, Zimbabwe, and Southern African Development Community, Maseru, Lesotho.

Savenije, H.H.G. (1995) 'New definitions for moisture recycling and the relationship with land-use changes in the Sahel' *Journal of Hydrology* 167: 57-78.

Savenije, H.H.G. (1996) 'The runoff coefficient as the key to moisture recycling' *Journal of Hydrology* 176: 219-225.

Schwarz, M. and M. Thompson (1990) *Divided we stand: Redefining politics, technology and social choice* Harvester Wheatsheaf, New York, USA.

Schwarz, H.E., J. Emel, W.J. Dickens, P. Rogers and J. Thompson (1990) 'Water quality and flows' In: B.L. Turner II, W.C. Clarck, R.W. Kates, J.F. Richards, J.T. Mathews and W.B. Meyer (eds.) *The earth as transformed by human action: global and regional changes in the biosphere over the past 300 years,* 253-270, Cambridge University Press, Cambridge, UK.

Serageldin, I. (1994) *Water supply, sanitation, and environmental sustainability: the financing challenge* World Bank, Washington, D.C., USA.

Serageldin, I. (1995) 'Water resources management: a new policy for a sustainable future' *Water International* 20: 15-21.

Shaw, E.M. (1994) *Hydrology in practice,* Third edition, Chapman & Hall, London, UK.

Shiklomanov, I.A. (1989) 'Climate and water resources' *Hydrological Sciences* 34(5): 495-529.

Shiklomanov, I.A. (1990) 'Global water resources' *Nature and Resources* 26(3): 34-43.

Shiklomanov, I.A. (1993) 'World fresh water resources' In: P.H. Gleick (ed.) *Water in crisis: A guide to the world's fresh water resources,* 13-24, Oxford University Press, New York, USA.

Shiklomanov, I.A. (ed.) (1997) *Assessment of water resources and water availability in the world: Scientific and technical report* State Hydrological Institute, St. Petersburg, Russia.

Showers, V. (1989) *World facts and figures* John Wiley and Sons, New York, USA.

Silver, C.S. and R.S. DeFries (1990) *One earth, one future: Our changing global environment* National Academy Press, Washington, D.C., USA.

Simon, J.L. (1980) 'Resources, population, environment: An oversupply of false bad news' *Science* 208: 1431-1437.

Simon, J.L. (1981) *The ultimate resource* Princeton University Press, Princeton, USA.

Slim, R.M. (1993) 'Turkey, Syria, Iraq: The Euphrates' In: G.O. Faure and J.Z. Rubin (eds.) *Culture and negotiation: The resolution of water disputes,* 135-155, Sage Publications, Newbury Park, California, USA.

Speidel, D.H. and A.F. Agnew (1988) 'The world water budget' In: D.H. Speidel, L.C. Ruedisili and A.F. Agnew (eds.) *Perspectives on water: Uses and abuses,* 27-36, Oxford University Press, Oxford, UK.

Stanners, D. and P. Bourdeau (eds.) (1995) *Europe's environment: The Dobris assessment* European Environment Agency, Copenhagen, Denmark.

Stigliani, W.M. (ed.) (1991) 'Chemical time bombs: Definition, concepts, and examples', Executive Report 16, International Institute for Applied Systems Analysis, Laxenburg, Austria.

Swart, R.J. and J.A. Bakkes (eds.) (1995) 'Scanning the global environment: A framework and methodology for integrated environmental reporting and assessment', Environment Assessment Technical Reports 95-01, United Nations Environment Programme, Nairobi, Kenya.

Szestay, K. (1982) 'River basin development and water management' *Water Quality Bulletin* 7: 155-162.

Tapfuma, V. (1995) 'The Batoka Gorge Hydroelectric Project: Design and potential environmental impacts' In: T. Matiza, S. Crafter and P. Dale (eds.) *Water resource use in the Zambezi basin: Proceedings of a workshop held at Kasane, Botswana,* 57-68, IUCN, Gland, Switzerland.

Thissen, W.A.H. (1978) 'Investigations into the Club of Rome's World3 model: Lessons for understanding complicated models' Ph.D. thesis, University of Technology, Eindhoven, the Netherlands.

Thom, A.S. and H.R. Oliver (1977) 'On Penman's equation for estimating regional evaporation' *Q.J.R. Meteorol. Soc.* 105: 345-357.

Thompson, M. (1988) 'Socially viable ideas of nature: A cultural hypothesis' In: E. Baark and U. Svedin (eds.) *Man, nature and technology: Essays on the role of ideological perceptions,* Macmillan Press, London, UK.

Thompson, M., R. Ellis and A. Wildawsky (1990) *Cultural theory* Westview Press, Boulder, USA.

Thompson, S.A. (1992) 'Simulation of climate change impacts on water balances in the Central United States' *Physical Geography* 13(1): 31-52.

Thornthwaite, C.W. (1948) 'An approach toward a rational classification of climate' *The Geographical Review* 38: 55-94.

Thornthwaite, C.W. and J.R. Mather (1955) 'The water balance', Publications in Climatology, Vol. VIII, No.1, Drexel Institute of Technology, Laboratory of Climatology, Centerton, New Jersey, USA.

Thornthwaite, C.W. and J.R. Mather (1957) 'Instructions and tables for computing potential evapotranspiration and the water balance', Publications in Climatology, Vol. X, No. 3, Drexel Institute of Technology, Laboratory of Climatology, Centerton, New Jersey,USA.

Tolba, M.K. and O.A. El-Kholy (eds.) (1992) *The world environment 1972-1992: Two decades of challenge* United Nations Environment Programme, Chapman & Hall, London, UK.

UN (1970) *Integrated river basin development* Second edition, United Nations, New York, USA.

UN (1978) *Register of international rivers* Centre for Natural Resources, Energy and Transport, Department of Economic and Social Affairs of the United Nations, Pergamon Press, Oxford, UK.

UN (1989) 'Handbook on social indicators', Studies in Methods, Series F, No.49, United Nations, Department of International Economic and Social Affairs, New York, USA.

UN (1990) 'Population prospects 1990', United Nations, New York, USA.

UN (1992a) 'Report of the United Nations Conference on Environment and Development', 3-14 June, Rio de Janeiro, Brazil.

UN (1992b) 'The United Nations energy statistics database (1992)', United Nations Statistical Division, New York, USA.

UN (1993) 'World population prospects: The 1992 revision', United Nations, New York, USA.

UN (1997a) 'Comprehensive assessment of the freshwater resources of the world: Report of the Secretary General', E/CN.17/1997/9, Commission on Sustainable Development, United Nations

UN (1997b) 'Critical trends: Global change and sustainable development', ST/ESA/255, Department for Policy Coordination and Sustainable Development, United Nations, New York, USA.

UNEP (1984) 'River basins of Africa database', United Nations Environment Programme, Nairobi, Kenya.

UNEP (1987) *Environmental data report 1987-88* Basil Blackwell, Oxford, UK.

UNEP (1991) *Freshwater pollution* UNEP/GEMS Environment Library No. 6, United Nations Environment Programme, Nairobi, Kenya.

UNEP (1993) *Environmental data report 1993-94* Blackwell Publishers, Oxford, UK.

UNEP (1995a) *Water quality of world river basins* UNEP Environment Library No. 14, United Nations Environment Programme, Nairobi, Kenya.

UNEP (1995b) 'Report of the data-definition workshop: Part of the Zambezi basin water resources sustainability assessment case study' Meeting Reports UNEP/EAP.MR.95-2, United Nations Environment Programme, Nairobi, Kenya.

UNEP (1996) 'Report of the workshop on the sustainability of freshwater resources in the Zambezi basin: Assessment approach, sustainability problems and policy options' Harare, Zimbabwe, 25-28 November 1996, Environment Information and Assessment Meeting Report UNEP/DEIA/MR.96-13, United Nations Environment Programme, Nairobi, Kenya.

UNESCO (1995) 'Discharge of selected rivers of Africa, Studies and reports in hydrology No.52', United Nations Educational, Scientific and Cultural Organization, Paris, France.

Van Amstel, A.R., G.F.W. Herngreen, C.S. Meyer, E.F. Schoorl-Groen and H.E. van de Veen (1988) 'Vijf visies op natuurbehoud en natuurontwikkeling: Knelpunten en perspectieven van deze visies in het licht van de huidige maatschappelijke ontwikkelingen', Publication No. 30, RMNO, Rijswijk, the Netherlands.

Van Asselt, M.B.A., A.H.W. Beusen and H.B.M. Hilderink (1996) 'Uncertainty in integrated assessment: A social scientific approach' *Environmental Modelling and Assessment* 1(1/2): 71-90.

Van Asselt, M.B.A. and J. Rotmans (1996) 'Uncertainty in perspective' *Global Environmental Change* 6(2): 121-157.

Van de Poel, I. (1991) 'Milieu en moraal, Een scriptie over ethische kwesties die spelen bij de mondiale milieuproblematiek' M.Sc. thesis, University of Twente, Enschede, the Netherlands.

Van der Leeden, F., F.L. Troise and D.K. Todd (1990) *The water encyclopedia* Second edition, Lewis Publishers, Chelsea, USA.

Van Deursen, W.P.A. (1995) 'Geographical information systems and dynamic models: Development and application of a prototype spatial modelling language' Ph.D. thesis, University of Utrecht, the Netherlands.

Van Deursen, W.P.A. and J.C.J. Kwadijk (1994) 'The impacts of climate change on the water balance of the Ganges-Brahmaputra and Yangtze basin', RA/94-160, Resource Analysis, Delft, the Netherlands.

Van Rijswijk, R.A. (1995) 'Calibration and validation of the AQUA model for the Ganges-Brahmaputra basin', M.Sc. thesis, Delft University of Technology, Delft, the Netherlands.

Van Wijk, W.R. and D.A. de Vries (1954) 'Evapotranspiration' *Netherlands Journal of Agricultural Science* 2: 105-119.

Vellinga, P. and S.P. Leatherman (1989) 'Sea level rise, consequences and policies' *Climatic change* 15: 175-189.

Vis, J.W.D. (1996) 'Experiments with and further development of the AQUA model for the Zambezi basin' M.Sc. thesis, University of Utrecht, the Netherlands.

Von Bertalanffy, L., C.G. Hempel, R.E. Bass and H. Jonas (1951) 'General system theory: A new approach to unity of science' *Human Biology* 23: 302-361.

Vörösmarty, C.J., B. Moore, A.L. Grace, M.P. Gildea, J.M. Melillo, B.J. Peterson, E.B. Rastetter and P.A. Steudler (1989) 'Continental scale models of water balance and fluvial transport: An application to South America' *Global Biogeochemical Cycles* 3(3): 241-265.

Vörösmarty, C.J. and B. Moore (1991) 'Modeling basin-scale hydrology in support of physical climate and global biogeochemical studies: An example using the Zambezi river' *Surveys in Geophysics* 12: 271-311.

Walker, J.C.G. (1977) *Evolution of the atmosphere* Macmillan, New York, USA.

Warrick, R.A. and J. Oerlemans (1990) 'Sea level rise' In: J.T. Houghton, G.J. Jenkins and J.J. Ephraums (eds.) *Climate change: The IPCC scientific assessment,* Cambridge University Press, Cambridge, UK.

Warrick, R.A., C.L. Provost, M.F. Meier, J. Oerlemans and P.L. Woodworth (1996) 'Changes in sea level' In: J.T. Houghton, L.G. Meira Filho, B.A. Callander, N. Harris, A. Kattenberg and K. Maskell (eds.) *Climate change 1995: The science of climate change, Contribution of WGI to the Second Assessment Report of the Intergovernmental Panel on Climate Change,* Cambridge University Press, Cambridge, UK.

WCED (1987) *Our common future* World Commission on Environment and Development, Oxford University Press, Oxford, UK.

Wetherald, R.T. and S. Manabe (1975) 'The effects of changing the solar constant on the climate of a generic circulation model' *Journal of Atmospheric Sciences* 32(11): 2044-2059.

WHO (1984) 'The International Drinking Water Supply and Sanitation Decade: Review of national baseline data (as at 31 December 1980)', Publication 85, World Health Organization, Geneva, Switzerland.

WHO (1993) *Guidelines for drinking-water quality, Volume 1: Recommendations* Second edition, World Health Organization, Geneva, Switzerland.

Wigley, T.M.L. and S.C.B. Raper (1995) 'An heuristic model for sea level rise due to the melting of small glaciers' *Geophysical Research Letters* 22(20): 2749-2752.

Williams, M. (1990) 'Forests' In: B.L. Turner II, W.C. Clarck, R.W. Kates, J.F. Richards, J.T. Mathews and W.B. Meyer (eds.) *The earth as transformed by human action: global and regional changes in the biosphere over the past 300 years,* 179-201, Cambridge University Press, Cambridge, UK.

Williamson, O. (1975) *Markets and hierarchies* Free Press, New York, USA.

Willmott, C.J., S.G. Ackleson, R.E. Davis, J.J. Feddema, K.M. Klink, D.R. Legates, J. O'Donnell and C.M. Rowe (1985) 'Statistics for the evaluation and comparison of models' *Journal of Geophysical Research* 90(C5): 8995-9005.

Wisserhof, J. (1994) *Matching research and policy in integrated water management* Delft University Press, Delft, the Netherlands.

WMO (1991) 'Hydrological models for water-resources system design and operation', Operational Hydrology Report No. 34, World Meteorological Organization, Geneva, Switzerland.

World Bank (1987) 'Community piped water supply systems in developing countries', Technical Paper No. 60, Washington, D.C., USA.

World Bank (1991) *World development report 1991* Oxford University Press, New York, USA.

World Bank (1993) 'Water resources management: A World Bank policy paper', International Bank for Reconstruction and Development, Washington, D.C., USA.

World Bank and UNDP (1990) 'Sub Saharan Africa Hydrological Assessment SADCC Countries, (a) Regional report, (b) Country report Angola, (c) Country report Botswana, (d) Country report Zambia, (e) Country report Zimbabwe, (f) Country report Tanzania, (g) Country report Malawi, (h) Country report Mozambique', Sir M. MacDonald & Partners, Cambridge, UK, and Hidroprojecto Consultores de Hidraulica e Salubridada, Lisbon, Portugal.

WRI (1992) *World resources 1992-93* Oxford University Press, New York, USA.

WRI (1994) *World resources 1994-95* Oxford University Press, New York, USA.
WRI (1996) *World resources 1996-97* Oxford University Press, New York, USA.
WRR (1994) *Duurzame risico's: Een blijvend gegeven* Wetenschappelijke Raad voor het Regeringsbeleid, Sdu Uitgeverij, The Hague, the Netherlands.
Yates, D. and K. Strzepek (1994) 'Potential evapotranspiration methods and their impact on the assessment of river basin runoff under climate change', Working Paper WP-94-46, International Institute for Applied Systems Analysis, Laxenburg, Austria.
Young, G.J., J.C.I. Dooge and J.C. Rodda (1994) *Global water resources issues* Cambridge University Press, Cambridge, UK.
Zweers, W. and W.T. de Groot (1987) 'Milieufilosofie en milieukunde: Een relatie in wording' *Milieu*(5): 145-149.

Summary

This book is the result of my investigation into the interaction between human beings and water in the long term. The notions behind the research are twofold. First, there is a growing awareness that water is not a resource which can be exploited limitlessly, nor a natural drain which can assimilate an unlimited amount of waste. Water on earth forms a natural cycle which can be disturbed by human intervention and this disturbance can have repercussions on society. One can think of the possible societal consequences of increasing water pollution and scarcity, declining groundwater tables, the construction of large-scale dams, deforestation, drainage of wetlands, climate change and sea-level rise. Changes in water cycles - both on a global and a regional level - often take place on a time scale of decades or centuries and are interconnected with societal and environmental changes to such an extent that they can only be understood through *integrated* research. This type of research focuses on the interaction between diverse processes, rather than on the nature of one particular process. A second notion behind this research is the growing awareness that the ability of people to predict the future is less than was thought, despite the advance in knowledge. This is because knowledge appears to provide some *understanding* of long-term processes, but it does not enable people to make meaningful *predictions*. This paradox of increasing knowledge and growing uncertainty requires an explorative approach, which regards uncertainty as inherent in knowledge and which explicitly shows that there is room for diverse assumptions and hypotheses. Such an approach is of crucial importance in the development of scenarios and the formulation of an effective development policy, because basic assumptions strongly influence the outcomes of these processes. For instance, in order to select the most effective type of investment, it is highly relevant to know whether one presupposes a mechanism in which increasing prosperity results in improved water supply and sanitation, or whether one presumes a reverse mechanism in which improved water supply and sanitation is a precondition for improved health and economic development.

This study aims to develop a method for integrated water assessment, so as to be able to analyse the coherence of different water-related developments and make results transparent to policy makers. The implications of uncertainties in basic assumptions should be expressed in an explicit way. The research has focused on the possible role of computer models. The main product is a simulation model for the analysis of the interaction between societal developments and changes in the water system. The model is generic, which means applicable on different spatial levels. As part of the study, this model has been applied on the global and on the river-basin level. In the latter case, it has been chosen to explore the basin of the Zambezi river in Southern Africa.

Method

A perspective approach has been chosen as the basis for developing scenarios. A perspective is a coherent perception (set of assumptions and hypotheses) with regard to the functioning of the world and the way people act. 'Perspectives on water' were developed by structuring the uncertainties and controversies in the water research field from four perspectives described in the cultural theory of Thompson *et. al.* (1990): the hierarchist, egalitarian, individualist and fatalist. The qualitative descriptions of the four perspectives on water have been translated into the model by implementing alternative equations and input data at the places in the model where the major uncertainties are located. In fact, one could speak of 'four models in one'. Both in the global and in the Zambezi study, the model has been calibrated separately for each of the perspectives, on the basis of historical data. Before running the model one can choose a particular type of context, a world-view and a management style. The context refers to exogenous developments which are input into the model, such as demographic developments, economic growth, etc. The world-view pertains to the functioning of the part of the physical world which is represented by the model. The management style relates to the response behaviour of people. Context, world-view and management style can all be chosen according to one of the perspectives.[1] Scenarios are constructed by choosing a certain combination of context, world-view and management style. Utopias are 'ideal' futures, in which context, world-view and management style all correspond to one particular perspective. Dystopias are scenarios in which this is not the case. Risks of different policy strategies are estimated by analysing the dystopian futures: what happens if a particular management style, which fits within a specific perspective, is applied to a world which behaves according to a different perspective. In this way, the risk concept is not defined from one perspective, but is understood at a level which exceeds the individual perspectives. Such a risk assessment can support the formulation of policy priorities which go beyond the preferences of separate perspectives.

---

[1] No specific fatalist world-view has been defined, because - in the perception of the fatalist - the world functions according to either the hierarchist, the egalitarian or the individualist world-view. A similar situation applies to the fatalistist context.

Results

The integrated approach has resulted in a number of interesting insights. A surprising conclusion from the global water study is that the loss of fresh ground water as a result of groundwater withdrawals might be a significant driving force behind sea-level rise, not only in the past but also in the future. This contradicts current opinion as embodied in the latest IPCC reports and deserves closer examination.

A conclusion of the Zambezi study is that meeting water demand in the Zambezi basin should not be a problem in an average hydrological year, either now or in future decades, but that the area is vulnerable to drought. This vulnerability increases considerably if water demand increases, particularly in upstream regions such as the parts of the basin situated in Angola, Zambia and Malawi. Downstream areas such as Mozambique are less vulnerable, due to the presence of Lake Kariba and the Cahora Bassa reservoir. However, these lakes are responsible for a considerable loss of water through evaporation: at present, evaporation from the two lakes is about twelve times as much as the evaporation resulting from water use in the Zambezi basin as a whole.

Both the global water study and the Zambezi study show that the interaction between mankind and water is gradually changing. This change is described as the 'water transition', consisting of three phases. In the first phase, a continuing development of new water resources takes place, to provide water to an increasing number of people and to increase yields in the agricultural sector. The second phase is characterized by growing water scarcity and competition between different water users, which slows down the rapid growth in water demand. In the third phase water scarcity has reached a level at which people are forced to use water more efficiently and to reduce pollution. Each phase requires its own type of water policy, which implies a shift from supply towards demand policy. Making this move can be a problem, but a greater problem can arise if disagreement exists about the exact current phase and thus about the necessity for change. One could say that developing countries are generally in the first or second phase of the transition and industrialized countries in the second or third phase. However, this notion is too simple, because different perspectives on the phase of development can be found in both developing and industrialized countries. The individualist perspective, which typically correlates with the first phase of the water transition, cannot only be found in developing countries, where ample possibilities for growth might be presumed, but also in industrialized countries, where continued growth is assumed to be possible through the use of new techniques and through increasing efficiency (e.g. desalination of sea water, stabilization of natural runoff through dams and artificial groundwater recharge, and concentration of water-intensive production processes in water-rich areas). Apart from the individualist perspective, the hierarchist perspective, which agrees with the second phase of the transition, can also be found all over the world. According to this view, people have to balance the desire for further growth with physical, societal, economic and technological limitations. The egalitarian perspective too, pertaining to the third phase of the transition, is represented everywhere to a greater or lesser extent. Furthermore, the large differences between the three perspectives leave room for a fourth perspective, viz. the fatalist perspective, according to which people cannot do other than await and cope with future developments as well as possible. Depending on the dominant perspective at a certain moment in a particular region, people will aim for a specific type of water policy, corresponding to the phase in which people think they are. From the individualist perspective, strong policy interference is not necessary, because there is a self-regulating mechanism: water demand and supply grow as a result of socio-economic development, which in turn is supported by an increasing exploitation of the water system. The most important aim of water policy should be to guarantee a proper functioning of the water price mechanism. According to the hierarchist perspective, increasing government interference is unavoidable, to be able to meet the different demands. For instance, governments would have to stimulate large-scale water resources projects and promote wastewater treatment. Water subsidies fit within the objective of adequate water supply for all. From the egalitarian perspective, it is considered most appropriate to charge full water costs to the consumer, even adding a water tax, to help slow down increasing water demand as much as possible. In addition a task of governments would be to increase people's awareness of the effects of profuse water use. In the fatalist perspective the type of water policy does not really matter, which means that there will be little active water policy.

A risk analysis of the different policy strategies shows that the fatalist management style generally carries the highest risks. Criteria which are used in this evaluation are for instance: what are - under varying context and world-view - the expected long-term consequences of the strategy on water scarcity, public water supply, sanitation, wastewater treatment, required expenditure in the water sector, decline of groundwater tables, and rise of the sea level. In the Zambezi basin, the egalitarian management style carries lower risks than the hierarchist or individualist styles. This can be understood through the egalitarian emphasis on demand policy. In the global water study, the differences between the hierarchist, individualist and egalitarian management styles are less pronounced. However, it has become clear that a number of elements from the egalitarian management style would deserve attention if one were to aim at excluding high risks. In particular the water-pricing policy preferred by the egalitarian fits within a strategy of risk avoidance. Subsidies for water would be allowed only if people would not be able to afford clean fresh water otherwise, which applies to public water supply to the poorest only.

Conclusion

This study shows that an integrated research approach can lead to fresh insights with regard to the interaction between socio-economic developments and changes within the water system. This approach can help in assessing the relative importance of the different processes involved in this interaction. The use of perspectives can be helpful in making the greatest uncertainties explicit and in estimating the risks of different policy strategies. Information about risks of policy can be relevant to policy makers in setting policy priorities.

Summary in Dutch - *Samenvatting*

Dit boek is de neerslag van mijn onderzoek naar de wisselwerking tussen mens en water op de lange termijn. De titel van het boek kan worden vertaald als: *Water in perspectief, Een integrale toekomstverkenning op basis van modellen*. De aanleiding voor het onderzoek is tweeledig.

Ten eerste is er het toenemende besef dat water niet een bron is waar onbegrensd uit kan worden geput of een natuurlijk riool waar ongelimiteerd in kan worden geloosd. Water op aarde neemt in feite deel aan een natuurlijke cyclus die verstoord kan worden door menselijk toedoen en deze verstoring kan weer repercussies hebben op de maatschappij. Men kan hierbij denken aan de mogelijke gevolgen van toenemende watervervuiling en -schaarste, dalende grondwaterspiegels, de aanleg van grootschalige dammen, ontbossing, het droogleggen van moerassige gebieden, klimaatsverandering en zeespiegelstijging. Veranderingen in waterkringlopen - zowel op mondiale als regionale schaal - zijn veelal pas merkbaar op een tijdschaal van decennia tot soms zelfs enkele eeuwen en zijn zo nauw verbonden met veranderingen in maatschappij en milieu in brede zin, dat ze alleen begrepen kunnen worden door *integraal* onderzoek. Dat is onderzoek waarbij vooral de relaties tussen uiteenlopende processen centraal staan, en niet zozeer de aspecten van één specifiek proces.

Een tweede aanleiding voor dit onderzoek wordt gevormd door het toenemende besef dat het menselijk vermogen om de toekomst te voorspellen - ondanks onze toenemende kennis - in feite kleiner is dan men dacht. Dit komt doordat kennis ons wel een zeker *begrip* van langetermijn-processen blijkt te geven, maar ons niet in staat stelt om zinvolle *voorspellingen* te doen. Deze paradox van toenemende kennis én onzekerheid vereist een exploratieve benadering die onzekerheid als inherent aan kennis beschouwt en die expliciet laat zien welke ruimte er is voor uiteenlopende aannamen en hypotheses. Voor het ontwikkelen van toekomstscenario's en het formuleren van een effectief ontwikkelingsbeleid is zo'n benadering van cruciaal belang omdat onzekerheden in basis-aannamen van grote invloed kunnen zijn. Om te weten waarin men het meest effectief kan investeren, is het bijvoorbeeld van belang of men uitgaat van een mechanisme waarbij een betere welvaart tot een betere watervoorziening en sanitatie leidt, ofwel dat men juist een omgekeerd mechanisme vooronderstelt, waarin een betere watervoorziening en sanitatie een voorwaarde vormt voor verbeteringen in volksgezondheid en economische groei.

Deze studie beoogt te komen tot een methode voor integraal water-onderzoek die het mogelijk maakt samenhangen tussen verschillende water-gerelateerde ontwikkelingen te analyseren en inzichtelijk te maken voor beleidsmakers, waarbij de implicaties van onzekerheden in basis-aannamen expliciet tot uitdrukking kunnen worden gebracht. De nadruk is gelegd op de rol die computermodellen hierbij kunnen spelen. Een belangrijk product van het onderzoek is een simulatiemodel voor de analyse van de interactie tussen maatschappelijke ontwikkelingen en veranderingen in het watersysteem. Het model is generiek van aard, dat wil zeggen toepasbaar op verschillende ruimtelijke schalen. Als onderdeel van deze studie is het model toegepast op mondiaal niveau en op stroomgebiedsniveau. In het laatste geval is gekozen voor het stroomgebied van de Zambezi-rivier in zuidelijk Afrika.

Methode
Voor het ontwikkelen van toekomstscenario's is een perspectief-benadering gekozen. Een perspectief is een samenhangende perceptie (stelsel van aannamen en hypothesen) van de wijze waarop de wereld functioneert en waarop mensen handelen. Er is in dit onderzoek voor gekozen om vier perspectieven te gebruiken: het hiërarchische, egalitaire, individualistische en fatalistische perspectief. Als basis voor deze perspectieven is uitgegaan van de culturele theorie van Thompson *e.a.* (1990). Een nadere uitwerking, gericht op water, is gemaakt door bestaande onzekerheden en controverses in het water-onderzoeksveld vanuit deze perspectieven te structureren. De kwalitatieve beschrijvingen van de vier perspectieven op water zijn naar het model vertaald door alternatieve vergelijkingen en invoergegevens te implementeren op die plaatsen waar de grootste onzekerheden voorkomen. Er is dus in wezen sprake van 'vier modellen in één'. Het model is - zowel in de mondiale als in de Zambezi-studie - gekalibreerd voor ieder perspectief afzonderlijk, op basis van historische gegevens. Voordat men het model laat rekenen, kan worden gekozen uit een zekere context, een wereldbeeld en een beleidsstijl. De context verwijst naar de exogene ontwikkelingen die als invoer van het model worden beschouwd, zoals bevolkingsontwikkeling, economische groei, enz. Een wereldbeeld is een bepaalde zienswijze op hoe de wereld functioneert. Een beleidsstijl heeft betrekking op het responsgedrag van mensen. Zowel context, wereldbeeld als beleidsstijl kan worden ingesteld volgens een van de perspectieven.[1] Toekomstscenario's worden opgebouwd door uit te gaan van een zekere combinatie van context, wereldbeeld en beleidsstijl. Utopia's zijn 'ideale' toekomsten waarin context, wereldbeeld en beleidsstijl alle met eenzelfde perspectief overeenkomen. Dystopia's zijn scenario's waarbij dit niet het geval is. Risico's van verschillende beleidsstrategieën worden geschat door een analyse van de dystopia's: wat gebeurt er als een zekere beleidsstijl, die past bij een

---

[1] Er is echter geen specifiek fatalistisch wereldbeeld gedefinieerd, omdat de wereld volgens de fatalist willekeurig functioneert volgens het hiërarchische, egalitaire ofwel individualistische wereldbeeld. Iets soortgelijks geldt voor de fatalistische context.

Samenvatting

bepaald perspectief, wordt toegepast binnen een wereld die zich volgens een ander perspectief gedraagt. Het risico-begrip wordt hierbij niet vanuit één bepaald perspectief gedefinieerd, maar wordt juist begrepen op een niveau dat de afzonderlijke perspectieven ontstijgt. Een dergelijke risico-analyse maakt het mogelijk beleidsprioriteiten te formuleren die uitstijgen boven de voorkeuren van de afzonderlijke perspectieven.

Resultaten

De integrale benadering heeft tot een aantal interessante inzichten geleid. Een verrassende conclusie uit de mondiale waterstudie is dat het verlies van zoet grondwater als gevolg van grondwateronttrekkingen een significante kracht achter zeespiegelstijging zou kunnen zijn, zowel in de afgelopen als in de toekomstige eeuw. Dit weerspreekt de huidige inzichten zoals verwoord door het IPCC en verdient nauwkeuriger onderzoek.

Een conclusie van de Zambezi-studie is dat de watervoorziening in het Zambezi-stroomgebied in een hydrologisch gemiddeld jaar geen problemen zou mogen geven, ook niet in de komende decennia, maar dat het gebied zeer kwetsbaar is voor droogte. Deze kwetsbaarheid wordt sterk vergroot bij een toenemende watervraag, vooral in de bovenstroomse regio's zoals de delen van het stroomgebied die in Angola, Zambia en Malawi vallen. Benedenstroomse gebieden zoals Mozambique zijn minder kwetsbaar door de aanwezigheid van het Karibameer en het Cahora Bassa reservoir. Deze meren zijn echter wel verantwoordelijk voor een groot verlies aan water door verdamping. Momenteel verdampt er uit de twee genoemde meren ongeveer twaalf keer zo veel als er verdampt als gevolg van watergebruik in het Zambezi-stroomgebied als geheel.

In zowel de mondiale waterstudie als de Zambezi-studie blijkt dat de aard van de wisselwerking tussen mens en water langzamerhand verandert. Deze verandering wordt beschreven als de 'watertransitie', bestaande uit drie fasen. In de eerste fase worden er voortdurend nieuwe watervoorraden aangesproken om een groeiend aantal mensen van water te voorzien en ter vergroting van de opbrengsten in de landbouw. De tweede fase wordt gekenmerkt door een toenemende waterschaarste en competitie tussen verschillende watergebruikers, hetgeen de snelle groei in watervraag afremt. In de derde fase heeft waterschaarste een niveau bereikt waarop mensen gedwongen zijn om efficiënter met water om te gaan en vervuiling te doen afnemen. Iedere fase vereist zijn eigen type waterbeleid, hetgeen betekent dat een ommezwaai moet worden gemaakt van aanbod- naar vraagbeleid. Het maken van een dergelijke ommezwaai kan een probleem vormen, maar een groter probleem kan ontstaan als er onenigheid bestaat over de fase waarin men verkeert en dus over de noodzaak van een ommezwaai. Men zou kunnen zeggen dat ontwikkelingslanden zich in het algemeen in de eerste of tweede fase van de transitie bevinden en geïndustrialiseerde landen in de tweede of derde fase. Dit is echter een te eenvoudige voorstelling van zaken, want zowel in ontwikkelingslanden als in geïndustrialiseerde landen blijken verschillende perspectieven op de fase van ontwikkeling te bestaan. Het individualistische perspectief, dat typisch hoort bij de eerste fase van de watertransitie, komt niet alleen voor in ontwikkelingslanden, waar men nog ruime mogelijkheden voor groei veronderstelt, maar ook in geïndustrialiseerde landen, waar toenemende groei mogelijk wordt geacht door gebruik van nieuwe technieken en een steeds toenemende efficiëntie (waarbij kan worden gedacht aan het ontzilten van zeewater, concentratie van water-intensieve productieprocessen in waterrijke gebieden, en stabilisatie van natuurlijke waterafvoer met behulp van dammen en grondwaterinfiltratie). Naast het individualistische perspectief komt ook het hiërarchische perspectief, dat past bij de tweede fase van de transitie, overal ter wereld wel voor. Volgens deze visie moet men zien te balanceren tussen de wens tot verdere groei en fysische, maatschappelijke, economische en technologische beperkingen. Ook het egalitaire perspectief, dat hoort bij de derde fase van de transitie, is overal wel in meer of mindere mate vertegenwoordigd. Tenslotte geven de grote verschillen tussen de drie perspectieven bestaansrecht aan een vierde perspectief, het fatalistische, volgens welke men niet anders kan doen dan afwachten en de toekomstige ontwikkelingen zo goed mogelijk het hoofd bieden. Afhankelijk van het dominante perspectief op een zeker moment in een zekere regio, zal men naar een bepaald type waterbeleid streven, dat is afgestemd op de fase waarin men denkt te verkeren. Volgens het individualistische perspectief is sterke overheidsbemoeienis overbodig, want er is een zichzelf regulerend mechanisme in werking: watervraag en -aanbod nemen toe als gevolg van de sociaal-economische ontwikkeling, welke op zijn beurt wordt ondersteund door een steeds intensiever gebruik van het watersysteem. De belangrijkste voorwaarde die aan waterbeleid wordt gesteld is dat een goed functionerend prijsmechanisme wordt gegarandeerd. In het hiërarchische perspectief wordt een toenemende overheidsbemoeienis onvermijdelijk geacht, om de verschillende belangen zo goed mogelijk te dienen. De overheid zou zich bijvoorbeeld moeten richten op de stimulering van nieuwe grootschalige waterprojecten en de promotie van afvalwaterbehandeling. Subsidie van water vormt een onderdeel van het streven naar een adequate watervoorziening aan alle gebruikers. Binnen het egalitaire perspectief past het volledig doorberekenen van waterkosten aan de gebruikers, met daarop zelfs een waterbelasting, om het toenemende watergebruik zoveel mogelijk te remmen. Bovendien zou de overheid het bewustzijn van mensen moeten vergroten ten aanzien van de effecten van overvloedig watergebruik. Volgens het fatalistische perspectief maakt het niet zoveel uit welk type waterbeleid wordt geformuleerd, hetgeen zal betekenen dat er weinig actief beleid wordt gevoerd.

Een risico-analyse van de verschillende beleidsstrategieën laat zien dat de fatalistische beleidsstijl in het algemeen het meest risicovol is. Criteria die bij deze beoordeling worden gehanteerd zijn bijvoorbeeld: wat zijn - onder variërende context en wisselend wereldbeeld - de verwachtte langetermijn-gevolgen van de strategie op waterschaarste, openbare watervoorziening, sanitatie, mate van afvalwaterbehandeling, benodigde uitgaven in de watersector, daling van grondwaterspiegels en stijging van de zeespiegel. Van de overige drie beleidsstijlen draagt, in het Zambezi-stroomgebied, de egalitaire beleidsstijl de minste risico's in zich, hetgeen kan worden begrepen uit de sterke nadruk op vraagbeleid. In de mondiale waterstudie blijken de verschillen tussen de hiërarchische, individualistische en egalitaire beleidsstijl minder geprononceerd, maar blijkt een aantal elementen uit de egalitaire beleidsstijl toch van groot belang als men al te grote risico's zou willen uitsluiten. Met name het egalitaire prijsbeleid past binnen een strategie van risico-mijding. Subsidie van water zou alleen plaats mogen vinden in die gevallen waarin mensen anders niet over schoon water zouden kunnen beschikken, hetgeen alleen voor openbare watervoorziening aan de armsten geldt.

Slot
Deze studie toont aan dat een integrale onderzoeksbenadering tot vernieuwende inzichten kan leiden als het gaat om de interactie tussen maatschappelijke ontwikkelingen en veranderingen binnen het watersysteem. De benadering kan helpen bij het inschatten van het relatieve belang van de verschillende processen die bij deze interactie een rol spelen. Het gebruik van perspectieven kan behulpzaam zijn bij het zichtbaar maken van de grootste onzekerheden en bij het inschatten van risico's van verschillende beleidsstrategieën. Informatie over risico's van beleid is voor beleidsmakers relevant voor het stellen van beleidsprioriteiten.

About the author

Arjen Hoekstra was born in Delft, an attractive old city in the Netherlands, on June 28, 1967. He completed his secondary education (VWO) at Hugo Grotius College, received his Master's degree in civil engineering, *cum laude*, at the Delft University of Technology and worked at Delft Hydraulics for two years. During his studies, he was chairman of the student association "Practische Studie" and councillor at the Faculty of Civil Engineering. His Master's thesis, about improving water monitoring networks in Indonesia, was written in that country. In 1991, he received an award from the Delft University Fund as the best civil engineering student of the year. In 1992, he participated in the Young Scientists' Summer Programme of the International Institute for Applied Systems Analysis (IIASA) in Laxenburg, Austria. In 1993 he moved to Utrecht and started the research which is reported in this book. He obtained employment at the Delft University of Technology, on detachment to the National Institute of Public Health and the Environment (RIVM), in Bilthoven. During the past four years at RIVM he has been a team member of the Global Dynamics and Sustainable Development Group. The main product of this group has been a book published by Cambridge University Press (Rotmans and De Vries, 1997). Arjen has now begun to work for the Dutch Organization for Scientific Research (NWO), on a project on the sustainability and environmental quality of trans-boundary river basins, carried out at the University of Technology in Delft and the Institute of Environmental Studies, Free University, in Amsterdam. Scattered over the past ten years, Arjen has cycled throughout Europe, from the Netherlands to Greece, Spain, and Poland, lived and hunted with pygmies in Gabon, and gone trekking in countries such as Indonesia, Turkey, Pakistan, India and Bolivia. Most winters, he goes for a week's camping in the Ardennes or Vosges, preferably when there is snow.

List of symbols

Symbol	Unit	Equation(s)	Explanation
a	-	3.23	Exponent in the equation for potential evaporation
A_{irr}	km²	3.4	Irrigated cropland area
A_{oc}	km²	3.17	Ocean area
A_{res}	km²	3.15	Artificial reservoir area
AF	-	3.56	Allocation factor
AF_{ind}	-	3.43, 3.56	Allocation factor industrial water supply
AF_{irr}	-	3.41, 3.56	Allocation factor irrigation water supply
AF_{liv}	-	3.42, 3.56	Allocation factor livestock water supply
AF_{pws}	-	3.40, 3.51, 3.56	Allocation factor public domestic water supply
AF_{san}	-	3.52, 3.56	Allocation factor sanitation
AF_{wwt}	-	3.53, 3.56	Allocation factor wastewater treatment
$AGWR$	kg/yr	3.35	Artificial groundwater recharge
AMT	kg/yr	3.19, 3.21, 3.22	Advective moisture transport from oceanic to terrestrial atmosphere
AMT_i	kg/yr	3.21	Advective moisture transport from oceanic to terrestrial atmosphere in the initial year
$APWL$	mm	3.25, 3.26	Accumulated potential water loss of the soil
c_{avg}	mg/l	3.36, 3.37	Average concentration
$c_{avg,nat}$	mg/l	3.36	Average concentration in natural waters
c_{dev}	mg/l	3.37	Standard deviation of the concentration distribution
$c_{st,A}$ to $c_{st,D}$	mg/l	3.37	Concentration standard (maximum concentration) for water quality class A to D
c_{ww}	mg/l	3.36	Average concentration in untreated wastewater
C_R	-	4.4	Runoff coefficient
$CAP_{coast}[Pr]$	US$	3.49	Coastal capital potentially flooded at probability Pr
CAP_{risk}	US$/yr	3.49	Coastal capital at risk
$COST_{desal}$	US$/kg	-	Desalination costs
$COST_{dyke}$	US$/m/km	-	Costs of dyke heightening
$COST_{hydr}$	US$/MW/yr	3.46	Hydropower costs
$COST_{san}$	US$/cap	-	Sanitation costs
$COST_{ws}$	US$/kg	4.10, 4.11	Average water supply costs
$COST_{ws,i}$	US$/kg	4.11	Average water supply costs in the initial year
$COST_{ws}[sect]$	US$/kg	3.50, 4.10	Water supply costs for sector $sect$
$COST_{wwt,dom}$	US$/kg	-	Wastewater treatment costs in the domestic sector

Symbol	Unit	Equation(s)	Explanation
$COST_{wwt,ind}$	US\$/kg	-	Wastewater treatment costs in the industrial sector
$COV_{pws,act}$	-	3.51	Actual public water supply coverage
$COV_{pws,dem}$	-	3.3, 3.40, 3.51	Demanded public water supply coverage
$COV_{pws,pol}$	-	3.51	Policy target for public water supply coverage
$COV_{san,act}$	-	3.52	Actual sanitation coverage
$COV_{san,dem}$	-	3.52	Demanded sanitation coverage
$COV_{san,pol}$	-	3.52	Policy target for sanitation coverage
$COV_{wwt,act}$	-	3.13, 3.53	Actual wastewater treatment coverage
$COV_{wwt,dem}$	-	3.53	Demanded wastewater treatment coverage
$COV_{wwt,pol}$	-	3.53	Policy target for wastewater treatment coverage
d	1/yr	3.8	Diffusion rate for the spread of new water-conserving technology
d_{res}	m	3.15	Average reservoir depth
$D[s,h]$	-	3.1	Distribution of a variable over socio-economic regions and hydrological areas
df	-	4.2	Dilution factor
DR	kg/yr	3.35, 3.38	Drainage of irrigation water
dt	yr	All differential equations	Time step
E_a	mm/month	3.24, 3.27, 4.5	Actual evaporation from a particular physical basic area
E_{fsw}	kg/yr	3.32	Evaporation from fresh surface water
E_{land}	kg/yr	3.22	Total land evaporation
E_{oc}	kg/yr	3.19, 3.20	Oceanic evaporation
$E_{oc,i}$	kg/yr	3.20	Oceanic evaporation in the initial year
E_p	mm/month	3.23 - 3.26, 4.5	Potential evaporation from a particular physical basic area
Eff_{act}	-	3.7, 3.8	Actual water-use efficiency
$Eff_{act,irr}$	-	3.12	Actual irrigation efficiency
Eff_{max}	-	3.8, 3.9	Maximum possible water-use efficiency
Eff_{min}	-	3.8	Minimum water-use efficiency
El_C	-	3.46	Cost elasticity of hydropower supply
El_G	-	3.7	Growth elasticity of water demand
El_P	-	3.7	Price elasticity of water demand
EXP_{act}	US\$/yr	3.55, 3.56, 4.15, 4.17	Actual expenditure
EXP_{dem}	US\$/yr	3.55, 3.56, 4.17	Demanded expenditure

List of symbols

Symbol	Unit	Equation(s)	Explanation
EXP_{max}	US$/yr	3.55	Maximum allowable expenditure
f	-	3.35	Part of reservoir storage contributing to stable runoff
F_{in}	kg/yr	3.16	Inflow of water into a store
F_{out}	kg/yr	3.16	Outflow of water from a store
$fcwu[sect]$	-	3.11	Fraction consumptive water use for sector $sect$
$fcwu_{irr}$	-	3.12	Fraction consumptive water use for irrigation
GBP	US$/yr	-	Gross basin product
GBP_{pc}	US$/yr	-	Gross basin product per capita
GNP	US$/yr	3.55, 4.15	Gross national product
GNP_{pc}	US$/yr	3.7, 3.51, 3.52, 3.53	Gross national product per capita
GWP	US$/yr	-	Gross world product
GWP_{pc}	US$/yr	-	Gross world product per capita
h	hours	3.23	Mean length of daylight
H_{crit}	m	3.47, 3.54	Critical water level (above which flooding will occur)
$H_{incr.dem}$	m	3.54	Increase in critical water level demanded
HG	GJ/yr	3.45	Hydropower generation
HGC	MW	3.14, 3.45, 3.46	Hydropower generation capacity
HGC_{pot}	MW	-	Potential hydropower generation capacity
I	-	3.23	Annual heat index
I_m	-	3.23	Monthly heat index
iaf	-	3.38	Inaccessible fraction of natural stable runoff
II	-	4.14	Overall water impact index
II_{ecol}	-	4.13, 4.14	Ecological impact index
II_{econ}	-	4.11, 4.14	Economic impact index
k_{fgw}	yr	3.30	Lag time of fresh ground water
k_{fsw}	yr	3.33, 8.4	Lag time of fresh surface water
lcf	-	3.23	Land cover factor
LGP	month	-	Length of growing period
LIV	-	3.5	Size of livestock
$loss_{evap}$	-	3.12	Evaporation fraction of irrigation loss
MI	-	4.5	Moisture index
N	-	3.23	Number of days in a month
n_{atm}	-	-	Number of atmospheric units
n_{clust}	-	-	Number of coastal clusters in the calculation of coastal impacts of sea-level rise

Symbol	Unit	Equation(s)	Explanation
n_{comp}	-	-	Number of spatial compartments
n_{dom}	-	-	Number of types of water supply in the calculation of domestic water demand
n_{exp}	-	4.15, 4.17	Number of items of expenditure
n_h	-	3.1	Number of hydrological areas (e.g. catchment areas)
n_{ice}	-	-	Number of ice sheets
n_{ind}	-	-	Number of industrial categories in the calculation of industrial water demand
n_{lct}	-	-	Number of land cover types in calculations of terrestrial hydrology
n_{livst}	-	-	Number of livestock categories in the calculation of livestock water demand
n_{pop}	-	-	Number of population categories in the calculation of domestic water demand
n_{prot}	-	-	Number of coastal protection types
n_s	-	3.1	Number of socio-economic regions (e.g. countries or country clusters)
n_{sect}	-	3.11, 4.10, 4.16	Number of water-demand sectors
n_{slr}	-	-	Number of sea-level rise components
n_{src}	-	-	Number of water source types
n_{store}	-	-	Number of water store types
n_{wqc}	-	-	Number of water quality classes
n_{wqv}	-	-	Number of water quality variables
p	-	3.29, 3.30	Exponent groundwater runoff
$p[h]$	-	3.1	Pressure variable for hydrological area h
$p[s]$	-	3.1	Pressure variable for socio-economic region s
P	mm/month	3.24 - 3.27, 4.12	Precipitation in a particular physical basic area
P_{land}	kg/yr	3.22, 4.4	Total precipitation on land
P_{net}	kg/yr	3.27, 3.28	Net precipitation on land
P_{oc}	kg/yr	3.19	Precipitation into oceans
P_{fsw}	kg/yr	3.32	Precipitation into fresh surface water
PI	-	4.3	Overall water pressure index
POP	people	3.3, 3.40	Population size
$POP_{coast}[Pr]$	people	3.48	Coastal population potentially flooded at probability Pr
POP_{rsk}	people/yr	3.48	Coastal population at risk
Pr	1/yr	3.48, 3.49	Flooding probability

List of symbols

Symbol	Unit	Equation(s)	Explanation
Pr_{crit}	1/yr	3.47, 3.48, 3.49	Critical flooding probability (probability that the critical water level is exceeded)
$Pr_{crit,max}$	1/yr	3.54	Maximum acceptable critical flooding probability
Q_A to Q_D	-	3.37, 4.8, 4.13	Fraction of a body of water within water quality class A to D
q	-	3.55	Maximum allowable expenditure as a fraction of gross national product
r	1/yr	3.7	Growth factor specific water demand
R_{del}	kg/yr	3.31, 3.32	Delayed surface runoff
R_{dir}	kg/yr	3.28, 3.32	Direct surface runoff
R_{fgw}	kg/yr	3.29, 3.31	Groundwater runoff
R_{firm}	kg/yr	8.2	Reservoir outflow needed to attain firm energy production
R_{perc}	kg/yr	3.28, 3.35, 3.38	Percolation into ground water
R_{riv0}	kg/yr	3.32, 3.33, 8.3	Undelayed river runoff
R_{riv}	kg/yr	3.33, 3.34, 3.36, 8.2, 8.3	Delayed river runoff
$R_{riv,min}$	kg/month	4.13	Minimum monthly river runoff (runoff in driest month of the year)
$R_{riv,min,i}$	kg/month	4.13	Minimum monthly river runoff in the initial year
$R_{riv,obs}[i]$	kg/month	8.1	Observed river runoff in month i
$R_{riv,obs,avg}$	kg/month	8.1	Average of observed monthly river runoff values in a year
$R_{riv,sim}[i]$	kg/month	8.1	Simulated river runoff in month i
$R_{riv,up}$	kg/yr	3.32	River runoff from upstream basins(s)
R_{stable}	kg/yr	3.35, 3.38	Stable runoff
R_{subs}	kg/yr	3.31, 3.34	Subsurface runoff
R_{tot}	kg/yr	3.34, 3.38, 4.4, 4.7	Total runoff
REI	-	4.12	Rain erosivity index
RI	-	4.18	Overall response index
$rpc[sect]$	-	3.50, 4.16	Ratio of water price to actual costs for water-demand sector *sect*
rpp	-	3.3, 3.40	Ratio of private to public water demand per capita
RSI	-	4.17, 4.18	Response satisfaction index
$s[h]$	-	3.1	State variable for hydrological area h
$s[s]$	-	3.1	State variable for socio-economic region s

Symbol	Unit	Equation(s)	Explanation
S	kg	3.16	Water storage
$S_{atm,land}$	kg	3.22	Water storage in terrestrial atmosphere
$S_{atm,oc}$	kg	3.19	Water storage in oceanic atmosphere
S_{cap}	mm	3.25, 4.6	Soil water-holding capacity
S_{fgw}	kg	3.29, 3.30	Fresh groundwater storage in zone of active water exchange
S_{fsw}	kg	3.33, 8.2, 8.3, 8.4	Fresh surface water storage
$S_{fsw,max}$	kg	8.2	Maximum fresh surface water storage (reservoir capacity)
$S_{fsw,min}$	kg	3.33, 8.2, 8.3, 8.4	Minimum fresh surface water storage (dead reservoir storage)
S_{glac}	kg	3.18	Water storage in glaciers
$S_{glac,i}$	kg	3.18	Water storage in glaciers in the initial year
$S_{glac,loss}$	kg	3.18	Loss of water storage in glaciers since the initial year
S_{ice}	kg	3.17	Storage of water in ice sheet
S_{res}	kg	3.14, 3.15, 3.35	Artificial reservoir storage
S_{soil}	mm	3.24, 3.25, 3.27, 4.6	Soil moisture content
$SCARC$	-	3.44	Water scarcity
SI	-	4.9	Overall water state index
slr	m	3.47, 3.54	Sea-level rise since the initial year
SMI	-	4.6	Soil moisture index
t	yr	All equations	Time
T	yr	-	Simulation period
T_{fsw}	yr	3.33, 8.3	Delay river runoff
T_{glac}	^0C	3.18	Critical temperature increase glaciers
$T_{glac,max}$	^0C	-	Maximum critical temperature increase glaciers
$T_{glac,min}$	^0C	-	Minimum critical temperature increase glaciers
T_{incr}	^0C	3.17, 3.18, 3.20, 3.21	Global mean surface air temperature increase
T_m	^0C	3.23	Mean monthly temperature
TD	-	3.9	Technological development
tmc	-	3.36	Transmission coefficient
uf	-	3.45	Utilization fraction of hydropower generation capacity ('load factor')
VA_{ind}	US$/yr	3.6	Value added in the industrial sector

List of symbols

Symbol	Unit	Equation(s)	Explanation
WAI	-	4.7, 4.9	Water availability index
WCI	-	4.16, 4.18	Water charge index
WD_{dom}	kg/yr	3.2, 3.3	Domestic water demand
$WD_{dom.pc}$	kg/yr/cap	3.3, 3.7, 3.40	Domestic water demand per capita (in the case of public supply)
WD_{exp}	kg/yr	3.2	Demand for water export
WD_{ind}	kg/yr	3.2, 3.6, 3.43	Industrial water demand
$WD_{ind.p\$}$	kg/yr/US\$	3.6, 3.7	Industrial water demand per dollar value added
WD_{irr}	kg/yr	3.2, 3.4, 3.41	Irrigation water demand
$WD_{irr.pha}$	kg/yr/ha	3.4, 3.7	Irrigation water demand per hectare
WD_{liv}	kg/yr	3.2, 3.5, 3.42	Livestock water demand
$WD_{liv.ph}$	kg/yr/head	3.5	Livestock water demand per head
WD_{spec}	kg/unit	3.7	Specific water demand: $WD_{dom.pc}$, $WD_{ind.p\$}$ or $WD_{irr.pha}$
WD_{tot}	kg/yr	3.2, 4.1	Total water demand
WDI	-	4.1, 4.3	Water demand index
WEI	-	4.15	Water expenditure index
WI	-	8.1	Willmott index (degree of correlation of simulated and observed hydrograph)
$WP[sect]$	US\$/kg	3.7, 3.50	Water price for sector *sect*
WPI	-	4.2, 4.3	Water pollution index
WQI	-	4.8, 4.9	Water quality index
$WS[sect]$	kg/yr	3.11, 4.10, 4.16	Water supply to sector *sect*
$WS[src]$	kg/yr	3.10	Water supply from source *src*
WS_{cons}	kg/yr	3.11, 3.21, 3.38, 3.44	Consumptive water use
WS_{dom}	kg/yr	3.39, 3.40	Domestic water supply
WS_{exp}	kg/yr	3.39	Water export out of the basin considered
WS_{fgw}	kg/yr	3.35, 3.38	Water supply from fresh ground water
WS_{fsw}	kg/yr	3.32	Water supply from fresh surface water
WS_{ind}	kg/yr	3.39, 3.43	Industrial water supply
WS_{irr}	kg/yr	3.39, 3.41	Irrigation water supply
WS_{liv}	kg/yr	3.39, 3.42	Livestock water supply
WS_{pot}	kg/yr	3.38, 3.44, 4.1, 4.2, 4.7	Potential water supply
WS_{tot}	kg/yr	3.10, 3.39, 3.44	Total water supply

Symbol	Unit	Equation(s)	Explanation
wsf[src]	-	3.10	Water source fraction (ratio of supply from source *src* to total water supply)
WW_{tot}	kg/yr	3.13, 3.32	Total wastewater disposal
WW_{tr}	kg/yr	3.13	Treated wastewater disposal
WW_{untr}	kg/yr	3.36, 4.2	Untreated wastewater disposal
x	-	8.3, 8.4	Exponent river runoff
α	mm/yr/°C	3.17	Ice sheet sensitivity
β	mm/yr	3.17	Initial imbalance ice sheet
γ_1, γ_2	-	3.47, 3.54	Constants in sea level - probability relationship
θ	MW/kg	3.14	Hydropower generation capacity per unit of reservoir storage
κ_{fgw}	yr	3.29, 3.30	Response factor fresh ground water
κ_{fsw}	yr	8.3, 8.4	Response factor fresh surface water
λ	1/°C	3.20	Sensitivity oceanic evaporation
μ	1/°C	3.21	Sensitivity advective moisture transport from oceanic to terrestrial atmosphere
ν	-	3.21	Loss of consumptive water use from the terrestrial to the oceanic atmosphere
τ	yr	3.18	Response time glaciers
ϕ	-	3.28	Ratio direct runoff / net precipitation
χ	-	3.31	Ratio delayed runoff / total groundwater runoff

Abbreviations

CSD	Commission on Sustainable Development
DANIDA	Danish International Development Assistance, Denmark
EEA	European Environment Agency, Copenhagen, Denmark
ESRI	Environmental Systems Research Centre
FAO	Food and Agriculture Organization, Rome, Italy
GBP	Gross Basin Product
GCM	General Circulation Model
GNP	Gross National Product
GWP	Gross World Product
ICWE	International Conference on Water and the Environment, Dublin, Ireland, 1992
IDWSSD	International Drinking Water Supply and Sanitation Decade (1981-1990)
IHD	International Hydrological Decade (1965-1974)
IIASA	International Institute for Applied Systems Analysis, Laxenburg, Austria
IPCC	Intergovernmental Panel on Climate Change
IUCN	The World Conservation Union (formerly the International Union for Conservation of Nature and Natural Resources), Gland, Switzerland
JICA	Japan International Cooperation Agency, Japan
OECD	Organisation for Economic Co-operation and Development, Paris, France
RIVM	National Institute of Public Health and the Environment, Bilthoven, the Netherlands
SADC	Southern African Development Community, Maseru, Lesotho
SADC-ELMS	Environment and Land Management Sector of SADC, Maseru, Lesotho
SADCC	Southern African Development Coordination Conference (later SADC), Maseru, Lesotho
SARDC	Southern African Research and Documentation Centre, Harare, Zimbabwe
SEI	Stockholm Environment Institute, Stockholm, Sweden
SHI	State Hydrological Institute, St. Petersburg, Russia
TARGETS	Tool to Assess Regional and Global Environmental and Health Targets for Sustainability
UN	United Nations, New York, USA
UNCED	United Nations Conference on Environment and Development, Rio de Janeiro, Brazil, 1992
UNDP	United Nations Development Programme, New York, USA
UNEP	United Nations Environment Programme, Nairobi, Kenya
UNESCO	United Nations Educational, Scientific and Cultural Organization, Paris, France
WCED	World Commission on Environment and Development
WHO	World Health Organization, Geneva, Switzerland
WMO	World Meteorological Organization, Geneva, Switzerland
WRI	World Resources Institute, Washington, D.C., USA
WWF	World Wide Fund for Nature (formerly World Wildlife Fund), Gland, Switzerland
ZACPLAN	Zambezi Action Plan
ZACPRO	Zambezi Action Plan Programme

Glossary

Aquifer. A layer of earth or rock containing ground water.

Calibration. Calibration of a model is the process of changing parameter values in order to correlate simulation results with observed data.

Climate sensitivity. The term climate sensitivity refers to the equilibrium change in surface air temperature following a unit change in radiative forcing and is expressed in $^{0}C/Wm^{-2}$. Radiative forcing is the perturbation to the energy balance of the earth-atmosphere system following, for example, a change in the atmospheric concentration of carbon dioxide (Houghton *et al*., 1996).

Consumptive water use. Consumptive (or irrecoverable) water use is the part of a freshwater withdrawal which is lost through evaporation. Consumptive water use is always smaller than total water use. Some authors also use the term *water consumption* (e.g. Shiklomanov, 1993), but this causes confusion among economists who then read this as *water supply*.

Context. In the simulation experiments in Chapters 7 and 9, the term 'context' refers to exogenous developments which are not part of the study and for which assumptions have to be made.

Cultural theory. A theory about socio-cultural viability which explains how ways of life maintain (or fail to maintain) themselves. As main references for this theory, which has its roots in anthropology, I use Thompson (1988), Thompson *et al*. (1990) and Schwarz and Thompson (1990).

Delay. The time difference between the inflow of an element into a store and its outflow.

Desalination. Production of fresh from salt or brackish water by removing the salt, using energy.

Dystopia. A future in which context, world-view and management style do not all correspond to the same perspective.

Evaporation. Potential evaporation from land is defined as the amount of water which will be lost from a surface completely covered with vegetation if there is sufficient water in the soil at all times for use by the vegetation (Thornthwaite and Mather, 1955). Actual evaporation from land is the volume of water actually lost, equal to or smaller than potential evaporation. In this study, the term evaporation does refer not only to direct evaporation from the soil, but includes transpiration by plants.

Explorative (exploratory) research. Research to explore the implications of varying assumptions and hypotheses.

Feedback. Negative feedback is a cyclic process which suppresses the signal that originally started the process. Positive feedback is a cyclic process reinforcing the original signal.

Flow. A flow is the amount of elements transferred from one store to another per unit of time.

Freshwater recharge. The freshwater recharge or freshwater renewal rate in an area is equal to the net precipitation in that area in a year.

Generic. A tool is generic if it can be applied under many different circumstances. In this study, the AQUA tool is called generic because it can be applied on different spatial levels (from river-basin to global level) and for different regions in the world.

Glacier. A mass of ice formed by accumulation of snow on higher ground. In this study the term 'glaciers' refers to all glaciers, ice caps and permanent snow cover on earth, excluding the ice sheets of Antarctica and Greenland.

Glacier sensitivity. This term is used here to refer to the water balance change in a glacier following a change in surface air temperature. The sensitivity is expressed as the contribution of the glacier to sea-level rise in mm/yr resulting from a $1^{0}C$ temperature increase. Defined in this way, one can speak of the *static* sensitivity of the glacier, ignoring changes in the configuration (Warrick *et al*., 1996).

Gross basin product. The concept of gross basin product (GBP) is used in the Zambezi study (Chapters 8-9) to refer to the output of goods and services for final use produced in the Zambezi basin in a particular year. It includes those parts of the gross national products of the Zambezi states which are generated within the Zambezi basin.

Gross national product. Gross national product (GNP) of a country is a measure of the total output of goods and services for final use generated by a national economy in a particular year, plus the net income from abroad. In other words, GNP is equal to the domestic and foreign value added to a national economy.

Gross world product. Gross world product (GWP) is the sum of gross national products in the world.

Groundwater recharge. Groundwater recharge or percolation is the process of water flowing from the unsaturated zone of the soil down to the saturated zone, the ground water.

Growth elasticity. The growth elasticity of demand for a commodity is a measure of the sensitivity of the demanded quantity to economic growth and is defined here as the percentage by which the quantity demanded changes for each 1 per cent change in gross national product per capita.

Hydrograph. A hydrograph or runoff curve shows the river discharge of a river as a function of time. A hydrograph is often drawn for one particular year, but it can also be drawn for an 'average' year.

Ice cap. A dome-shaped glacier usually covering a highland near a water divide (Houghton *et al.*, 1996).

Ice sheet. The IPCC defines an ice sheet as a glacier of more than 50×10^3 km^2 in area forming a continuous cover over a land surface or resting on a continental shelf (Houghton *et al.*, 1996). According to this definition, the earth has two ice sheets: Antarctica and Greenland.

Ice sheet sensitivity. A parameter to describe the response of an ice sheet's water balance to a change in surface air temperature. Cf. glacier sensitivity.

Indicator. An instrument to communicate key information about a system in a simplified way to policy makers and the public. Although an indicator is not necessarily a figure, the indicators described in this study are all quantitative in nature.

Integrated water assessment. A process of combining and interpreting knowledge from diverse scientific disciplines to allow a better understanding of the long-term interaction between water and development, explicitly distinguishing the role of water policy.

Integrated water management. The concept of integrated water management is often used loosely to refer to the process of integrated water assessment, the process of policy formulation, the resulting integrated water policy, the process of policy implementation, and the phase of operation and maintenance.

Integrated water policy. Comprehensive policy with regard to water, including water supply policy, water demand policy and policy in respect of societal activities which somehow relate to or affect water. Integrated water policy flows from the insights obtained through integrated water assessment.

Lag time. The lag time (renewal, residence, circulation time) of a store is the ratio of its storage to the total flow emerging from it. Not to be confused with *delay*.

Management style. A particular coherent perception of how people act and respond to their environment. In this study, the term generally refers to a style of *water* management.

Model. A representation of a part of reality. A model is not the same as a *system*, which is a part of reality itself. A model is thus a description of a particular system. In this study, the term generally refers to a computer model, but a computer model is of course only one of many possibilities to describe a system.

Net precipitation. Net precipitation is defined as precipitation minus evaporation and soil moisture change in a particular area and period, and is the part of precipitation available for runoff. Over a period of

Glossary

one year the net soil moisture change can generally be ignored, so that net precipitation equals precipitation minus evaporation.

Parameter. A constant in an equation. Strictly speaking, a parameter has one constant value, but in this study a broader interpretation is used and some constant curves (such as cost curves) and time-dependent response variables (such as water source fractions) are also classed as parameters.

Percolation. See *Groundwater recharge*.

Perspective. A perspective is understood as a coherent perception of how the world functions and how people act. In this study, I use four of the five perspectives described in the cultural theory: the hierarchist, egalitarian, individualist and fatalist.

Policy analysis. An applied discipline which uses multiple methods of inquiry and argument to produce and transform policy-relevant information which may be utilized in political settings to resolve public problems (Dunn, 1981).

Potential water re-use. The volume of potential water re-use is the part of a freshwater withdrawal which can possibly be re-used. It is the counterpart of *consumptive water use*.

Potential water supply. Defined in this study as the maximum amount of fresh water which can reasonably be withdrawn from natural water sources annually. Opinions on what is reasonable differ (see Chapters 2 and 5). If an area under consideration has water coming in from upstream, a distinction is made between potential water supply from internal sources and potential supply from external sources (see Chapters 8 and 9).

Prediction. A prediction is understood as an unconditional statement about a future development, which can be evaluated later on the basis of the real development. In the field of this research, it is not useful to speak of 'prediction' (see Section 3.2).

Price elasticity. The price elasticity of demand for a commodity is a measure of the sensitivity of the demanded quantity to a change in the price of the commodity and is defined as the percentage by which the quantity demanded changes for each 1 per cent change in the price.

Public water supply coverage. The fraction of a population with public water supply.

Renewal time. See *Lag time*.

Risk. The term risk is used to refer to the chance of undesirable consequences. In Section 7.6, two levels of risks are introduced. First-level risks are risks as perceived within a particular world-view. Second-level risks are the risks which emerge if one begins to question a world-view and considers the implications of alternative world-views.

River basin. A river basin or catchment area is an area from which net precipitation flows off through one particular water course. A large river basin is composed of several smaller sub-basins, because each tributary of the main river has its own basin.

Runoff. Runoff refers to water flow through gravity. The precipitation minus the evaporation and soil moisture change in an area is available for runoff. This net precipitation is divided into *direct surface runoff* and *groundwater runoff*. Part of the groundwater runoff reaches the earth's surface again within the same area and becomes *delayed surface runoff*; the other part becomes *subsurface runoff*. Direct and delayed surface runoff together form *river runoff*. River runoff and subsurface runoff together form *total runoff* from the area. *Stable runoff* is the part of the total runoff which is available throughout the year, often assumed to be equal to the groundwater runoff. In this study, stable runoff is precisely defined as the sum of natural groundwater recharge, artificial groundwater recharge, drainage of irrigation water and the stable runoff contribution from artificial surface reservoirs, but minus groundwater withdrawals. *Natural stable runoff* refers to natural groundwater recharge only.

Sanitation coverage. The fraction of a population with access to proper sanitation facilities.

Scenario. The term 'scenario' is used for various purposes by different authors. In the broadest sense, a scenario is an imagined sequence of future events. Policy analysts and modellers sometimes apply the term to refer to a future development which is used as *input* for some analysis. At other times, the concept refers to the *output* of an analysis. If the concept is used in its first meaning, I speak of *exogenous* or *input* scenarios, to stress that they are not part of the study. Some authors use the term scenario to refer to a particular set of measures or a strategy, but this has been avoided in this study. The term 'scenario approach' is used in this study to refer to a method of dealing with uncertainties (see Sections 1.2 and 5.1). In this context, a scenario is understood as a development under a particular set of assumptions.

Sensitivity analysis. In a sensitivity analysis of a model, one analyses how and to which extent uncertainties in the input values (for example parameter values) result in uncertainties in the model results.

Sink. A store receiving a flow of elements.

Soil water-holding capacity. The water-holding or saturation capacity of a soil is the amount of water in the soil available for plant uptake if the soil is at field capacity (i.e. the difference between field capacity and wilting point) and is generally expressed in mm or as a percentage of the upper zone of the soil (the first metre below ground level).

Source. A store from which a flow of elements originates.

Specific water demand. The term specific water demand or water-use intensity is used to express water demand per unit. Specific domestic water demand for example is the per capita demand, specific irrigation demand the per hectare demand and industrial water demand the demand per dollar of value added or per ton of manufactured goods.

Steady state. A store is in a steady state if the sum of the outflows is equal to the sum of the inflows.

Store. A spatially or otherwise delimited entity which contains a certain quantity of elements.

Sustainable development. The most cited definition of sustainable development is that in the influential report *Our Common Future* of the World Commission on Environment and Development: a development which "meets the needs of the present without compromising the ability of future generations to meet their own needs." (WCED, 1987).

System. A part of reality, conceived as a coherent whole of entities (elements, components). An *open* system is a system which is connected to, and interacts with, its environment. A *closed* system is a system which does not take in from, or give out to, its environment. In practice closed systems do not exist, but for certain purposes it may be reasonable to regard a system as closed. This does not, however, apply in the context of this thesis.

Transmission coefficient. A term used in this study to refer to the ratio of the input of a polluting substance into the surface water of a river basin to the outflow of this substance from the basin through river runoff.

Utopia. A future in which context, world-view and management style correspond to the same perspective.

Validation. Validation of a model can be understood in many different ways (see Section 6.4). In validating AQUA, I focus on its heuristic value, i.e. its value in guiding further study.

Virtual water. Water 'hidden' in an agricultural or industrial product, defined as the quantity of water used in the production process. Whereas farmers use water *directly* for irrigating their fields, consumers of food use this water *indirectly*. International trade in virtual water takes place in the form of international trade in water-intensive products.

Wastewater treatment coverage. The fraction of the total wastewater flow which is treated before discharge into the environment.

Water availability. This term refers to either the availability of freshwater stocks, the availability of freshwater flows or the availability of water in an even wider sense (e.g. including salt water), depending on the context. The concept of *potential water supply* is introduced in this study as a formal and more strict definition of water availability.

Water demand. Water demand can be loosely defined as the volume of water which has to be withdrawn in response to some human purpose. Depending on the context, the term is used in a more specific sense, namely water need or desire, or in its economic meaning (see Sections 2.2, 5.3 and 5.4).

Water literacy. Knowledge of people with regard to the effects of water use on the environment, the possibilities of water conservation, etc.

Water policy. In this study, water policy is understood as any government (local, national or international) plan of action with regard to water use, water management or any other water-related issue. According to this definition, water policy may also include climate policy or land-use policy.

Water policy analysis. Examination of current or future water-related problems and of possible government measures to solve these problems.

Water pricing. In most places in the world, the actual water supply cost per litre does not correspond to the water tariff, i.e. the price paid by the user of the water. Water-pricing policy generally means an increase in prices by lowering subsidies.

Water re-use. Water re-use or recycling means that water which has been used once is used again (for the same or another purpose). Before water is re-used, it generally undergoes some kind of treatment.

Water scarcity. Water scarcity refers to the quantity of water demanded for human purposes compared to the potential water supply. Opinions differ on the proper measure of water scarcity (see Chapters 2 and 5).

Water source fraction. The ratio of the supply from a certain water source (e.g. fresh surface water or renewable fresh ground water) to total water supply.

Water supply. Water supply is here defined as the volume of water withdrawn from any natural water source for human purpose. In this study, 'water supply' is used interchangeably with the terms 'water use' and 'water withdrawal'. 'Water supply' is here not used to refer to 'water availability' or 'potential water supply' (unlike for example in Kulshreshtha, 1993), because this is confusing for economists. A distinction is made between domestic, agricultural and industrial water supply. Domestic water supply refers to water supply to households, municipalities, commercial establishments and public services. Agricultural water supply includes water supply for irrigation and livestock. Industrial water supply refers to water supply to different industrial sectors, and includes groundwater withdrawals by mining industries and water use for thermoelectric power generation. The use of water for generating hydropower does not fall within my definition of industrial water supply, as 'instream water use' does not require any withdrawal.

Water transition. The 'water transition' hypothesis is formulated in this study to explain the changing interaction between water and development in terms of three phases: exponential growth of water demand, balancing the desirable and the possible and stabilization (see Section 7.7).

Water use. See Water supply.

Water-use efficiency. Irrigation efficiency is defined as the fraction of the total water withdrawal which actually benefits the crop. The remainder consists of water losses through evaporation and groundwater recharge. The maximum possible irrigation efficiency has a natural upper limit of 100 per cent. For domestic and industrial water use, efficiency is a relative concept, which means that an efficiency value has meaning only if compared to a previous efficiency value. Assuming the maximum possible efficiency in an initial year at 100 per cent implies that the maximum possible efficiency in later years can exceed 100 per cent. If actual efficiency reaches maximum efficiency, this can be understood as the absence of water losses; further water conservation will inevitably result in a reduced performance of the activity which depends on the water.

Water-use intensity. See *Specific water demand.*

Water withdrawal. See *Water supply.*

World-view. A particular coherent perception of how the world functions. In this study, the term is generally used to refer to a particular perception with regard to water issues.

Series **Environmental Studies**

Bas Arts
The Political Influence of Global NGOs
Case Studies on the Climate and Biodiversity Conventions

Arjen Y. Hoekstra
Perspectives on Water
An Integrated Model-based Exploration of the Future

Barbara Hogenboom
Mexico and the NAFTA Environment Debate
The Transnational Politics of Economic Integration

Ans Kolk
Forests in International Environmental Politics
International Organisations, NGOs and the Brazilian Amazon

Arthur P.J. Mol
The Refinement of Production
Ecological Modernization Theory and The Chemical Industry

Nico Nelissen, Jan van der Straaten and Leon Klinkers (eds.)
Classics in Environmental Studies
An Overview of Classic Texts in Environmental Studies

Martijntje Smits (ed.)
Polymer Products and Waste Management
A Multidisciplinary Approach